Advances in Materials
Problem Solving With the
Electron Microscope

MATERIALS RESEARCH SOCIETY
SYMPOSIUM PROCEEDINGS VOLUME 589

Advances in Materials Problem Solving With the Electron Microscope

Symposium held November 30–December 3, 1999, Boston, Massachusetts, U.S.A.

EDITORS:

Jim Bentley
Oak Ridge National Laboratory
Oak Ridge, Tennessee, U.S.A.

Charles Allen
Argonne National Laboratory
Argonne, Illinois, U.S.A.

Uli Dahmen
Lawrence Berkeley National Laboratory
Berkeley, California, U.S.A.

Ivan Petrov
University of Illinois
Urbana, Illinois, U.S.A.

Materials Research Society
Warrendale, Pennsylvania

CAMBRIDGE UNIVERSITY PRESS
Cambridge, New York, Melbourne, Madrid, Cape Town,
Singapore, São Paulo, Delhi, Mexico City

Cambridge University Press
32 Avenue of the Americas, New York NY 10013-2473, USA

Published in the United States of America by Cambridge University Press, New York

www.cambridge.org
Information on this title: www.cambridge.org/9781107413351

Materials Research Society
506 Keystone Drive, Warrendale, PA 15086
http://www.mrs.org

First published 2001
First paperback edition 2013

Single article reprints from this publication are available through
University Microfilms Inc., 300 North Zeeb Road, Ann Arbor, MI 48106

CODEN: MRSPDH

ISBN 978-1-107-41335-1 Paperback

CONTENTS

*Invited Paper

MICROELECTRONIC MATERIALS

PARTIALLY ORDERED AND
NANOPHASE MATERIALS

INTERFACES IN METALS
AND CERAMICS

*Invited Paper

PREFACE

Symposium Q, "Advances in Materials Problem Solving With the Electron Microscope," was held November 30-December 3 at the 1999 MRS Fall Meeting in Boston, Massachusetts. More than 100 papers were presented in ten sessions including four poster sessions. The sessions were well attended and the discussions were lively.

The symposium was motivated by the remarkable advances that continue to be made in electron microscope instrumentation and techniques for applications to materials science. Characterization problems can now be tackled that were beyond reach just a few years ago. The advances include quantitative high resolution imaging, atomic-resolution Z-contrast imaging, elemental mapping by energy-filtered TEM or spectrum imaging, atomic resolution EELS for composition and bonding, quantitative CBED, site occupancy determination by ALCHEMI, electron holography, EBSP in the SEM for phase identification and orientation imaging microscopy, low-voltage microanalysis of bulk specimens, and in-situ experiments of dynamic phenomena. The aim of the symposium was to emphasize how these recent developments in electron microscopy are being used to solve materials problems.

For the most part, the sessions were organized to feature different groups of materials or microstructural components rather than electron microscope techniques or instrumentation. Attendees heard discussions of low-energy electron microscopy of surfaces, crystallography, defects, specimen preparation, and interfaces in metals and ceramics. Technological applications ranged over magnetic materials, microelectronic materials, partially ordered and nanophase materials, polymers, ceramics, metallic alloys, concrete, biomaterials, and glasses.

The symposium organizers are grateful to the following for assisting in chairing sessions: I.M. Anderson, N.D. Browning, V.P. Dravid, C.P. Flynn, R.J. Gottschall, R. Hull, K.M. Krishnan, M.R. Libera, J. Mayer, J.R. Michael, D.J. Miller, I.M. Robertson, M. Sarikaya, R. Sinclair, D.J. Smith, A. Thust, R.D. Twesten, G.C. Weatherly, and Y. Zhu. Many of these colleagues also presented invited talks.

Symposium support was provided by Argonne National Laboratory, Lawrence Berkeley National Laboratory, Oak Ridge National Laboratory, the University of Illinois, JEOL USA, Inc. and Fischione Instruments, Inc. The symposium organizers and the Materials Research Society gratefully acknowledge their support. Finally, special thanks go to the helpful MRS staff.

Jim Bentley
Charles Allen
Uli Dahmen
Ivan Petrov

June 2001

MATERIALS RESEARCH SOCIETY SYMPOSIUM PROCEEDINGS

MATERIALS RESEARCH SOCIETY SYMPOSIUM PROCEEDINGS

Prior Materials Research Society Symposium Proceedings available by contacting Materials Research Society

Magnetic Materials

MICROSTRUCTURAL CHARACTERIZATION OF LONGITUDINAL MAGNETIC RECORDING MEDIA

ROBERT SINCLAIR, DONG-WON PARK, CLAUS HABERMEIER and KAI MA
Department of Materials Science and Engineering, Stanford University, Stanford, CA 94305-2205, bobsinc@stanford.edu

ABSTRACT

The optimization of disc manufacturing conditions is required to increase the storage capacities of magnetic recording media, which is strongly related to both magnetic properties and microstructural features. Analyzing the microstructure requires transmission electron microscopy (TEM), since the small grain sizes of the media prevent other tools from characterizing them. This paper discusses several fascinating characteristics of TEM in understanding and analyzing the properties of the recording media.

INTRODUCTION

One of the most remarkable high-technology industries at the present time concerns computer hard disc manufacturing. For the last several years, increases in the areal density of magnetic recording have occurred at a rate of 60% or more per annum, which is achieved by advances both in the media and in the recording heads. At the time of writing, products with 36 Gbits/in^2 have been demonstrated, and the industry goal of 100 Gbits/in^2 is clearly in sight. This article focuses on the magnetic medium itself.

The important magnetic properties usually comprise coercivity, squareness of the magnetization hysteresis loop and the signal-to-noise ratio of the recording. All are manipulated by the disc processing conditions, which at the materials level influence the underlying microstructure. Structural parameters which are thought to play a role in determining properties include crystal size and orientation, defect density, phase identity, grain boundary segregation or separation etc. As the grain sizes are typically in the range 10-20 nm, only transmission electron microscopy (TEM) has the capability to analyze the microstructure in a detailed fashion. In this paper, we highlight how TEM can be applied to establish the structure-property relationships and illustrate the difficulties associated with addressing this critical problem.

BACKGROUND

The magnetically active material currently used in hard disc technology is a thin film of a cobalt-chromium-X alloy (X being one or more minor additional elements), in the hexagonal close packed (HCP) crystal structure. For longitudinal magnetic recording, the magnetically "easy" c-axis is induced to lie in the plane of the film (i.e. the basal planes are standing proud with respect to the thin film surface). This is achieved by suitable epitaxial growth, the most common arrangement utilizing a body-centered cubic (BCC) underlayer of chromium or a chromium alloy. These films are grown sequentially by sputtering onto either nickel-phosphorus plated aluminum or glass substrate discs. A final carbon overcoat is deposited to protect the film during use.

The most commonly employed processing conditions bring about (200) oriented Cr polycrystalline films, with the cobalt alloy growing with (1120) planes parallel to the surface (i.e. the c-axis must crystallographically lie in the plane of the film). Loss of this epitaxy, for instance by an interfacial reaction, results in a vertical c-axis with concomitant degradation in the longitudinal recording performance. Two orientations of the cobalt crystals are possible, with the c-axis paral-

3

lel to either the [011] or [01$\bar{1}$] Cr directions. This results in a cobalt grain size usually smaller than that of the chromium, with orthogonal orientations growing on the same underlying Cr grains (the so-called "bicrystal structure") as shown in Fig.1. Further explanations of this detail can be found elsewhere [1,2]. Discs described here were manufactured in a standard fashion at HMT Technology, Seagate Technology and Komag Corporation.

Specimens for TEM analysis are prepared by conventional means. 3 mm diameter discs may be cut from the (larger) hard discs, mechanically dimpled from one side to less than 10 μm thickness and finally ion-beam milled to perforation. The final ion-milling step can also be refined so that either the cobalt or the chromium film could be examined preferentially. Bright field, dark field, high-resolution images and their associated diffraction patterns were obtained in regular TEM's (Philips EM430 or CM20), and nanoprobe analysis for X-ray energy dispersive spectroscopy was achieved in a field-emission gun TEM. Energy filtered imaging was carried out at Oak Ridge National Laboratory.

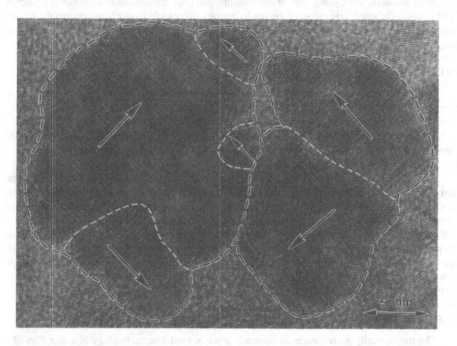

Figure 1. A high-resolution TEM image showing the bi-crystal grain structure of a CoCrPtTa alloy media. Arrows point along the c-axis directions.

RESULTS

Clearly there are many possible avenues for altering the microstructure to achieve superior recording performance. One primary goal from the magnetic point-of-view is to bring about the sharpest possible "bit transition regions", where the induced magnetization switches from parallel to anti-parallel. Because of the random nature of the cobalt crystal orientations in the plane of the film, this requires very small grain sizes (e.g. 10 nm), with some degree of magnetic decoupling of adjacent grains. Thus the analysis of grain size and the degree of grain separation is of major technological concern at the present time. There are various scientific difficulties associated with such analyses by TEM, and this presents a major emphasis of the present article.

Grain Sizes

The determination of average grain size has long been a task for metallography. But when the dimensions involved are in the 10-20 nm range, this is not a straightforward venture at all. In order to utilize a computer analysis such as the NIH program, all of the grain boundaries in the area of interest must be identified [3]. In conventional bright and dark field images only a small fraction of the grains are diffracting strongly at any one time. Moreover, when adjacent grains also diffract to a similar extent, the presence of a grain boundary may not be detected in a single image. Figure 2 illustrates this problem by showing the subtle change of appearance with very small specimen tilts. Therefore in order to achieve a somewhat reliable analysis, a large number of complementary images from the same area is required, with very detailed documentation of each grain, which requires significant investment of time.

Furthermore, when a bicrystal structure is present, it is not possible to distinguish the two variants by conventional imaging. High-resolution pictures are required. As the latter involve superior specimens, more careful microscope adjustment, higher magnifications, and so fewer grains in the image, developing sufficient statistics becomes a key issue.

Measurement of the grain size itself is not so obvious. The NIH program, when it can be used, converts the grain area into its equivalent representative dimension. But when individual measuring is necessary, the length across the grain is to be determined. We generally use the approach to estimate the longer axis and its orthogonal dimension, and take the average. An alternative which has also been employed is to determine the grain size parallel and perpendicular to the c-axis (from high-resolution images), and to quote both numbers [4]. Note that the linear intercept method commonly employed in optical microscopy cannot be used here because very few grain boundaries can be seen in the images. In reality any sensible approach is acceptable, but perhaps it is useful to specify the method in each case.

Of course, one of the major concerns with TEM analyses concerns the statistical nature of the data, especially compared to bulk techniques such as X-ray diffraction. In one study comparing data from high-resolution images with those from grazing incidence X-ray line broadening [5], both the absolute values from each technique and their trend as a function of underlayer material were in reasonable agreement. However, in our case, the microstructure is at a very small scale. Clearly the magnetic properties are reproducible across the disk (otherwise the manufacturing process would be suspect!), and the written bit sizes are not that much larger than the grains themselves (e.g. 63 nm at 400 kfci). Therefore, it is commonly found that samples taken from different areas on the same, or even an alternative, disc yield similar data.

In our experience, it is generally not necessary to analyze a large number of samples for reliable results. This issue has been considered in some detail by Carpenter et al. [6]. One perhaps surprising result is that the TEM grain size average for 100 grains in polycrystalline Al is within two percent of the value for 1000 or even 2500 grains (see Fig. 10 in Ref. 6). We obtain a similar finding (e.g. Fig. 3). Accordingly, a quantitative analysis does not require as many grains as might be expected - we generally use about 200 grains to allow same safety level.

One interesting point is associated with the subjectivity of the analysis. Different researchers, even those with significant experience, may adopt slightly different criteria in deciding what constitutes a grain, or in evaluating the dimension of "grain size". Both the grain size distribution and average may vary from one person to another (e.g. Fig. 4). We recommend, therefore, that one researcher only determines the grain sizes in a particular comparative set, and that the individual criteria and approach be clearly stated.

5

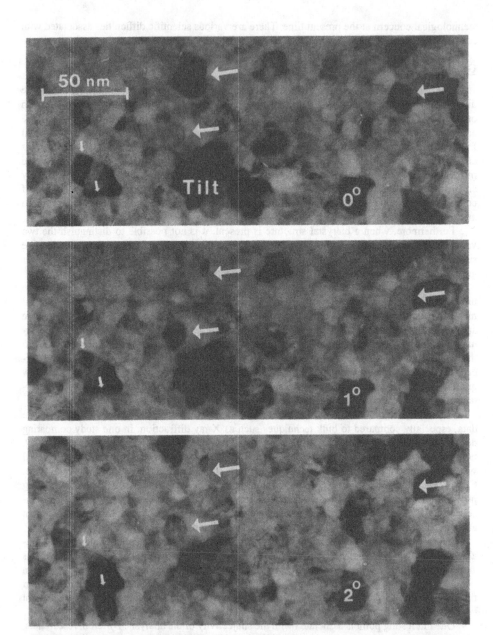

Figure 2. Bright field (BF) images of a CoCrPtTa medium grown on CrMo with 1° specimen tilts. The underlayer orientation is <112> which does not induce the bi-crystal structure, and so the "BF grain size" is reliable. Only small changes in tilt bring about significant differences in the appearance of diffracting grains (arrowed).

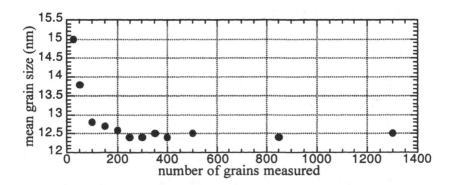

Figure 3. The variation of "average grain size" with increasing number of grains in the statistical population, showing that a reasonable average is obtained with about one hundred grains.

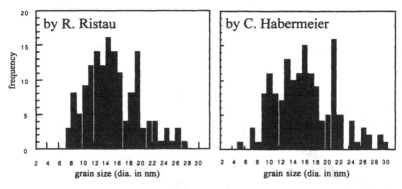

Figure 4. Grain size distributions obtained from the same micrographs by two different researchers. Note that the averages were determined to be 15.3 and 16.5 nm, respectively.

Grain Size Distributions

In addition to the grain size itself, the variation in grain size, or rather the "grain size distribution", is an important microstructural parameter. Technologists aim to achieve as narrow a distribution as possible, but manipulation of this feature by processing is naturally limited.

Display of the data is often achieved using a standard histogram (e.g. Fig. 5) [7,8], but its nature is altered by the "bin size" populations. We find that a cumulative percentage curve is superior, plotted either linearly with increasing grain size, or as a function of the logarithm of grain size. The latter allows simple distinction of mathematical descriptions of the distribution. For instance, for the former a Gaussian distribution is a linear plot and a log-normal is curved, and vice versa for the log-log graph (e.g. Fig. 6). In all our data, on Co-Cr-Ta and Co-Cr-Pt alloys and

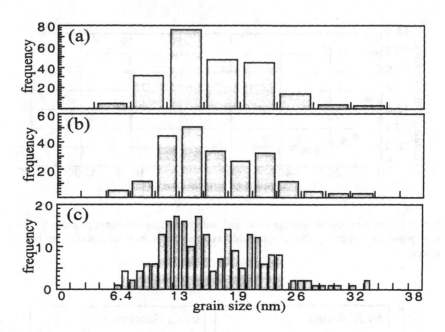

Figure 5. The variation of standard histograms with "bin size" for a Cr underlayer: (a) 10, (b) 15 and (c) 50 bins.

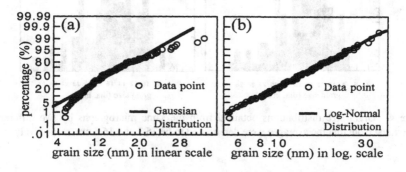

Figure 6. A comparison of cumulative percentage curves versus (a) linear or (b) logarithmic increase in grain size for CoCrPtTa media. Matching the data with standard mathematical distributions (e.g. Gaussian or Log-normal) allows assessment of the latter.

on Cr and Cr alloy underlayers, we find that a log-normal distribution is the best fit, although researchers at Hitachi have developed a "modified" Gaussian for their data on the Co-Cr-Pt alloy media [7]. Of some surprise here is that again we would anticipate that a very large sampling would be required to establish the distribution. This turns out not to be the case, as shown by Fig. 7. On the logarithmic plot the distribution from a small, randomly chosen number of grains (22) remains rather constant even up to a sampling an order-of-magnitude larger. This indicates that an

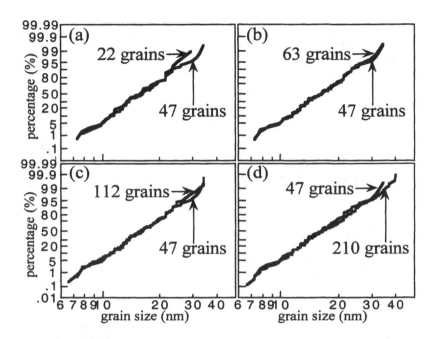

Figure 7. The effect of increasing grain population on the cumulative percentage curves, showing that a reasonable distribution can be established from even a small number of grains for this type of display (Cr underlayer).

analysis of a few hundred grains is more than adequate to establish the nature of the grain size distribution when displayed this way, although of course "more is better".

However, while the graphical data yield useful numbers, it is worthwhile remembering that the original images still remain useful in evaluating the microstructure, as shown in Fig. 8 for a Cr underlayer analysis.

Defects

One of the central paradigms of materials research is that "defects influence properties". Therefore it should be no surprise that this same statement should be applied to recording media, not necessarily always discriminatively. Dislocations appear to play no major role as the grains themselves are so small. However, with the very low stacking fault energy of cobalt and its alloys, significant stacking fault densities are present which vary with, at the least, alloy content. At the faults, the stacking sequence is changed locally to an FCC arrangement. As the FCC cobalt phase has different magnetic properties from those of the HCP phase, some effect on recording might be anticipated.

A detailed study of the effects of stacking faults is not easy. Really, only high resolution imaging is capable of unequivocally identifying the faults, and so the development of reasonable statistics becomes an issue. One such attempt was made by Ishikawa et al. [9]. By varying the alloy content of Co-Cr-Pt and Co-Cr-Ta alloys (15% Cr, with up to 8% Pt and 6% Ta), it was found that the stacking fault density increased (from about 0.3 nm^{-1} up to 0.5 nm^{-1} for the former,

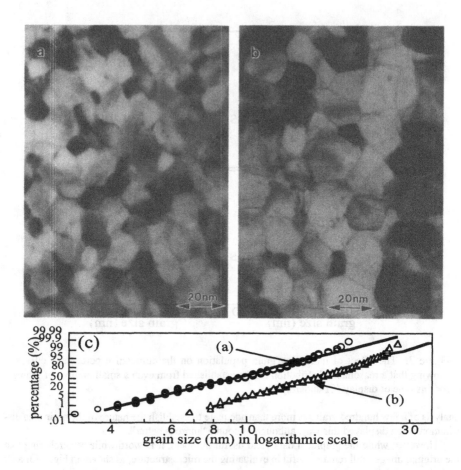

Figure 8. A comparison of the microstructure and cumulative percentage curves for Cr under-layers deposited with different film thickness. The images reveal the microstructural differences just as well as the graphical data.

and close to 0.4 nm^{-1} for the latter). However the coercivity and anisotropy field approximately doubled for the Co-Cr-Pt alloys, but rather decreased by about 20% for the Co-Cr-Ta alloys. Thus there was no clear concomitant changes of magnetic properties with stacking fault density or alloy content in general. One can only conclude at this stage that the role of stacking faults remains unclear, and that some general trend (e.g. either increasing or decreasing stacking fault populations) does not necessarily influence the magnetic performance one way or the other.

Grain Boundaries and Segregation

The grain boundaries themselves represent perhaps the most interesting microstructural variable. Not only are there random high angle and low angle boundaries (where cobalt grains, grown on adjacent, random chromium grains, meet), but also the 90° boundaries in bicrystal domains (see Fig. 1) and even 0° boundaries (where two parallel cobalt crystals, nucleated on the same underlying chromium grain, meet). The latter can be recognized in high-resolution images by a slight offset of the lattice fringes and by discontinuous stacking faults. One would anticipate decreasing grain boundary energy associated with this sequence, and also decreasing effect on the magnetism.

Several groups have shown that sputtering at slightly elevated temperatures (e.g. up to 250°C) brings about chromium segregation to the grain boundaries. The conventional means of study involves compositional profiling by energy dispersive X-ray analysis utilizing nano-scale electron probes in a field-emission gun TEM. For instance Tang et al. [10] showed that in a Co-Cr-Ta alloy (10% Cr, 6% Ta), an average grain boundary content of 18% Cr was achieved for 250°C sputtering, with associated Cr depletion in the near-grain boundary region. Interestingly the third element (Ta in this case) does not show any significant segregation. Cobalt alloys with this high level of alloying become paramagnetic, and so it is thought that the segregation leads to magnetic decoupling of the grains. This is manifested in TEM Lorentz images in a refinement of the vortex structure of demagnetized samples [10], and in the recording properties by improved signal-to-noise ratio. Wittig and coworkers are making significant progress in associating the degree of segregation with the nature of the grain boundaries as described above, which also requires detailed painstaking work [11].

Perhaps even more impressive is the application of energy filtered imaging for this type of study, as first demonstrated by the Hitachi group [12]. Cobalt jump ratio images (i.e. those formed by the ratio of the electron signal at and adjacent to the cobalt $L_{2,3}$ energy loss edge) clearly show a cobalt depletion delineating the grain boundaries. Likewise the Cr jump ratio images are complementary, with the grain boundaries highlighted by the enhanced Cr signal [11]. Converting these impressive pictures to quantitative form involves careful background subtraction. By using a four-window method, Bentley et al. have shown that the degree of segregation determined by this electron energy loss method is consistent with that established by X-ray energy dispersive spectroscopy [13]. This is indeed an important development in fully characterizing the magnetic media microstructure.

SUMMARY AND CONCLUSIONS

Transmission electron microscopy is an indispensable tool for the microstructural characterization of magnetic longitudinal recording media. Grain size averages and their distribution require careful application of imaging, often involving high resolution methods, but a reasonable statistical analysis may be obtained from only one or two hundred grains. Knowledge of grain boundary segregation can be supplied by energy dispersive and now energy filtering techniques, and the results appear to be closely equivalent. This information can be combined for micromagnetic modeling of the media, and for providing the manufacturing engineers with data as to the parameters important to improving the recording performance. In short, this is a fascinating application of TEM in a highly competitive contemporary technology.

ACKNOWLEDGMENTS

This work would not have been possible without the support and collaboration of several industrial colleagues, and without the contributions of former students and researchers in our group. In particular we appreciate the contributions of T. Yamashita and G. Bertero (Komag Corp.), G. Rauch, R. Ranjan and R. Ristau (Seagate Technology), S. Malhotra, B. Lal and M. Russak (HMT Technology), and M. Doerner (IBM).

REFERENCES

1. T. P. Nolan, R. Sinclair, R. Ranjan and T. Yamishita, IEEE Trans. Magn. **29**, pp. 292-299 (1993).
2. M. Mirzamaani, C. V. Jahnes and M. A. Russak, J. Appl. Phys. **69**, pp. 5169-5171 (1991).
3. R. A. Ristau, Ph. D. Dissertation, Lehigh University, 1998.
4. S. McKinlay, N. Fussing and R. Sinclair, IEEE Trans. Magn. **32**, pp. 3587-3589 (1996).
5. S. McKinlay, Ph. D. Dissertation, Stanford University (in preparation).
6. D. T. Carpenter, J. M. Rickman and K. Barmak, J. Appl. Phys. **84**, pp. 5843-5854 (1998).
7. Y. Uesaka, Y. Takahashi, Y. Nakatani, N. Hayashi and H. Fukushima, J. Magn. Magn. Mater. **174**, pp. 203-218 (1997).
8. H. S. Chang, K. H. Shin, T. D. Lee and J. K. Park, IEEE Trans. Magn. **31**, pp. 2731-2733 (1995).
9. A. Ishikawa and R. Sinclair, IEEE Trans. Magn. **32**, pp. 3603-3607 (1996).
10. K. Tang, M. E. Schabes, C. A. Ross, L. He, R. Ranjan, T. Yamishita and R. Sinclair, IEEE Trans. Magn. **33**, pp. 4074-4076 (1997).
11. J. E. Wittig, T. P. Nolan, R. Sinclair and J. Bentley, Mater. Res. Soc. Symp. Proc. **517**, pp. 211-216 (1998).
12. K. Kimoto, T. Hirano and K. Usami, J. Elect. Micro., **44**, pp. 86-90 (1995).
13. J. Bentley, J. E. Wittig and T. P. Nolan, Microscopy and Microanalysis, **5**(Suppl 2) pp. 634-635 (1999).

ELECTRON HOLOGRAPHY OF NANOSTRUCTURED MAGNETIC MATERIALS

R. E. DUNIN-BORKOWSKI[a,b], B. KARDYNAL[c,d], M. R. MCCARTNEY[a],
M. R. SCHEINFEIN[e,f], DAVID J. SMITH[a,e]
[a] Center for Solid State Science, Arizona State University, Tempe, AZ 85287-1704
[b] Now at: Department of Materials, University of Oxford, Parks Road, Oxford OX1 3PH, UK
[c] Center For Solid State Electronics Research, Arizona State University, Tempe, AZ 85287-6206
[d] Now at: Clarendon Laboratory, University of Oxford, Parks Road, Oxford OX1 3PU, UK
[e] Department of Physics and Astronomy, Arizona State University, Tempe, AZ 85287-1504
[f] Now at: FEI, 7451 NW Evergreen Parkway, Hillsboro, OR 97124

ABSTRACT

Off-axis electron holography and micromagnetic calculations that involve solutions to the Landau-Lifshitz-Gilbert equations are used to study magnetization reversal processes in lithographically patterned submicron-sized Co and Co/Au/Ni magnetic elements.

INTRODUCTION

A detailed understanding of magnetic nanostructures is essential for their utilization in information storage applications such as high density recording media and read heads. Such applications require reproducible magnetic domain structures and a good understanding of the interactions between neighboring elements. Here, we use off-axis electron holography in the transmission electron microscope (TEM) [1] to study the magnetic microstructure of submicron-sized elements that have been patterned lithographically onto electron transparent Si_3N_4 windows. We then compare our results with solutions to the Landau-Lifshitz-Gilbert equations [2]. The microscope geometry for off-axis electron holography and the approach used to obtain the phase of the holographic interference fringes, which is sensitive to the in-plane component of the magnetic induction integrated in the incident beam direction, are shown in Figs. 1a and 1b, respectively. Further details about the application of the technique can be found elsewhere [3].

Off-axis electron holograms were recorded at 200 kV using a Philips CM200-FEG TEM equipped with a field-emission electron source, an electrostatic (rotatable) biprism located in the selected-area aperture plane and a 1024×1024 pixel Gatan 794 multi-scan CCD camera. An additional Lorentz minilens ($C_S = 8m$ and 1.2 nm line resolution at 200 kV), located in the bore of the objective lens pole-piece, allowed images to be obtained with the main objective lens switched off and the sample located in almost field-free conditions. The objective lens could also be excited slightly and the sample tilted by up to ±30° in order to apply a known in-plane magnetic field, allowing magnetization processes to be followed *in situ* through hysteresis cycles. Reference holograms were acquired from the adjacent silicon nitride to remove artifacts caused by local irregularities of the image/recording system, and the mean inner potential contribution to the holographic phase was always subtracted to obtain the magnetic contribution of primary interest [3].

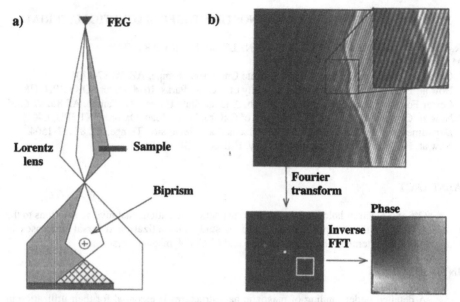

Fig.1. a) Setup used to generate off-axis electron holograms. Field emission source provides coherent illumination and electrostatic biprism causes overlap of object and reference waves. Lorentz lens allows imaging in close-to-field-free conditions. b) Off-axis electron hologram from thin crystal with enlargement showing interference fringes within sample. Fourier transform of hologram and phase obtained after inverse Fourier transform of sideband are also shown.

RESULTS

We begin by comparing holographic data from two 30nm-thick Co nanostructures with simulations. Figure 2a shows a representative hologram taken from one end of an ordered, close proximity (170nm) array of elements (Fig. 2b), whose intended cross-sectional geometry is given in Fig. 2c. The measured magnetic contribution to the holographic phase for a hysteresis loop is shown in Fig. 2d for an in-plane field of between ±1930 Oe and an out-of-plane field of 3600 Oe. The holographic phase contours follow lines of constant magnetization, and their separation is proportional to the in-plane component of the magnetic induction integrated in the incident beam direction. Computed results are shown in Fig. 2e for initial conditions that were selected to best match the data [5]. The vortex helicities match those measured and almost all of the S-shaped domain structures are reproduced. However, the fact that differences in the starting state are of importance in the formation of subsequent domain structures is illustrated by Fig. 2f, in which the vortex unrolls in the opposite sense in the second half of the cycle after changing the initial conditions slightly. Simulations also showed that the domain structures are sensitive to the out-of-plane component of the applied field and to interactions between neighboring elements [5].

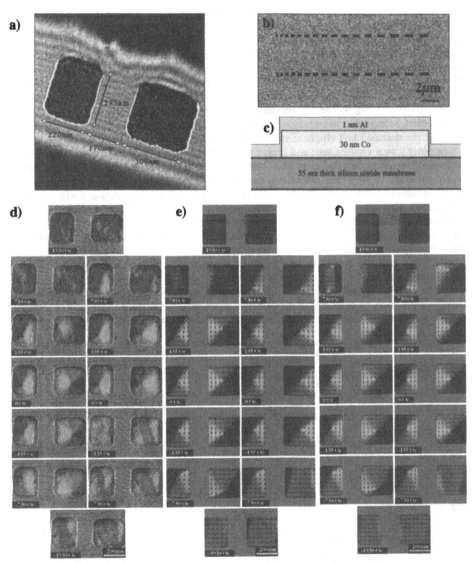

Fig.2. a) Off-axis electron hologram of 30 nm thick Co elements which form extreme left end of top linear array shown in b); c) Intended cross-sectional geometry of each element; d) Magnetic contributions to holographic phases during hysteresis cycle for 3600 Oe out-of-plane field and applied in-plane fields indicated. Phase contour spacing is 0.21π radians; e) Simulation for initial S-state (large cell) and C-state (small cell) at start of cycle, and 3600 Oe field into page; f) As for e) but with very slight change to starting state and to curvature of corners of each element.

15

Magnetic interactions between two thin, closely-separated ferromagnetic layers within individual lithographically defined spin valve or tunnel junction structures can also influence their switching mode and coercive field. Here, we examined Co/Au/Ni trilayers patterned into diamonds, ellipses and rectangular bars, which were separated laterally in order to reduce inter-element interactions. Figure 3a shows a hologram of two rectangular bars, which form the left end of the middle array shown in Fig. 3b. The intended cross-sectional geometry of each element is given in Fig. 3c. Results for the second largest element in each row are tabulated in three sets of four columns in Fig. 4. The left column in each set shows the magnetic contributions to the experimental holographic phases recorded during a complete hysteresis cycle. The switching fields of the diamond and elliptical shaped elements are smaller than for the rectangular bar, which never forms a vortex state and is also too narrow to form end-domains (which govern the reversal of larger rectangular elements). Instead, the phase contours curve at their ends by a maximum angle of ~45° just before magnetization reversal. The experimental phase contours have two distinct spacings in each element (narrower at higher applied fields and wider close to remanence). This observation provides direct evidence for the presence of ferromagnetic and antiferromagnetic coupling between the Ni and Co, which can be understood with reference to the simulations shown in the remaining columns of Fig. 4. The columns labeled 'Co' and 'Ni' track the magnetization states of the individual magnetic layers within each element, while those labeled 'Total' show the computed holographic phase shifts, which can be compared directly with the experimental data. Changes in the 'Total' contour spacing are apparent between fields at which the Ni layer has reversed but the Co layer is still unchanged, and similar behavior is seen clearly in the form of steps in the experimental hysteresis cycles shown in Fig. 5, which are plotted in the form of the change in the magnetic contribution to the phase across each element.

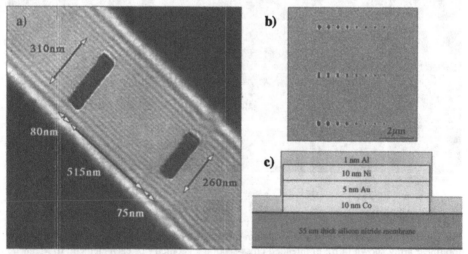

Fig.3. Off-axis electron hologram of two patterned bar-shaped elements of nominal layer sequence 10 nm Co/ 5 nm Au/ 10 nm Ni, also shown at left hand end of linear arrays of diamonds, bars and ellipses in b); c) Schematic diagram of intended cross-sectional geometry.

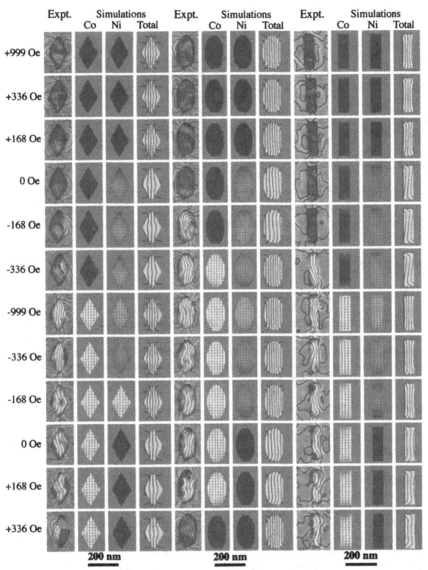

Fig.4. Experimental hysteresis cycles for second largest Co/Au/Ni diamonds, ellipses and bars in Fig. 3b, for average out-of-plane field of 3600 Oe and in-plane fields indicated applied along major axis of each shape. Contour spacing is 0.064 π radians. Simulations show magnetization in Co and Ni layers separately, as well as computed phase contours for direct comparison with data.

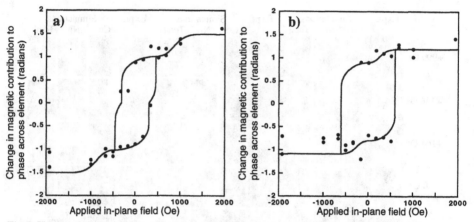

Fig.5. Hysteresis loops deduced from magnetic contributions to holographic phases of elements shown in Fig. 3 for a) largest diamonds and b) largest rectangular bars.

CONCLUSIONS

Magnetization reversal processes in submicron-sized magnetic elements have been followed using off-axis electron holography in the transmission electron microscope. Comparisons with micromagnetic simulations showed that the reproducibility of an element's domain structure in successive cycles is affected both by the out-of-plane component of the applied field and by the exact details of its starting magnetic state. Closely-separated magnetic layers within individual elements were observed to couple to each other antiferromagnetically, and narrow rectangular bars were found to reverse without the formation of end domains.

ACKNOWLEDGMENTS

We thank J. Speidell for the SiN membranes. The work was partly supported by an IBM subcontract on the DARPA Advanced MRAM Project (Contract No. MDA-972-96-C-0014).

REFERENCES

1. A. Tonomura, Ultramicroscopy **47**, 419 (1992).
2. LLG Micromagnetics Simulator is available commercially. See www.dancris.com/~llg.
3. David J. Smith and M.R. McCartney, in *Introduction to Electron Holography*, edited by E. Volkl, L.F. Allard and D.C. Joy (Plenum, New York, 1998).
4. R.E. Dunin-Borkowski, M.R. McCartney, B. Kardynal and David J. Smith, J. Appl. Phys. **84**, 374 (1998).
5. R.E. Dunin-Borkowski, M.R. McCartney, B. Kardynal, David J. Smith and M.R. Scheinfein, Appl. Phys. Lett. **75**, 2641 (1999).

FLUX MAPPING AND MAGNETIC BEHAVIOR OF GRAIN BOUNDARIES IN Nd-Fe-B MAGNETS

V.V. VOLKOV AND YIMEI ZHU
Dept.of Applied Science, Brookhaven National Laboratory, Upton, NY 11973, volkov@bnl.gov

ABSTRACT

Advanced Fresnel- & Foucault-Lorentz microscopy were applied to analyze magnetic behavior of the grain boundaries in Nd-Fe-B hard magnets. *In-situ* TEM magnetizing experiments combined with these imaging methods revealed the process of magnetization reversal in polycrystalline sintered and die-upset Nd-Fe-B under various magnetic fields. Fine details of magnetic flux distribution, derived from the magnetic interferograms created by phase-coherent Foucault imaging, provide a quantitative description of the local variation of magnetic flux. Our study suggests that the grain boundaries play an important multi-functional role in the reversal of magnetization, by acting as (a) pinning centers of domain walls, (b) centers of nucleation of reversal domains, and (c) sinks or sources for migrating magnetostatic charges and/or dipoles. They also ensure a smooth transition for irreversible remagnetization in polycrystalline samples.

INTRODUCTION

Anisotropic Nd-Fe-B hard magnets have received considerable scientific attention because of their importance in the technology of permanent magnets. The range of the applications is expected to grow rapidly as their properties and cost-effectiveness are improved. Although the high remanence of anisotropic permanent magnets is known to be closely related to grain texture and alignment, the coercivity mechanisms and their relation to microstructure such as grain size, grain boundary (GB) structure and admixtures of secondary phases are still elusive. Therefore, a nanoscale characterization of GB properties and their magnetic behavior serves an important step towards the understanding of coercivity mechanisms in anisotropic hard magnets.

Different TEM methods of magnetic imaging, based on Lorentz microscopy principles, have been developed recently to analyze magnetic materials. However, only a few of them are quantitative and can be applied in real time, which is crucial for addressing magnetization reversal mechanisms in Nd-Fe-B magnets. A unique tool may be the novel phase-coherent Foucault (PCF) microscopy [1, 2]. This method does not require any special attachment such as a bi-prism in electron holography. The PCF-images can be acquired in any TEM with a coherent source. They display in-plane local distribution of magnetic flux. If the variation of sample thickness is negligible, the PCF-images can be easily converted to a local induction map.

Another purpose of our work was to demonstrate how *in-situ* TEM magnetizing experiments reveal GB magnetic behavior in Nd-Fe-B magnets under a variable external field. We demonstrate that dynamic magnetic imaging, combined with microstructural analysis, can yield a better understanding on how magnetic behavior of GBs might be related to coercivity of Nd-Fe-B magnets.

EXPERIMENTAL

Two types of anisotropic hard magnets were used: sintered $Nd_{15}Fe_{78.5}B_{6.5}$ and die-upset $Nd_{13.75}Fe_{80.25}B_6$. The nominal compositions of the samples had a slight excess of Nd over the stoichiometry of $Nd_2Fe_{14}B$ and both had high energy-products. Their fabrication procedures were published in [3, 4] and [5-8], respectively. *In-situ* TEM was carried out using a JEM FEG

19

3000F microscope under the free-lens control mode. The internal magnetic field (0.02~3T) of the TEM was carefully calibrated by SQUID [9] and Hall probe [10] measurements.

RESULTS AND DISCUSSION

A. Magnetic flux mapping

It is essential that any method used for image analysis of local domain structure of magnetic material is simple, reliable and quantitative. The novel approach based on magnetic interferograms, created by PCF-imaging, complies with most of these demands. Recently, PCF microscopy was applied for induction mapping of closure domains in soft Fe- and Co-films [1, 2]. In the present work, we modified this method to analyze the hard magnets. Fig.1 illustrates the principles of magnetic-flux observation and PCF-interferograms. We assumed that all the interactions of electron waves with a sample could be explained by their interactions with electromagnetic fields. Then, from the Schrödinger equation the phase shift ($\Delta\theta$) between the two electron beams $\Psi = R \exp(i\theta)$, where $\theta = \varphi / \hbar$, can be defined as [1]

$$\Delta\theta = \Delta\varphi / \hbar = \left(\sqrt{2me} / \hbar\right)\oint \sqrt{V} ds \ - \ (e/\hbar)\oint \mathbf{A} ds = \left(\sqrt{2me} / \hbar\right)\oint \sqrt{V} ds \ - \ (e/\hbar)\int \mathbf{B} ds, \quad (1)$$

Here, m and e are the electron mass and charge, and $\hbar = h / 2\pi$ is Planck's constant. V and **A** are the scalar and vector potentials (div **A** = 0), representing the electrostatic and magnetic field contributions. The contour integral of **A** is taken for a closed path along two electron trajectories, and the integral of **B** is performed for the normal component of flux density (B_n) over the surface enclosed by the two electron paths.

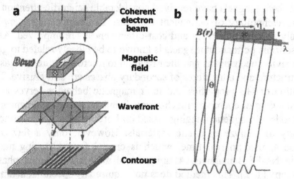

Fig. 1 (a) Principle of magnetic-flux observation, and (b) phase-coherent Foucault imaging.

In a FEGTEM, the fringe pattern (interferogram) formed by the phase shift between intersecting coherent electron waves can be well described by the Eq.(1), which can be reduced to the following one, assuming a uniform film thickness (t) and absence of electrostatic stray field:

$$\Delta\varphi / \hbar = \left(\sqrt{2meV}/2\hbar\right)\cdot(V_0/V)\cdot t + (e/\hbar) \int \mathbf{B} \cdot ds \approx C\cdot V_0 \cdot t + (e/\hbar)\cdot B_n \cdot \Delta S \quad (2)$$

where V_0 is the inner potential. It is assumed here that $V_0 \approx 20V$, $V_0 \ll V = 300$ kV and $C = \sqrt{me/2V}$ is a constant. The first term in Eq.(2) accounts for the contribution of film thickness to phase shift, the second is known as the Aharonov-Bohm shift [11]. Assuming that B-field is available only in the film, the integral in Eq.(2) transforms to $\Delta\Phi(x,y) = B_n\Delta S$, where $\Delta S = (r \cdot t)$, as shown in Fig. 1b. When two electron beams enclose an elementary flux $d\Phi =$

h/e, their phase shift in Eqns.(1-2) due to the magnetic field reaches $\Delta\theta = 2\pi$, which gives the maximum fringe intensity in the interferogram.

A similar result can be derived from a classical approach by taking into account the deflection angle θ of the electron beam due to the action of Lorentz force by magnetic domains:

$$\sin\theta \approx \theta = e \cdot B_n \cdot t / mv = e \cdot B_n \cdot t \cdot \lambda / h \qquad (3)$$

where v and λ are the velocity and wavelength. The Lorentz deflection angle of the electron beam is very small $\sin\theta \approx \theta = 4 \cdot 10^{-5}$ rad in our TEM experiments (electron wavelength is $\lambda = 0.02$ Å at 300 kV). The second necessary equation we derive from the interference maxima condition (Fig. 1b) as

$$\eta \cdot \sin\theta = \lambda \qquad (4)$$

where $\eta(x,y)$ is the fringe spacing, corresponding to a local 2π phase-shift between the coherent electron beams. It follows from (3) and (4) that for a single fringe spacing η we find

$$D_\eta(x,y) / t = 1/(\eta \cdot t) = (e/h) \cdot B_n(x,y) \qquad (5)$$

where $D_\eta = 1/\eta$ is a fringe density (number of fringes per unit surface area). Further from Eq.(5), for the n-th interference maximum with coordinate $r(x,y) = n \cdot \eta$ (n-integer) the expression for magnetic flux transforms simply to

$$\Delta\Phi = B_n \cdot \Delta S = B_n \cdot (n\eta \cdot t) = B_n \cdot (n\frac{h}{e \cdot B_n \cdot t} \cdot t) = n \cdot \frac{h}{e} \qquad (6)$$

where h/e – is a single quantum of magnetic flux, corresponding to each fringe spacing.

Eqns.(5-6) open a very effective, easy way for interpretation and nanometer scale mapping of magnetic flux $\Phi(x,y)$ on the basis of PCF-images, as we can visualize a single fluxon $d\Phi = h/e = 4 \cdot 10^{-15}$ Wb (Eq.6) flowing between two adjacent contour lines. A PCF-image thus provides direct quantitative flux mapping $\Phi(x,y)$. The vector of magnetic flux density $B_n(x,y)$ follows the local fringe line direction and its local amplitude is defined by the ratio of local fringe density to a film thickness as $D_\eta(x,y)/t = 1/(\eta t) = (e/h)B_n(x,y)$ (Eq.(5)). Hence, the fringe density map normalized to film thickness $D_\eta(x,y)/t(x,y)$ gives a direct quantitative $B_n(x,y)$ inductance mapping .

Fig. 2 shows the experimental in-focus PCF-image of $Nd_2Fe_{14}B$. It contains much more magnetic and structural information than an out-of-focus Fresnel image. Under a properly collimated illumination angle of $4*10^{-7}$ rad, the spatial coherence length of ≤ 5 µm ("the width of interferogram") can be achieved at 300 kV. The black and white contrast of magnetic domains here is similar to one created by non-coherent Foucault imaging. It shows the opposite components of their magnetization. As mentioned above, the interferogram in Fig. 2 directly represents a map of magnetic flux $\Phi_1(x,y)$ for the white-contrast domains, provided the thickness variations of the film are sufficiently small ($\Delta t < 60$ nm) to cause an additional 2π-phase shift in the fringe pattern. A similar interferogram, $\Phi_2(x,y)$, can be obtained for the black domains after switching the illumination conditions to complementary ones. Superposition of $\Phi_1 + \Phi_2 = \Phi(x,y)$ creates a complete map of magnetic flux in a sample. The fringe spacing, directly measured from Fig. 2, yields $\eta = 43$ nm. By taking into account the magnetic moment of a single domain in $Nd_2Fe_{14}B$ ($B_n \approx I_s = 1.6$ T) we get from Eq.(5) a reasonable estimated film thickness $t \approx 60$ nm. It suggests that from 1π phase-shift criteria the spatial resolution of typical magnetic mapping by the PCF-technique is about ~20nm and may be a function only of the film

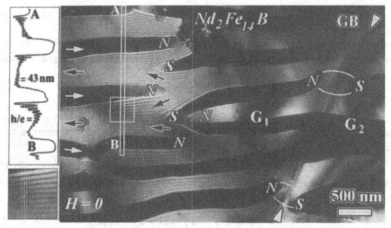

Fig. 2 Phase-coherent Foucault image (magnetic interferogram) of NdFeB-magnet near the grain boundary area. The density of fringes and their directions are directly associated with magnetic flux density $B_n(x,y)$ distribution. The insets show the intensity profile along A-B line.

thickness. We note that fringes in Fig. 2 disappear at the domain tips, where the magnetic flux leaves the sample and creates the magnetic poles (N/S charges). These charges concentrate near the GB like simple N/S dipoles or form metastable multipole's configurations ...N/S/N/S... running across the domain structure. In both cases, the magnetostatic energy is decreased. As a result, a complex magnetic-field distribution appears near the GB (Fig. 2). Notice that for real closure domains, the number of fringes (n) within any fringe contour should remain constant because of the absence of magnetic flux dissipation.

B. In-situ TEM magnetizing experiments

The nanoscale magnetic behavior of grain boundaries and the magnetization reversal in die-upset and sintered magnets were found to be complex and ruled by different mechanisms.

Die-upset (DU) magnets. Many general features of the grain boundary magnetic behavior can be derived directly from magnetic-sensitive conventional Foucault images as shown in Fig3.

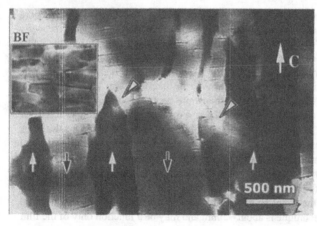

Fig.3 Non-coherent Foucault image of large "interacting /cluster" domains, running across the platelet-like grain texture in DU-magnet along the easy (die-upset) direction marked with c-arrow, showing remanence state of DU-magnet after saturation along hard direction. The local polarization of magnetic domains is marked by small black and white arrows, and the pinning centers by arrowheads. Inset is a bright-field image.

Notice that the tips of domains with opposite magnetization are often pinned by defects or locally misaligned grains of Nd-Fe-B hard magnetic phase. A few very small reversal domains were observed in the remanent state after magnetization of DU-magnet along the easy magnetic (or die-upset) direction. Their presence is associated with local imperfections, such as secondary phase inclusions or large-angle GBs, similar to those shown in Fig.3. *In-situ* TEM magnetizing experiments confirmed that small reversal (or negative) domains, associated with non-aligned grains, continue to grow slowly under the increasing negative field until they nucleate a c-axis aligned reversal domain at the nearest GB. The magnetization reversal in a DU-magnet then transforms rapidly to an irreversible process, until only the pinning of the domain walls at GBs controls the further expansion of reversal domains as shown by successive images (Fig. 4a) captured on video and recorded at different points of the magnetic hysteresis loop (Fig.4b). Thus, the leading mechanism in magnetization reversal of DU-magnets seems to be the nucleation of reversal domains, preferentially at misaligned GBs and interfaces or sample surface where the demagnetizing field is the largest. Fig.4 shows also an example of strong pinning/trapping center (Fig.4a, Frame-3) not vanished at strong negative field $H << -H_c$ (Fig.4b) and its structure model (Fig.4c). This center was found to be responsible for the nucleation of positive domains at $H \geq +H_c$ in the 4-th quadrant of the hysteresis loop (frames 4-6, Fig.4a), starting from negative magnetization. The presence of a nonmagnetic "pocket" phase at the GBs [12,13] is the necessary condition for "strong pinning" by such a defect center. It reduces the high density of magnetic charges (Fig.4c) by spreading them around this GB-buffer layer.

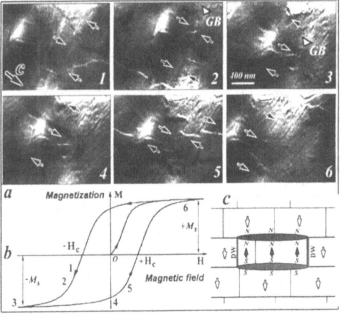

Fig.4 Successive Fresnel images (a) of domain structure in DU-magnet taken at different points (1-6) of the hysteresis loop (b) of the Nd-Fe-B - hard magnet.

Magnetization directions of local domains are marked by small arrows and the fine line-contrast perpendicular to these arrows are the grain boundaries.

(c) Structural model showing the origin of strong pinning/trapping center (fig. a: Frame-3) not vanished at strong negative field $H \leq -H_c$ (fig. b: point-3).

Sintered (SI) magnets. Our *in-situ* TEM observations revealed that the irreversible stage of remagnetization in SI-magnets is often associated with GBs. Fig.5 serves an example showing a domain-splitting mechanism responsible for the reverse domain nucleation near the GBs. It is an essentially irreversible process - the nucleation of negative domains takes place via several sudden splitting of positive domains (Fig.5). Here, a positive domain (marked with black arrow) splits at the GB into "positive-negative-positive" domains under a moderate negative magnetic field. The newly formed domain configuration encloses a pair of new 180°-domain walls, which facilitates further remagnetization of magnet by a simple slow motion of the domain walls.

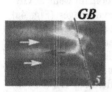

Fig.5 Successive images (5-8) recorded from video showing reversal domain nucleation at the GB in Nd-Fe-B SI-magnet via splitting of domain with opposite magnetization.

On the other hand, this process may be considered as a cascade-like discharge of magnetic poles (or charges) at the GBs. The domain-splitting mechanism can well explain some experimental observations. For instance, the fact that a high concentration of secondary phases in SI-magnets cannot greatly improve the coercivity is because the classical mechanism of domain wall pinning by defects is no longer the only mechanism for magnets with a high density of domain walls.

ACKNOWLEDGEMENT

This research was supported by the U.S. Department of Energy, Division of Materials Sciences, Office of Basic Energy Science, under the Contract No. DE-AC02-98CH10886.

REFERENCES

[1] A. Tonomura, Surf. Sci. Rep., **20**, 317 (1994).
[2] J.N. Chapman, A.B. Johnston, L.J. Heyderman, S. McVitie and W.A.P. Nicholson, IEEE Trans. Magn., **30-6**, 4479 (1994); A.B. Johnston, J.N. Chapman, B. Khamsehpour and C.D.W. Wilkinson, J. Phys. D: Appl. Phys. **29**, 1419 (1996).
[3] M. Sagawa, S. Fujimura, N. Togawa, H. Yamamoto, Y. Matsuura, J. Appl. Phys. **55**, 2083 (1984).
[4] J. Hu , Y. Liu , M. Yin , Y. Wang , B. Hu, Z. Wang, J. of Alloys and Comp., **288**, 226 (1999).
[5] R. W. Lee, Appl. Phys. Lett. **46**, 790 (1985).
[6] R. W. Lee, E. G. Brewer, N. A. Schaffel, IEEE Trans. Magn. MAG-21, 1958 (1985).
[7] C. D. Fuerst, E. G. Brewer, J. Appl. Phys. **73**, 5751 (1993).
[8] V.V. Volkov, Y. Zhu, J. Appl. Phys. **85**, 3254 (1999).
[9] V.V. Volkov, D. C. Crew, Y. Zhu, L. H. Lewis, Proc. "Microscopy and Microanalysis", Vol. 5, Suppl. 2, Portland, Oregon, Aug.1-5, 1999, p. 46-47.
[10] V.V. Volkov, Y. Zhu. Unpublished.
[11] Y. Aharonov, D. Bohm, Phys. Rev. **115**, 485 (1959).
[12] R. K. Mishra, J. Appl. Phys. **62**, 967 (1987); R. K. Mishra, Mater. Sci. Eng. **B 7**, 297 (1991).
[13] V.V. Volkov and Y. Zhu, submitted to J. Magn. Magn. Mater., Nov., 1999.

Crystallography and Defects

LEEM INVESTIGATIONS OF BCC METALS GROWN HETEROEPITAXIALLY ON SAPPHIRE.

W. Swiech, M. Ondrejcek, R.S. Appleton, C.S. Durfee, M. Mundschau and C.P. Flynn
University of Illinois at Urbana-Champaign, 104 S. Goodwin, Urbana, IL 61801, USA

ABSTRACT

We describe studies of refractory metals Mo and Nb, grown heteroepitaxially in the (011) orientation on sapphire (11$\bar{2}$0), using low energy electron microscopy (LEEM). A wide variety of structural and dynamical phenomena are observed and recorded as video sequences. These are organized here in five categories as follows. First are surface impurity phases identified by LEED. It is an important point, established here for Mo, that almost ideally clean surfaces with mainly single stepped terraces, can be prepared on thin films for use in surface science. Second, surface reconstruction phenomena can also be identified by diffraction. This is illustrated here by a reconstruction of the Nb surface, with two equivalent domains that form a stripe phase, owing to competing long and short range interactions. Detailed studies of nucleation, fluctuations and the equilibrium of these variants are described. The third area is surface topography, illustrated here by almost ideal stepped surfaces of Mo and Nb at high temperatures, and by nanofaceting by step edge coalescence to form {110} facets on Nb at lower temperatures. The phase diagram for nanofaceting is discussed. Fourth are processes at the internal Al_2O_3-metal interface, as interfacial dislocations, misorientation steps and defects become visible in the LEEM image through the displacement fields they cause. The final category includes bulk process where the observed dynamics of threading screw and edge dislocations and accompanying slip traces reveal their release at high temperatures from Cottrell atmospheres. The kinetics are tracked as surface diffusion smooths dislocation slip traces, and the interactions among dislocations and surface steps are measured.

INTRODUCTION

Low energy electron microscopes have been successfully developed by Bauer [1] and Tromp [2] but still rather few operational machines exist worldwide. Impressive inroads have been made in pioneering the study of single crystal metal surfaces [3-5] and semiconductor surfaces [6-9], including surface topography [10-12] and reconstruction [13-14], bulk defects [15-16] and buried nuclei of strain [17]. The evolution of structure during annealing and growth in ultrahigh vacuum has been the focus of considerable interest [18-20]. The power of the technique is enhanced by the ability to perform both diffraction and bright and dark field studies of structure in ultrahigh vacuum and over a wide range of temperatures [10].

The work reported here has a separate direction. We are concerned with the heteroepitaxy of refractory bcc metals like Mo and Nb on sapphire. In addition to fundamental interest, two practical interests motivate exploration of these systems. The context is that sapphire substrates are readily available on a variety of planes and with selected vicinal miscut. On these surfaces the refractory metals can be made to grow as excellent single crystals in a variety of selected orientations [21]. The first opportunity is to employ the surfaces of heteroepitaxial deposits as samples for use in surface science experiments. Their potential advantages over bulk crystals are clearly defined: it is much simpler and cheaper to obtain a new epitaxial film on commercial sapphire than a new metal single crystal, and, in some cases, the film in a variety of orientations can be available already in ultrahigh vacuum, uncontaminated, and without need for bulk crystal preparation followed by sputtering and annealing cycles. The second main motivation for understanding refractory metals on sapphire is to enhance the practical use of sapphire as an important highway for heteroepitaxial synthesis. Many elements such as rare earths and noble metals react with sapphire at high synthesis temperatures, replacing Al, forming compounds like spinels, or entering as mobile interstitial ions. Refractory metals that do not react with the sapphire substrate provide stable, thermally durable new surfaces, which offer selection from a variety of alternative symmetries, thus providing templates for subsequent growth of heteroepitaxial deposits

27

that include compounds like L_{12} Cu_3Au and Laves phases (eg $DyFe_2$) [22-25] in addition to pure metals and alloys.

Refractory buffers of this type have been widely grown but incompletely characterized. RHEED studies are notably misleading as monitors of surface structure, ex situ transmission electron microscopy of defect structure risks structural degradation from intrusive sample preparation, and the Fourier transform involved in x-ray diffraction studies of structure largely obliterates the spatially resolved information required to understand epitaxial behavior. To a significant extent the use of LEEM overcomes these problems. The purpose of this paper is to illustrate the application of LEEM to the characterization of heteroepitaxial structure, using the example of refractory metals grown on sapphire.

In the experiments summarized here, five main areas of information about single crystal refractory metal thin films are explored. First is the investigation of crystal structure on the outer metal interface with the ambient vacuum, including the identification of contaminants present on the initial surface. In this area the text highlights the surface chemistry of Mo as surface impurity phases are removed to leave an almost ideal, clean surface with single steps, for use in surface experiments. Second is the investigation of surface reconstructions and superstructures. This we illustrate by a reconstruction stripe phase of Nb (011), that is believed to originate in competing long and short range interactions, and that exhibits an important sensitivity to the epitaxial state, notably strain. The third area concerns surface topography where a bewildering variety of phenomena appear, including at high temperature almost ideally stepped surfaces, with step configurations and fluctuations determined by line energy and stiffness; at lower temperatures terraces are observed to ripen on length scales up to tens of microns, with nanofaceting caused by step coalescence, strongly dependent on miscut. Here the phase diagram for nanofaceting of Nb (011) is described. The fourth area concerns the metal-Al_2O_3 interface. It turns out that interfacial steps and dislocations both can be imaged in LEEM owing to the displacement fields they cause, and that the observed phenomena look very different for the cases of Mo and Nb. The final area is bulk evolution under annealing where the motion of threading screw dislocations with slip traces and edge dislocation released from Cottrell atmospheres at high temperature are both clearly visible and can be followed in real time video sequences. Both exhibit interesting interactions with surface steps that have contours which smooth dynamically by surface diffusion. These five areas are summarized in five sections, explaining the results, and the paper finshes with a summary.

EQUIPMENT

The LEEM employed here was built by Tromp at IBM Yorktown Heights. It has been fitted with evaporation sources for epitaxial growth while the surface is under observation, and also more elaborate capabilities in a separate but connected chamber. Methods have been developed that permit LEEM (bright and dark field) and LEED observations on samples grown on insulating or conducting substrates, and maintained at temperatures above 1700K [26]. Samples of refractory metals on sapphire were grown by well-practiced methods that yield single crystal films of very high quality, by methods that are detailed elsewhere [21, 22, 24, 25]. Images obtained in LEEM are sensitive to features of experimental conditions that are too numerous for detailed mention in this summary, so reference should be made to the original reports for complete information.

PREPARATION OF CLEAN Mo (011)

The diffraction capabilities of the LEEM make it possible to identify and image areas that exhibit foreign atomic periodicities such as surface phases. In the present research, samples were prepared elsewhere in an MBE chamber, and carried through air to the LEEM where they were transferred into uhv using an introduction assembly. The initial materials thus comprised single crystal films contaminated by exposure to the atmosphere and subsequently brought once more into ultrahigh vacuum.

It proved possible to clean the Mo surfaces to produce a 1 x 1 diffraction pattern. Surface carbide and oxide phases were readily recognizable from earlier studies on bulk crystals [27-29]. Figures 1(a) and (b) show two-variant patterns for the carbide and oxide respectively, and Fig 1(c) shows the clean surface at 1090K after flashing to 1580K. Dark field images taken with carbide spots reveal the domain structures in Figs 1(d) and (e). The two hexagonal patterns are rotated by 10.5° from each other, and by ± 5.25° from Mo [100], as expected for Kurdjumov-Sachs orientations of the carbides. Only for the surfaces that initially are most contaminated by carbide is the coverage complete. More commonly the carbide is found in elongated islands (Fig 1(f)) oriented along the two <111> directions of Mo (011). From these variants originate the two diffraction pattern in Fig 1(a). They are lost after heating above 1660K, leaving the (1 x 1) surface of Fig 1(c) with the almost perfect appearance shown in Fig 1(g). For further details of phases observed during the cleaning of Mo (011) the reader is referred to the original report [26].

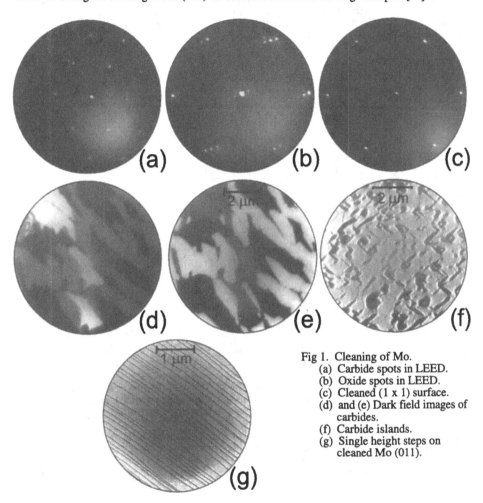

Fig 1. Cleaning of Mo.
(a) Carbide spots in LEED.
(b) Oxide spots in LEED.
(c) Cleaned (1 x 1) surface.
(d) and (e) Dark field images of carbides.
(f) Carbide islands.
(g) Single height steps on cleaned Mo (011).

RECONSTRUCTION OF Nb (011)-O, AND ITS STRIPE PHASE

The LEED capabilities allow surface reconstructions to be identified in the LEEM, and the equilibrium and evolution of their spatial arrangements to be explored in real time. This is illustrated by the O-induced reconstruction of Nb (011) grown on sapphire (11$\bar{2}$0). A small O-coverage is known [30, 31] to induce two symmetry-related variants of a surface reconstruction. We find by LEEM that the reconstructions are highly sensitive to epitaxial strain, they disappear from the O-covered surface above $T_c = 1500K$, they form a stripe phase in the range 1470-1500K, and at lower temperature symmetry is broken so that one variant is dominant. Just below the stripe phase the minority variant is thermally activated.

After thermal processing of a 500 nm film to 1700K in the LEEM studies, the Nb surface at 1420K exhibits the LEED pattern shown in Fig 2(a), which comprised superposed patterns from two symmetry-related variants that are mirror images in the [100] line. This is the known oxygen-induced effect . The stripe phase is illustrated by five images in Figs 1(b)-(f). The stripe pattern imaged by the LEEM in dark field, using a diffraction spot from one variant, is shown for two temperatures in Figs 2(b) and (c). The

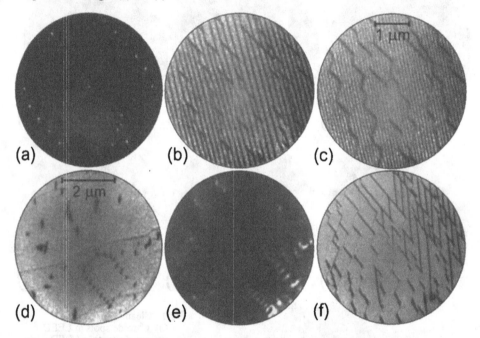

Fig 2. (a) Nb LEED at 1420K showing two variants. (b) and (c) Stripe phase in dark field at 1475K and 1485K. (d) and (e) Strain-induced variants below and above the temperature range of the stripe phase. (f) Thermally activated structure below the stripe phase.

stripe period is seen to shrink as T_c is approached. Figure 2(d) shows the favored variant everywhere at low temperature except where the minority variant occurs on defects, in this case edge dislocations arranged in a low angle grain boundary, and Fig 2(e) is the same region, but above T_c, when variants induced by strain fields mark the locations of the edge dislocations. In the thermally activated regime the minority variant often occurs in elaborate patterns, nucleating on step edges, as illustrated for a 50nm thick film in Figure 2(f).

The temperature dependence of the stripe period is shown in Fig 3(a) for two films with different thicknesses. Thicker films evidently display wider stripes. The mechanisms that contribute to the observed behavior are indicated in Fig 3(b). We believe that the stripes originate in the competition between long range elastic interactions and the short range repulsion associated with the interface between variants, much as in the Landau stripe phases of Type II superconductors; the general phenomenology of competing interactions has been reviewed[32]. In

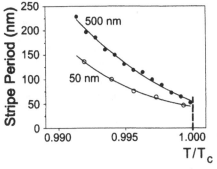

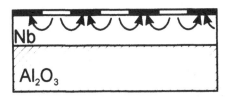

Fig 3. (a) The dependence of stripe width on reduced temperature T/T_c for two films. (b) Sketch of stripe phase indicating elastic interactions between variants limited by the stiff sapphire substrate.

the present case the attraction probably originates in a periodic shear that accommodates the differently oriented surface stresses of the two variants. The dependence on film thickness probably arises from the way the stiff sapphire modifies the shear field at its interface with the clamped film, since the stripe period is comparable with the thickness and the periodic shear field must die exponentially with depth below the surface. The explanation of the broken symmetry may lie in the anisotropic thermal expansion of the sapphire (1120) surface, which is correctly oriented to distinguish between variants. We mention that the stripe behavior is insensitive to added hydrogen and oxygen, other than in excess of 10^{-6} torr.

SURFACE NANOTOPOGRAPHY

LEEM contrast arises by interference caused by variations of surface height, and by diffraction from variations of atomic spacing. Consequently surface features like steps, facets and defects can be imaged at a resolution ~ 10 nm which may often be adequate for the spacings of interest. Here we describe behavior of surfaces in the absence of defects, deferring until Sections 6 and 7 observations related to dislocations, slip traces, etc. At the highest temperatures available in the LEEM, Mo and Nb films both exhibit almost ideal surface structures, with well-spaced monatomic steps. The step energies are fairly isotropic. Figure 4(a), for example, shows step-loops shrinking with reasonably circular shape even though their path is perturbed by dislocations. Below about 1600K the two metals exhibit differing behaviors in that the interacting step edges on the Mo surface tend to form step bunches only in the presence of oxygen contamination, whereas on Nb the steps coalesce to form {110} nanofacets. The comments here are confined to Nb and the phase diagram for nanofaceting of Nb (011).

When imaged in LEEM the high temperature surfaces are in vigorous motion on the length scale of the LEEM resolution. This arises from equilibrium populations of adatoms and advacancies on the terraces [33], and the statistical fluctuations of step edge location that accompany their emission and reabsorption. Both the line energy and the line stiffness of the steps can be deduced quantitatively from the Fourier transform of the fluctuating profiles [34]. Measurements to determine these quantities are currently in progress and will be reported in due course. At temperatures below about 1600K an attractive short ranged interaction overcomes the entropic repulsion [35] of steps on Nb (011) and causes them to coalesce. Nb is distinct in that its {110} surfaces have unusually low surface energy [36], and we have discovered that, as a consequence, the evolution of the vicinal Nb (011) surface is dominated by steps that coalesce to form {110} nanofacets [37,38]. The atoms of the bcc (011) surface are located in Fig 4(b), and Fig 4(c) shows three of the six {110} planes that intersect (011). Two intersect (011) at 90° along [01$\bar{1}$], and four at 60° along the close packed directions. These are the likely planes for {110} nanofaceting.

We have studied the evolution of vicinal Nb (011) surfaces miscut along [01$\bar{1}$] and [100] by as little as 0.1°; smaller miscuts are hard to create. When annealed below about 1600K, the step edges are seen to coalesce to form single lines along the facet directions identified above. The

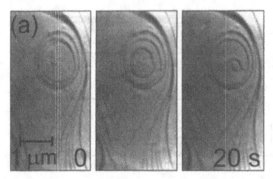

Fig. 4. (a) A sequence of images showing step loops shrinking while maintaining a reasonably circular shape. This indicates that the step energy is fairly isotropic. (b) Sketch showing atomic positions and principal directions of Nb (011). (c) Facet planes that lie at 60° and 90° to (01$\bar{1}$) are indicted, together with the lines along which they intersect the surface.

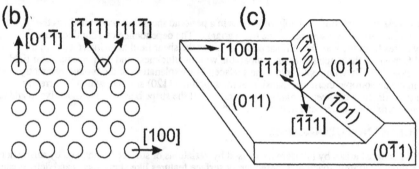

resulting images are shown in Figure 5. Figs 5(a) and 5(b) present images of surfaces miscut along [01$\bar{1}$] that have achieved facet heights of about 5 and 20 steps, respectively. The facets are not straight, which is consistent with the fact from symmetry that their facet energy can vary only in second order with azimuthal angle. The lower images show two alternative appearances of the surface miscut along [100]. These correspond to different facet lengths compared to the terrace width. Short facets produce terraces with serrated edges whereas long facets make for terraces with parallelogram shapes. The sketches associated with the images offer plausible interpretations in terms of nanofacets. By scanning tunneling microscopy we have confirmed that selected facets are steeper than 30°, even with broadening because of the tip shape. This is consistent with the inference from LEEM orientations that they are facets, as identified in Fig 4(c).

A wealth of additional details of nanofaceting behavior are available in LEEM images. In Fig 6(a), a change of surface gradient causes facets to dissociate into their constituent steps. This allows stable facets containing 3, 4, 5, etc steps to be identified. However, in the two visible cases in which two steps become entangled, they fail to follow the exact facet direction. This constitutes evidence that at least three steps must coalesce for a 60° facet to be stable. Presumably the positive line energies where the facet joins the terraces are large enough to make two steps markedly less stable. In addition, we observe that a temperature range exists in which dissociated steps and facets alternate to form a new 'periodically faceted' phase, imaged in Fig 6(b). This arrangement is favored over both dissociated steps and facets for a limited temperature range because the facets generally do not follow the contour lines of surface height. By balancing the relative lengths of step and facet segments, the structure is able to follow the local contour lines. These new states appear both on heating and cooling, and accordingly are believed to belong to the constrained equilibrium of the surface. This assignment has been confirmed by theoretical models [38]. The behavior observed here has been summarized in a speculative phase diagram for nanofaceting [37] presented below as Fig 6(c). In this diagram, facets act like compounds and dissociated steps like solutions, in an otherwise normal binary phase structure. The presence of eutectics at which sawtooth nanofacets transform directly into dissociated steps has been reproduced by theory [38].

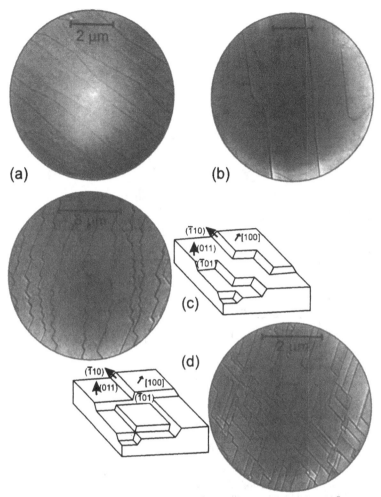

Fig 5. (a) and (b) show 90° nanofacets on surfaces miscut by 0.1° along [01$\bar{1}$], with heights of about 5 steps and 20 steps respectively. (c) and (d) show two configurations of 60° nanofacets that differ in the ratio af facet width to terrace width (see inset sketches).

INTERFACIAL PROCESSES

Defects at interfaces can be imaged in LEEM by means of the displacements their strain fields cause at the front surface of the sample[17,39]. An example in Fig 7(a) shows dislocations at the interface of a Mo sample 50 nm thick grown on sapphire. The apparent bulge of the front surface is about the same width as the sample is thick, which reflects the radial portion of the defect-induced displacement field. Defects of opposite sign give reversed contrast, and the contrast is reversed because the sample has been tilted between the two images shown. Weaker nuclei of strain can still be imaged. Figure 7(b) contains two images of a Mo film 50 nm thick. One has steps at fairly uniform spacings, while in the second image the steps are perturbed by a pinning center to reveal a broad terrace. Weak intensity modulations are visible on this terrace, and once identified, it is easy to detect them over much of the area of both images. The modulations are parallel to the steps on the front surface, and exhibit closely the same spacing. The modulation is

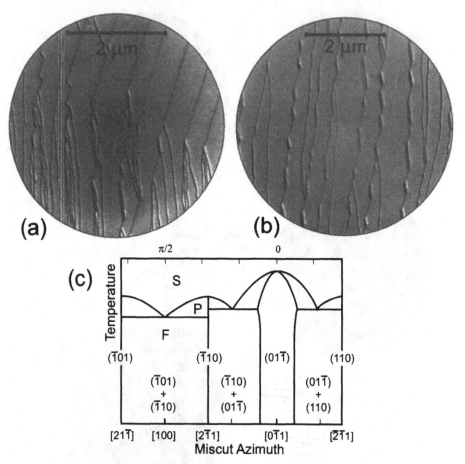

Fig 6. Nanofacets and periodic states. (a) Stable facets form from three or more step edges but not from just two steps. (b) The periodically faceted states shown form upon both cooling and heating, and correspond to equilibrium structures. (c) The suggested phase diagram for nanofaceting that is consistent with the LEEM observations.

the strain field at the front surface arising from vicinal miscut at the buried interface. The difference between the sapphire and Mo step heights at their interface causes strains with precisely the same periodicity as that of the steps on the front surface. At the Nb interface with sapphire the dislocations show much less geometrically rigid structure. Figures 7(c) reveal interfacial dislocations more as locally stretched metal with broad structures that occasionally terminate at a threading screw dislocation (marked by an arrow) which passes to the front surface where it initiates a new surface step edge.

DEFECT PROCESSES IN THE BULK

The thermal evolutions at the front surface and the interface are often linked by the threading edge and screw dislocations which are driven to slip in order to reduce epitaxial strain. These processes are clearly visible in LEEM, and may be observed at temperature as evolution is

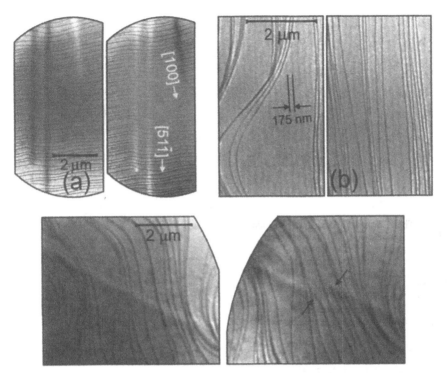

Fig. 7. (a) Edge dislocations of both signs at the Mo-Al$_2$O$_3$ interface, illustrating also contrast reversal by sample tilt (second image). (b) Misorientation steps at the Mo-Al$_2$O$_3$ interface are visible as a contrast modulation with the same mean spacing as surface steps, particularly at the wide terrace. (c) Interfacial dislocations at the Nb-Al$_2$O$_3$ interface evolve with time.

initiated[40]. Example are dislocations that are freed from the Cottrell atmospheres that pin them, or temperature changes that induce additional epitaxial strain. A record of the resulting slip is written in slip traces on the front surface, as illustrated in Fig 8(a). As time elapses, the diffusion of adatoms smooths the initially sharp intersection of slip traces with terraces, so that the time evolution of the relaxation process is recorded in the slip trace profiles. In fact, LEEM images can simultaneously detect the progress of the interfacial dislocation, the surface slip trace and the threading screw dislocation. In Fig 8(b) the dislocation is visible as a mark at the terminus of an added step edge, while the underlying interfacial edge dislocation appears as a broad hump beneath the slip trace. The geometry of the coupled defects is clarified by the sketch in Fig 8(c).

It is possible to track the evolution of the structure as the screw dislocation slips to accommodate strain. Video records of the progress reveal detailed aspects of the process which are often apparent in real time during the actual observations but which can be analyzed quantitatively during later examination of the video pictures. An example is provided in Figure 9, where slip of the structure in Fig 8(b) is tracked as a function of time lapse. During this interval the screw dislocation passes through many of the step bunches that are visible on the surface in Fib 8(b). It turns out the screw dislocation interacts with the step bunches and that the progress of slip is delayed as the screw lingers at a number of the surface steps features. This is illustrated in Fig 9, which is a record of screw velocity as a function of position, relative to the surface features. The important point for emphasis here is that dips in the slip rate shown in the figure frequently occur where the screw dislocation passes through a surface step bunch. These points of coincidence are

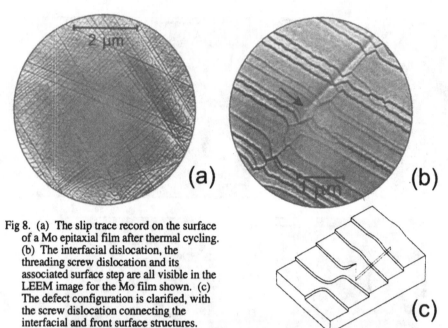

Fig 8. (a) The slip trace record on the surface of a Mo epitaxial film after thermal cycling. (b) The interfacial dislocation, the threading screw dislocation and its associated surface step are all visible in the LEEM image for the Mo film shown. (c) The defect configuration is clarified, with the screw dislocation connecting the interfacial and front surface structures.

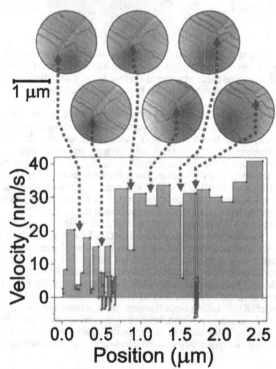

Fig 9. A graph of velocity as a function of position shows, for the defect in Figure 8, that the dislocation slip at 1440K is often hindered as the screw passes through a surface step bunch. The surface configurations at several prominent velocity minima are imaged above

documented above the main figure by LEEM images that locate the screw dislocation (arrows) relative to the steps. Thus the figure reveals, for the first time, how the interaction with surface step bunches delays the time evolution of a slip system as it relieves epitaxial strain in a thin single crystal film.

SUMMARY

Here we report initial LEEM studies of refractory metal films grown epitaxially on sapphire substrates. Significant results are reported in (i) the preparation of ideal systems for surface studies, (ii) in the investigation of surface reconstructions and their strain dependence, (iii) in surface nanotopography, of which Nb nanofaceting is described here, (iv) in the in situ examination of interfacial defects, and (v) in tracking bulk defect processes. A topic neglected here for reasons of space is the dynamical evolution of epitaxial systems, which can be pursued in real time and recorded on video tape. In light of the results reported here only for Mo and Nb samples miscut by 0.1° on a single principal sapphire orientation, it seems clear that the study of heteroepitaxy by LEEM offers boundless future opportunities.

ACKNOWLEDGMENTS

This research was supported in part by DOE grant DEFG02-96ER45439. The LEEM was purchased with funds from DOE and the University of Illinois, together with an equipment grant from NSF. The LEEM is maintained by the Material Research Laboratory.

REFERENCES

1. W. Telieps and E. Bauer, Ultramicroscopy **17**, 99 (1985); L.H. Veneklasen, Rev. Sci. Instrum. 63, 5513 (1992).

2. R.M. Tromp and M.C. Reuter, Ultramicroscopy **36**, 99 (1991); R.M. Tromp, M. Mankos, M.C. Reuter, A.W. Ellis and M. Copel, Surf. Rev. Lett. **5**, 1189 (1998).

3. W. Telieps, M. Mundschau and E. Bauer, Surf. Sci. **225**, 87 (1990).

4. M. Mundschau, E. Bauer and W. Swiech, Phil. Mag. A **59**, 217 (1989).

5. M.S. Altman, Surf. Sci. **344**, 65 (1995).

6. M. Mundschau, E. Bauer, W. Telieps and W. Swiech, Phil. Mag. A **61**, 257 (1990).

7. R.M. Tromp and M.C. Reuter, Phys. Rev. Lett. **68**, 820 (1992).

8. J.B. Hannon, N.C. Bartelt, B.S. Swartzentruber, J.C. Hamilton and G.L. Kellogg, Phys. Rev. Lett. **79**, 4226 (1997).

9. P. Sutter, E. Mateeva, J.S. Sullivan and M.G. Lagally, Thin Solid Films **336**, 262 (1998).

10. M. Mundschau, E. Bauer, W. Telieps and W. Swiech, Surf. Sci. **223**, 413 (1989).

11. K. Pelhos, J.B. Hannon, G.L. Kellog and T.E. Madey, Surf. Sci. **432**, 115 (1999).

12. F.-J. Meyer zu Heringdorf, D. Kahler, M. Horn-von Hoegen, T. Schmidt, E. Bauer, M. Copel and H. Minoda, Surf. Rev. Lett. **5**, 1167 (1998).

13. M.S. Altman and E. Bauer, Surf. Sci. **347**, 65 (1996).

14. M.S. Altman and E. Bauer, Surf. Sci. **344**, 51 (1995).

15. E. Bauer, Rep. Prog. Phys. **57**, 895 (1994).

16 . M. Mundschau, E. Bauer and W. Swiech, Surf. Sci. **203**, 412 (1988).

17. R.M. Tromp, A.W. Denier van der Gon, F.K. LeGoues and M.C. Reuter, Phys. Rev. Lett. **71**, 3299 (1993).

18. M. Mundschau, E. Bauer, W. Telieps and W. Swiech, Surf. Sci. **213**, 381 (1989).

19. W. Swiech, E. Bauer and M. Mundschau, Surf. Sci. **253**, 283 (1991).

20. A.W. Denier van der Gon, R.M. Tromp and M.C. Reuter, Thin Solid Films **236**, 140 (1993).

21. S.M. Durbin, J.A. Cunningham, M.E. Mochel and C.P. Flynn, J. Phys. F **11**, L223 (1981); S.M. Durbin, J.A. Cunningham, C.P. Flynn, J. Phys. F **12**, L75 (1982).

22. G.L. Zhou, S. Bonham and C.P. Flynn, J. Phys.: Condens. Matter **9**, L671 (1997).

23. C.P. Flynn and M.H. Yang in Epitaxial Oxide Thin Films, edited by J.S. Speck, D.K. Fork, R.M. Wolf and T. Shiosaki (Mater. Res. Soc. Proc. **401**, Pittsburgh, PA, 1996) p. 1.

24. M. Huth and C.P. Flynn, J. Appl. Phys. **83**, 7261 (1998).

25. C.P. Flynn and M.B. Salamon, in *Handbook of Physics and Chemistry of Rare Earths*, Vol. **22**, edited by K. Gschneider and L. Eyring, (Elsevier, Amsterdam, 1996), p. 1.

26. W. Swiech, M. Mundschau and C.P. Flynn, Surf. Sci. **437**, 61 (1999).

27. T.W. Haas and J.T. Grant, Appl. Phys. Lett. **16**, 172 (1970).

28. T.W. Haas, J.T. Grant and G.J. Dooley III, J. Appl. Phys. **43**, 1853 (1973).

29. E. Bauer and H. Poppa, Surf. Sci. **127**, 243 (1983).

30. R. Pantel, M. Bujor and J. Bardolle, Surf. Sci. **62**, 589 (1977).

31. R. Franchy, T.U. Bartke and P. Gassmann, Surf. Sci. **366**, 60 (1996).

32. R.P. Huebener, Magnetic Flux Structures in Superconductors, (Springer, Berlin, 1979).

33. N.C. Bartelt and R.M. Tromp, Phys. Rev. B **54**, 11731 (1996).

34. N.C. Bartelt, J.L. Goldberg, T.L. Einstein, E.D. Williams, J.C. Heyraud and J.J. Métois, Phys. Rev. B **48**, 15453 (1993).

35. M. Seul and D. Andelman, Science **267**, 476 (1995).

36. L. Vitas, A.V. Ruban, H.L. Skriver and J. Kollar, Surf. Sci. **411**, 186 (1998).

37. C.P. Flynn, W. Swiech, R.S. Appleton and M. Ondrejcek, Phys. Rev. B (in press).

38. C.P. Flynn and W. Swiech, Phys. Rev. Lett. **83**, 3482 (1999).

39. W. Swiech, M. Mundschau and C.P. Flynn, Appl. Phys. Lett. **74**, 2626 (1999).

40. M. Mundschau, W. Swiech, C.S. Durfee and C.P. Flynn, Surf. Sci. Lett. **440**, L831 (1999).

ELECTRON BACKSCATTER DIFFRACTION: A POWERFUL TOOL FOR PHASE IDENTIFICATION IN THE SEM

J. R. MICHAEL, R. P. GOEHNER
Sandia National Laboratories, Albuquerque, NM 87185-1405, jrmicha@sandia.gov

ABSTRACT

EBSD in the SEM has been developed into a tool that can provide identification of unknown crystalline phases with a spatial resolution that is better than one micrometer. This technique has been applied to a wide range of materials. Use of the HOLZ rings in the EBSD patterns has enabled the reduced unit cell to be determined from unindexed EBSD patterns. This paper introduces EBSD for phase identification and illustrates the technique with examples from metal joining and particle analysis. Reduced unit cell determination from EBSD patterns is then discussed.

INTRODUCTION

The identification of unknown micrometer-sized phases in the scanning electron microscope (SEM) has been limited by a lack of a robust and simple way to obtain crystallographic information about the unknown while observing the microstructrure or morphology of the specimen. Electron backscatter diffraction (EBSD) in the SEM has become an established technique for the determination of the orientation of individual crystallites and has recently been developed into a tool that can provide identification of unknown crystalline phases.[1] Previous to the development of EBSD for phase identification, the only applicable technique for the identification of micrometer-sized phases was electron diffraction in the transmission electron microscope (TEM). Convergent beam and selected area electron diffraction (CBED and SAED) can provide information from submicrometer sized phases, but requires the preparation of electron transparent samples which can be very difficult and time intensive. Diffraction in the TEM is better suited to the identification of smaller sample areas due to the limited electron transparent areas of thin samples, while phase identification using EBSD is best suited to larger (bigger than 0.1 micrometers) sample areas due to the spatial resolution of the technique. Thus, phase identification using EBSD is a complementary technique to TEM methods.[2]

Previous attempts at using EBSD for phase identification were based on the recognition of symmetry elements in the patterns. When all of the symmetry elements for a given crystal were determined, the point group of the crystal could be determined. It was found that EBSD could identify 27 of the 32 possible point groups. This approach was quite difficult and required a very good understanding of crystallography. Also, high quality patterns were required and these were obtained by using photographic film in the SEM.[3,4]

39

This paper will discuss the progress that has been made in using EBSD patterns in the SEM for the identification of unknown phases. The use of a large crystallographic database in conjunction with the chemistry of the sample has now become a standard and relatively simple technique for phase identification. Recent work has shown that it is possible to determine reduced unit cells from the EBSD patterns with no a priori information about the sample. Examples of phase identification of bulk materials and particulate materials will be discussed. Finally, the determination of reduced unit cells from EBSD patterns will be discussed.

EXPERIMENTAL

In order to overcome the disadvantages of using photographic film a charge coupled device (CCD) based camera was developed. This camera consists of a single crystal yttrium aluminum garnet (YAG) phosphor that was fiber optically coupled to a cooled 1024 x1024 slow-scan scientific-grade CCD. The fiber optic used was a 2.5:1 tapered optic. For most applications the CCD was binned to 512 x 512 resulting in exposure times of 1 to 10 seconds. The camera has been described in detail previously.[1]

The CCD camera was installed on a JEOL 6400 SEM. The specimens were mounted on a pre-tilted specimen holder at 70.5 ° from the horizontal. The exact sample tilt is not critical as we are not measuring the sample orientation. EBSD patterns were usually obtained with an accelerating voltage of 20 kV. Sample preparation varied depending on the sample. Polished sections of metals and ceramics were prepared using standard metallographic techniques followed by a light etch to remove any deformed surface layers. EBSD patterns were obtained from many samples with little or no specimen preparation. Single crystal samples of minerals were mounted in the SEM with a crystal facet oriented toward the EBSD

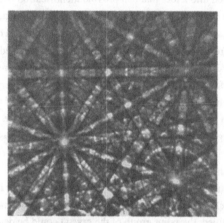

Figure 1. EBSD pattern obtained from a 5 μm RuO_2 crystal at 20 kV.

camera. The only specimen preparation required was cleaning with a suitable solvent. An example of a typical pattern is shown in Figure 1. Flat fielding is the only image processing procedure that is normally required for phase identification. The raw image acquired with a focussed electron probe contains the crystallographic information from the specimen and artifacts related to the backscattered electron distribution and the camera. A second image, called the flat field image, is obtained while the electron beam is scanned over a large number of grains. This effectively removes the crystallographic information from the pattern, but preserves the image artifacts. The final high quality pattern is then obtained by dividing or normalizing the raw pattern by the flat field image. Large single crystals present a problem as it is difficult to obtain a flat field image that contains no crystallographic information. In these cases it is useful to have a fine grained sample of nearly the same average atomic number as the unknown. It is then possible to obtain a flat field from the fine grained material and use it for flat fielding.[1]

RESULTS AND DISCUSSION

Phase Identification Procedure

The identification of unknown phases is accomplished in the following manner. First an EBSD pattern is obtained from the region of interest. The qualitative chemistry of the area must also be determined using either energy dispersive spectrometry (EDS) or wavelength dispersive spectrometry (WDS). This is important as the chemistry of the sample is one of the parameters used to search the crystallographic database.

The EBSD pattern is then analyzed. The Kikuchi line pairs are identified through the use of the Hough transform. The Hough transform simply transforms lines in real space to points in Hough space that can be located automatically. Once the positions of the Kikuchi lines are identified, the widths of the lines are determined. The width of the Kikuchi line pairs are inversely proportional to the spacing of the atomic planes. Once the plane spacings for a number of lines in the pattern are determined, a reduced unit cell volume is calculated. In most cases a subcell volume is obtained. The reduced unit cell volume along with the sample chemistry is then used to search a crystallographic database. We have used the Powder Diffraction File (PDF) produced by the International Center for Diffraction Data (ICDD). This database currently contains over 100,000 inorganic compounds. The typical search that includes both the unit cell volume and the chemistry requires a few seconds and will usually return at most 20 possible matches. It is then necessary to index the pattern using the database information. If a consistent set of indices for the Kikuchi lines is found, the pattern is then simulated using the database crystallographic structure. A comparison of the simulation and the experimental pattern is usually sufficient to determine the identification of the phase.[5]

41

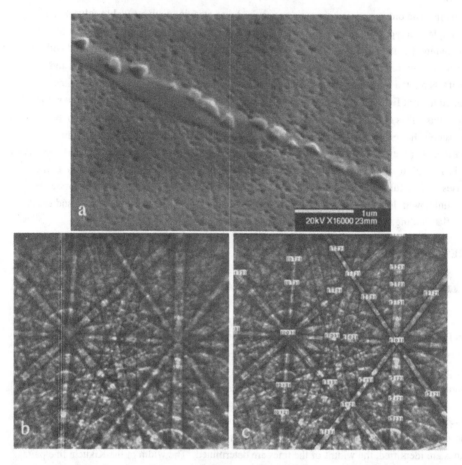

Figure 2. Identification of precipitates found at grain boundaries in a welded Ta alloy. a) SE M image of a grain boundary, b) EBSD pattern acquired from precipitate phase, c)EBSD pattern indexed as HfO_2.

Examples of Phase Identification Studies

Application of EBSD Phase Identification to Welding Research

The physical properties of welds are often dependent on the phases that form during the welding process. Often these phases can be quite small and have been traditionally identified by SAED in the TEM. Figure 2 shows a grain boundary in a Ta alloy weld. The grain boundary is decorated with small precipitates (about 0.2 μm in size) that were shown to contain mostly Hf. Figure 2b is an EBSD pattern obtained from one of the Hf-containing

Figure 3. Identification of austenite and ferrite on a fracture surface in a welded austenitic alloy. a) SEM image of the fracture surface, b) EBSD pattern from area labeled F in 3a, c)EBSD pattern from area labeled A in 3a.

precipitates. The pattern was analyzed and the database searched using the calculated reduced unit cell volume and the possible chemistry of Hf and C, N or O. This search of the database resulted in 5 possible matches. The only match that could index the EBSD pattern was monoclinic HfO_2. The experimental pattern overlaid with the simulation is shown in Figure 2c. There is excellent agreement between the experimental pattern and the simulation. The presence of HfO_2 in the weld region indicates that during the welding process there was inadequate shielding of the weld resulting in the formation of the oxides.

Figure 4. Identification of individual Pb-containing particles on a carbon substrate. a) SEM image of 0.5 μm particle, b) EBSD pattern obtained at 20 kV. c) EBSD pattern indexed as PbO_2 (Plattnerite, tetragonal).

EBSD studies need not be conducted on flat polished samples. Figure 3a is an example of a hot crack that formed during the welding of an austenitic alloy. There are two phases present on the fracture surface and these are indicated by arrows. EBSD patterns were obtained from the two phases. The chemistry of the regions were very similar in that they both contained Fe, Ni and Cr. The EBSD patterns shown in Figure 3b and 3c were analyzed and used to identify the globular phases at the grain boundaries as ferrite and the remainder of the fracture surface as austenite.[6]

Application of EBSD to Particulate Identification

The identity of small particles on substrates is very difficult to determine using SEM imaging and x-ray microanalysis. X-ray microanalysis of particles is very difficult as a result of the small particle size and the uneven surfaces of the particle. The small size results in the electron beam interacting with the substrate as well as the particle and the uneven or rough surfaces make quantitative x-ray microanalysis very difficult if not impossible. EBSD is a useful technique for the identification of particulate materials because patterns may be obtained from individual particles. Figure 4a is an SEM image of a small particle collected on carbon tape. X-ray analysis showed the particle to contain Pb and possibly O.

Figure 4b is the EBSD pattern obtained from the particle shown in 4a. There are at least 4 known lead oxide compounds. The phase identification algorithm identified this particle as PbO_2, a tetragonal phase. Figure 4c is the experimental pattern overlaid with the simulation based on the database information for PbO_2. The agreement is quite good indicating that the particle has been identified as Plattnerite, tetragonal PbO_2 .[7]

Structure Determination From EBSD Patterns

EBSD patterns are formed by the elastic scattering of inelastically scattered electrons. The patterns appear as though the electrons that contribute to the pattern diverge from a point source of radiation within the sample and are therefore termed divergent beam diffraction patterns. EBSD patters are related by reciprocity to channeling patterns or rocking beam patterns. Many of the features observed in EBSD patterns are also observed in convergent beam diffraction (CBED) patterns in the TEM, although there is only an approximate

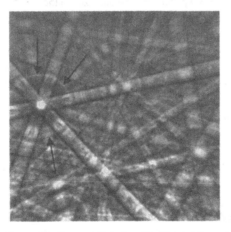

Figure 5. EBSD pattern from a Cr_7C_3 precipitate in a ferritic stainless steel. A prominent HOLZ ring is indicated by the arrows.

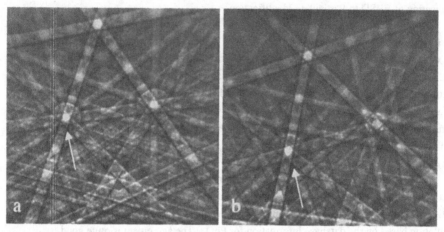

Figure 6. EBSD patterns obtained from two polytypes of SiC. The HOLZ rings analyzed are indicated by arrows. a) EBSD pattern from 6H polytype. b) EBSD patten from 15R polytype.

relationship between EBSD patterns and CBED patterns because the geometry is different in the two cases. In many EBSD patterns there are rings visible around zone axes. These rings have been shown recently to be higher order Laue zone (HOLZ) rings. These HOLZ rings correspond to an envelope of excess lines paired with deficiency lines excited close to the zone axis, but where the reciprocal lattice point is not in the zero layer.[8] HOLZ rings have been used in CBED to determine the reciprocal lattice layer spacing. The accuracy is no better than 1 – 2 % due to lens distortions introduced by the use of short camera lengths. Recent work has shown that HOLZ rings in EBSD patterns can be analyzed in the exact same way as in CBED patterns. The only modification is that the diameter of the rings are most easily measured in terms of angle. [8]

Figure 5 is an EBSD pattern from a Cr_7C_3 carbide in a steel alloy. The [001] HOLZ ring is indicated by the arrows. The measurement of this HOLZ ring results in a real lattice spacing of 0.453 nm in the [001] direction that compares well with the value calculated from the crystal structure of 0.453 nm. The accuracy of spacings determined from HOLZ rings is much greater than measurements made from the widths of the Kikuchi line pairs. This technique has been demonstrated in a large number of materials and all crystal systems with an accuracy of 0.1 to 2%. The only problem occurs for higher atomic number materials where the accuracy of the technique is rather poor. This is most probably due to the rather short extinction distances in higher atomic number materials. An extrapolation technique has been developed to correct for this problem.[8]

One use of HOLZ rings in EBSD is to discriminate different polytypes. Figure 6a and b are EBSD patterns obtained from two polytypes of SiC. Note the HOLZ rings indicated by the arrows in Figures 6a and b. The measured value for the HOLZ ring indicated in figure

6a is 1.858 nm which is excellent agreement with the expected value of 1.846 nm for this zone in the 6H polytype of SiC. The HOLZ ring indicated in Figure 6b results in a calculated spacing of 1.545 nm which is in excellent agreement with the expected value of 1.538 nm for this zone in the 15R polytype. This example demonstrates that the HOLZ rings in EBSD patterns may be useful for identifying polytypes.[8]

The spacings determined from the HOLZ rings have other uses. If the spacings from three non-coplanar zone axes are measured it is possible to calculate the primitive unit cell for the specimen. The cell reduction algorithms are based on the fact that any three non-coplanar prime lattice vectors will produce an arbitrary unit cell that in a unique way represents the Bravais lattice in which it is embedded. Every Bravais lattice contains an infinite number

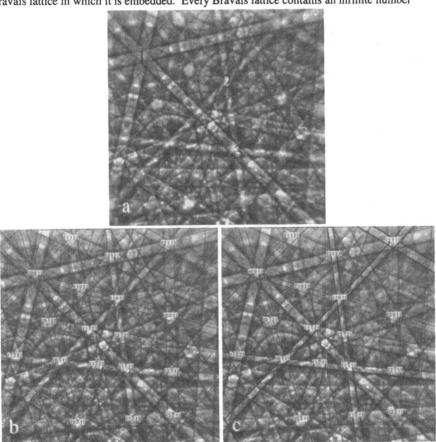

Figure 7. a) EBSD pattern from Mo_2C precipitate in a cast iron. b) EBSD pattern indexed with primitive unit cell calculated directly from the pattern. c) EBSD pattern indexed with information from the crystallographic database.

of primitive cells that are different in shape and size, but are equivalent in generating the Bravais lattice. Therefore, from three interatomic plane spacing measurements and the angles between the planes an arbitrary unit cell can be determined. This arbitrary primitive cell is then used as input to the primitive unit cell algorithm for the determination of the primitive unit cell of the phase. This approach has been demonstrated previously where primitive unit cells were generated from unindexed CBED patterns in the TEM.[9]

From EBSD patterns arbitrary primitive unit cells can also be constructed by combining three lattice vectors determined from the Kikuchi line pairs. However, the accuracy of plane spacings determined from the Kikuchi line pairs is relatively poor, whereas reciprocal lattice layer spacings determined by HOLZ ring analysis are much more accurate as shown above. The use of HOLZ ring measurements results in sufficient lattice spacing accuracy to determine primitive unit cells and has been demonstrated in a number of crystals.

Figure 7a is an EBSD pattern from a Mo_2C precipitate in a cast iron. An arbitrary primitive unit cell was determined from the numbered zones on the pattern. The real lattice spacings determined were: zone 1, 0.48 nm, zone 2, 0.56 nm zone 3, 0.71 nm. The angles between the perpendiculars to the HOLZ rings were measured and are: zone $1-2 = 32.0°$, zone $2-3 = 24.3°$ and zone $1-3 = 47.3°$. These angles and layer spacings define an arbitrary primitive cell of a = 0.71 nm , b = 0.56 nm , c = 0.48 nm , $\alpha = 32.0°$, $\beta = 47.3°$ and $\gamma = 24.3°$. Cell reduction is then used to determine the primitive unit cell. The resulting primitive unit cell is a=0.296 nm, b = 0.298 nm , c = 0.48 nm , $\alpha = 88.2°$, $\beta= 89.1°$ and $\gamma = 118.7°$ and a unit cell volume of 37.0 $Å^3$. This compares well with the actual cell parameters for Mo_2C of a=0.3012 nm, b = 0.3012 nm , c = 0.4735 nm , $\alpha = 90°$, $\beta= 90°$ and $\gamma = 120.0°$ with a unit cell volume of 37.1$Å^3$. Figure 7b shows the experimental pattern overlaid with the simulation based on the calculated reduced unit cell and 7c shows the experimental pattern overlaid with the simulation calculated from the database. There is good agreement between the simulation and the experiment in both cases. Thus, it is possible to start with an unindexed EBSD pattern, calculate the primitive unit cell from the unindexed pattern and then use the calculated primitive unit cell to index the pattern. Future developments of this technique could use the calculated primitive unit cell from the EBSD pattern as an additional condition to search a suitable crystallographic database. Also, the determination of the reduced unit cell is useful when phases that are not in the database are encountered.

CONCLUSIONS

EBSD in the SEM now provides a robust and relatively simple way to identify unknown crystalline phases from their crystallography and chemistry. Procedures have been developed that permit automated identification of unknown phases in the SEM. The technique has been applied to unprepared bulk samples, polished sections, fracture surfaces and particles. Recent work shows that reduced unit cells may be directly determined from unindexed EBSD patterns providing another way to search crystallographic databases for matches. Phase identification in the SEM using EBSD is certainly a powerful tool for materials characterization.

ACKNOWLEDGEMENTS

This work was supported by the United States Department of Energy under contract DE-ACO4-94AL8500. Sandia is a multiprogram laboratory operated by Sandia Corporation, a Lockheed Martin Company, for the United States Department of Energy.

REFERENCES

1. R. P. Goehner and J. R. Michael, J. of Res. of the Nat. Institute of Standards and Technology, **101**, 1996, pp. 301.
2. J. W. Steeds, Convergent Beam Electron Diffraction, in: Introduction to Analytical Electron Microscopy, edited by J. J. Hren, J. I. Goldstein and D. C. Joy, (Plenum Press, New York, 1979) p. 399.
3. K. Z. Baba-Kishi and D. J. Dingley, Scanning, **11**,1989, p. 305.
4. K. Z. Baba-Kishi, Scanning, **20**,1998, p. 117.
5. J. R. Michael, M. E. Schlienger and R. P. Goehner, Microscopy and Microanalysis, **3 Supplement**, 1997, p. 879.
6. C.V. Robino, J.R. Michael, M.C. Maguire, , Welding Journal Research Supplement,**77**, 1998, pp. 446-s-457-s.
7. J. Small and J. Michael, Microscopy and Microanalysis, **5 Supplement 2**, 1999, p. 226.
8. J. R. Michael and J. A. Eades, Ultramicroscopy, in press.
9. Y. Lepage, Microscopy Research and Technique, **21**, 1992, p. 158.

MEASURING THE THIN FILM STRAIN TENSOR NEAR ALUMINUM GRAIN BOUNDARIES VIA A NEW IMAGE PROCESSING APPROACH TO CBED HOLZ PATTERNS

J. INOUE*, A. F. SCHWARTZMAN**, and L. B. FREUND**
*Department of Civil Engineering, University of Tokyo, Bunkyo, Tokyo 113-8656 JAPAN
inoue@ohriki.t.u-tokyo.ac.jp
** Division of Engineering, Brown University, Providence, RI 02912

ABSTRACT

HOLZ lines formed in a CBED pattern provide the most accurate means to measure a local strain tensor with high spatial resolution over mesoscopic length scales. With the advent of energy-filtering in a field-emission TEM, the precision of this measurement increases by filtering out the inelastically scattered electrons. This paper presents an alternate approach to obtaining the same increased precision by image processing of CBED patterns formed in a conventional LaB$_6$ microscope. This technique results in the determination of the full strain tensor within $\pm 0.01\%$. It is based on developing a Wiener filter for CBED patterns, deconvoluting the point spread function of the CCD camera, using the Hough transform to measure distances between HOLZ line intersections, and subtracting out an experimentally determined projector lens distortion. The present technique has been used to measure the strain tensor for the two types of grain boundaries found in MBE grown Al thin films on Si which have the mazed bicrystal microstructure.

INTRODUCTION

Owing to the sensitivity of the position of higher-order Laue zone (HOLZ) lines in a convergent beam electron diffraction (CBED) pattern to small changes in the lattice parameters and the ability to focus an electron beam to a small region, CBED has the potential to be a more accurate lattice parameter measurement technique with better spatial resolution than conventional ones, such as X-ray diffraction. The application of the technique to strain measurement, however, has been limited. There are several reasons for the infrequent use of CBED for strain measurement:

1. unlike X-ray and neutron beams, an electron beam is strongly and multiply scattered by a crystal, so that HOLZ lines are shifted from a position which is determined simply by the Bragg law, in particular, for high-symmetry zone axes,

2. inelastically scattered electrons, which have a modified wavelength, can affect the positions of HOLZ lines and broaden them, and

3. images are distorted because of the asymmetry of the projector lens.

Traditionally the measurement has been done by visual comparison between a simulated pattern created by using a kinematic assumption and the experimental pattern. However, several methods have been proposed recently to overcome these problems. An adjusted high voltage method, in which the deviation of dispersion surfaces from the kinematic dispersion surface is included in the acceleration voltage, has been suggested by Lin et al [1].

Mat. Res. Soc. Symp. Proc. Vol. 589 © 2001 Materials Research Society

Zuo proposed an automated lattice parameter measurement in which the distance between HOLZ lines intersections is used as a quantifying parameter of a HOLZ pattern [2]. Streiffer extended the automated measurement method by introducing an omega energy filter which removes most of the inelastically scattered electrons, and by including a dynamical correction term, which is evaluated by Block-wave theory, to kinematically determined HOLZ line positions [3].

In this paper, we propose a method in which a full strain tensor is determined with an increased accuracy. The method is mainly based on a newly developed image processing approach in addition to the conventional strain measurement technique. The applicability of the present method is demonstrated by measuring the strain tensor for the two types of grain boundaries found in a mazed bicrystalline aluminum film grown by MBE on a (100) silicon wafer. A JEOL 2010 microscope was used for this experiment.

DESCRIPTION OF THE MEASUREMENT METHOD

In order to enhance HOLZ lines in a raw image of a central disk of a CBED pattern, which is obtained via a JEOL 2010 with a slow scan CCD camera, two types of filters are introduced. The first filter is the Wiener filter which eliminates random noise from electronic devices, and the second is the deconvolution by a point spread function which reduces the background noise. The noise, which decreases linearly as the frequency increases, is assumed for the Wiener filter to be a random electrical white noise, as shown in Fig. 1. The point spread function is obtained directly by imaging the transmitted beam

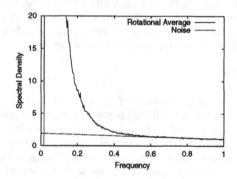

Figure 1: Rotational average of a spectrogram from a raw image of a HOLZ pattern obtained with a GATAN Slow Scan CCD attached to JEOL 2010 microscope.

in an SAD pattern from the same area as the strain measurement. Fig. 2 shows (a) a raw image of HOLZ lines in the central disk and (b) the processed image after applying the Wiener filter and the deconvolution of the point spread function. The image enhancement is clearly demonstrated. After applying these filters, each HOLZ line is extracted by the Hough transform[4].

In order to achieve an improvement in accuracy of the strain measurement, a method which eliminates image distortion induced by asymmetry of the projector lens is developed. Considering the size of a CBED disk on the TEM screen, a uniform distortion is assumed inside the disk. The projector lens distortion is determined by finding a deformation gradient which minimizes the standard deviation of a characteristic length, such as the distance

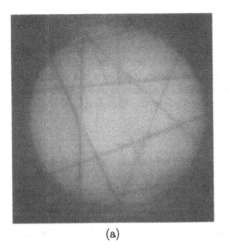
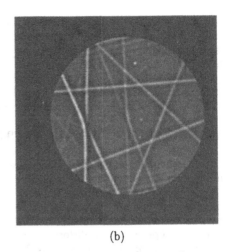

(a)   (b)

Figure 2: Effect of applying filters as described in the text. The HOLZ pattern is from an aluminum film with [110] normal orientation. The zone axis is located between [331] and [441]. Image (a) is an original raw image obtained directly from a slow scan CCD, while (b) is an image deconvoluted with a point spread function and Wiener filter.

between intersections of HOLZ lines which is measured from randomly rotated single aluminum crystals. The idea is that, since the original lattice parameter is constant, the original characteristic length before distortion should also be constant. Although the size of the crystal used in the experiment was $2 \times 2 \, \mu m$ and a uniform strain can be expected, the measurement was conducted within a region measuring $50 \times 50 \, nm$. Table 1 shows the measurement error after eliminating the projector lens distortion. Without subtracting the effect of the projector lens distortion, the standard deviation is more than 0.05% strain when a sample is randomly rotated.

Component	Average (strain %)	Standard Deviation (strain %)
ϵ_{xx}	-0.028	0.005
ϵ_{yy}	0.002	0.0048
ϵ_{xy}	0.010	0.0084

Table 1: Measurement error after subtracting projector lens distortion determined experimentally using an annealed bulk aluminum specimen. The x and y axes are selected so as to be parallel to $[1\bar{1}0]$ and $[001]$, respectively, while the z axis is parallel to the foil normal, $[110]$.

Each component of the strain tensor is determined from the distances between HOLZ line intersections in a central disk of a CBED pattern, so as to satisfy a theoretically obtained mapping function between distances between HOLZ line intersections and strain components. In the present study, the position of a HOLZ line is assumed to be a linear function of strain tensor components for a sufficiently small deformation:

$$d_i = a_{ijk}\epsilon_{jk} + c_i \tag{1}$$

where d_i and ϵ_{ij} are the distance between intersections and strain components, respectively, while a_{ijk} and c_i denote the linear mapping coefficients and intercepts, respectively. The linear mapping function for a certain zone axis is numerically given by applying dynamical diffraction theory. The absorptive atomic form factor, which is given by Bird and King [5], is used. The Debye-Waller factor is referred to Shears and Shelley [6]. From the dynamical simulation, the standard deviation from the linear mapping function is determined to be 0.2 pixel, which is less than the measurement error expected for the Hough transform.

APPLICATION OF THE TECHNIQUE

Our technique is applied to the measurement of a strain field around a grain boundary in a bicrystalline aluminum film which was grown on a silicon substrate in the molecular beam epitaxy facility at Brown University. Normal vectors of the silicon substrate and the aluminum film are [100] and [110], respectively. In this bicrystalline aluminum film, there are two dominant orientations, which are aligned at 90 degrees to each other. For the calibration of projector lens distortion and the acceleration voltage, a bulk silicon specimen and bulk aluminum specimen were used. The latter was prepared by annealing at 500 °C for about one hour in a ultra-high vacuum chamber. The TEM samples were mechanically dimpled to 15 μm and ion milled to a thickness transparent to an electron beam. The acceleration voltage was set to 160 kV, since a sufficient number of HOLZ lines with good contrast are obtained at a relatively low tilt angle for the bicrystaline aluminum sample.

From both of the preliminary experiments, on aluminum and silicon, it is concluded that the gradient of distortion in the image on the screen is described by

$$F = \begin{bmatrix} 1.0 & 0.0 \\ 0.0145 & 1.0145 \end{bmatrix}$$

In the present study, linearity is also assumed between acceleration voltage and distances between HOLZ line intersections. The actual acceleration voltage of our TEM was determined by the same procedure as used for measuring the strain tensor. In this case, the acceleration voltage was chosen as a free parameter in the linear mapping function. From the experiment, the acceleration voltage is found to be unstable up to 60 minutes after the filament current is turned on. The acceleration voltage of the TEM at the stabilized state is estimated to be $162.320 \pm 0.025 keV$; a small oscillation of the estimated acceleration voltage was still observed.

Strain fields in the vicinity of several grain boundaries were measured, an example of which is shown in Fig. 3(a). In order to estimate the strain, and then to infer the stress field, conditions of finite deformation due to states of plane stress are assumed, which are appropriate in light of the magnitude of the measured strain and the geometry of the foil. As a result of these assumptions, the strain components ϵ_{xx}, ϵ_{yy}, and ϵ_{xy} are the free parameters to be determined. Fig. 4 shows the strain measured along a straight line which is perpendicular to the grain boundary; the latter is straight and $200nm$ long. In this example, the [110] direction of the crystal on one side of the grain boundary is perpendicular to the grain boundary, and is parallel to the grain boundary on the other side. The inferred value of normal stress acting on the grain boundary is around 0.02 GPa on both sides and the corresponding shear stress is around 0.01 GPa; both of them are found to be continuous along the grain boundary, consistent with the requirement of equilibrium. The stress component parallel to the grain boundary is not continuous, which indicates the

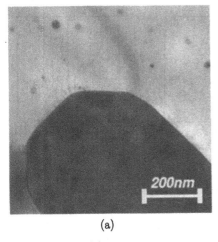

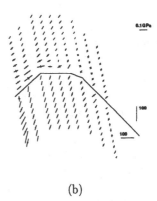

(a)   (b)

Figure 3: Image (a) is an electron micrograph of the grain boundary, and (b) is the inferred stress field obtained by the present technique. Each cross represents the principal stress axes of the measurement point. The length of each line is proportional to the magnitude of the corresponding principal stress. The solid line shows the position of the grain boundary. The 100 planes in each grain are also denoted.

existence of a huge mismatch strain at the grain boundary. Contaminant particles on the surface will affect the beam condition striking the aluminum surface, thus creating local spikes in the measured strain field, as seen in Fig. 4.

CONCLUSIONS

A thin film strain field measurement technique is developed, based on analysis of the HOLZ pattern formed in the central disk. A raw image of a HOLZ pattern is processed by applying the Wiener filter and a point spread function as a deconvolution kernel. Each HOLZ line is extracted by the Hough transformation. A linear mapping relation is assumed between strain and distance between intersections of HOLZ lines. By introducing a dynamical diffraction simulation to achieve a linear mapping function, dynamical effects influencing the accuracy of the measurement are reduced.

From the experimental study of residual stress in an annealed bulk aluminum sample and also a bulk silicon sample, it is established that the strain is determined with an error of less than 0.01%.

The technique is applied to the measurement of a stress field around grain boundaries in a textured bicrystalline aluminum foil. The results of the experiments demonstrate the feasibility of the technique to study local strain in thin films.

Acknowledgements

The research support of the Brown University Materials Research Science and Engineering Center, funded by the National Science Foundation under Award DMR-9632524, and of the Institute for Materials Research and Engineering, Singapore, is gratefully acknowledged.

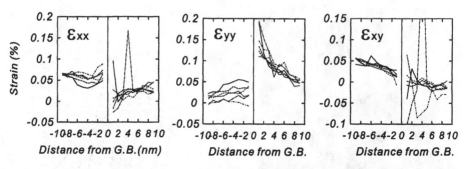

Figure 4: Example of strain ϵ_{ij} plotted along a line perpendicular to a grain boundary across which the crystals are rotated 90 degree with respect to each other. For simplicity, the error bars are omitted, but the absolute value of the error is 0.01%.

References

[1] Y. Lin, D. Bird, and R. Vincent, Ultramicroscopy **27**, 233 (1989).

[2] J. Zuo, Ultramicroscopy **41**, 211 (1992).

[3] S. Streiffer, S. Bader, C. Deininger, and M. R. J.Mayer, in *Polycrystalline Thin Films: Structure, Texture, Properties and Applications, Mater. Res. Soc. Symp. Proc.* **343**, edited by K. Barmuk *et al.* (Materials Research Society, Pittsburgh, PA, 1994), pp. 615–620.

[4] S. Krämer and J. Mayer, Journal of Microscopy **194, Pt 1**, 2 (1999).

[5] D. Bird and Q. King, Acta Crystallographica **A46**, 202 (1990).

[6] V. Shears and S. Shelley, Acta Crystallographica **A47**, 441 (1991).

CONTROLLED ENVIRONMENT TRANSMISSION ELECTRON MICROSCOPY

I. M. ROBERTSON
Department of Materials Science and Engineering, University of Illinois, 1304 W. Green St.
Urbana IL 61801, ianr@uiuc.edu

ABSTRACT

The basic design features of a controlled environment transmission electron microscope and the details of the one at the University of Illinois are described. Examples of how this instrument has been used to determine fundamental mechanisms of hydrogen embrittlement in metals are presented.

INTRODUCTION

The use of the transmission electron microscope to investigate the dynamics of gas-solid interactions dates back to the earliest electron microscopes. For example, Marton [1] in 1935 had a gaseous environment in the microscope to observe hydrated biological samples and to study contamination rates. Since this early application, specimen holders and objective pole-pieces have been modified to control the gas environment and pressure without adversely affecting the performance of the instrument.

In modifications to sample holders, the gas inlet and outlet are contained within the body of the holder [2]. The gas leak rate into the column is restricted by either electron transparent windows or small apertures which are attached to the sample holder and are positioned above and below the sample. To change samples the windows or apertures must be removed. This is one of the main drawbacks of this system as the integrity of the windows can be compromised during this procedure. In window-limited holders, the strength and thickness of the window material determines the maximum gas pressure and, in principal, pressures as high as atmospheric pressure can be attained, although the thickness of window material needed to sustain such pressures may not be compatible with instrument performance. As the electron beam passes through the window material, additional information is superimposed on the image and diffraction pattern of the sample. In this regard, amorphous materials are preferred over crystalline ones. In general, window-limited environmental cells are restricted to room temperature, as the combination of temperature and environment may compromise the integrity of the window. Multi–layered windows have, however, been made to allow elevated temperature studies to be performed [2,3]. Window-limited environmental cell specimen holders have been constructed and used to study gas-solid [3] and wet chemistry [4,5] reactions in-situ in the transmission electron microscope.

The other approach is to use apertures within the objective pole-piece to restrict the flow of gas from the sample region into the microscope column. In the simplest design, one aperture is placed above the sample, which restricts gas flow toward the gun, and another below the sample, which restricts flow toward the camera. In other words, the apertures separate the higher pressure volume around the sample, p_1, from the microscope column pressure, p_2. The gas pressure that can be maintained around the sample is determined by the gas leak rate through the apertures, Q_i, and the pumping speed in their vicinity [6]. The leak rate through an aperture, Q_i, is

$$Q_i = C_i (p_1 - p_2)$$ (1)

In equation (1), C_i is the gas conductance, p_1 is the pressure between the apertures of the cell and p_2 is the pressure in the column just outside the cell apertures. The expression used for the conductance depends on the nature of the gas flow (molecular or viscous) through the aperture. Molecular flow conditions exist when the mean-free path of the gas is greater than the aperture diameter and viscous flow when it is smaller than the aperture diameter. For a 100-micrometer diameter aperture, it can be shown that the change from molecular to viscous flow occurs at a gas pressure of about 5 torr, with molecular flow occurring below 5 torr and viscous flow above. Assuming that viscous flow conditions are established, the leak rate is given by the expression:

$$Q_v = \left\{ \frac{\sqrt{\left(\frac{\nu}{\nu - 1} \right)\left(\frac{2RT}{M} \right)}}{\left(1 - \frac{p_2}{p_1} \right)} \right\} \left(\frac{p_1}{p_2} \right)^{\frac{1}{\gamma}} \left[1 - \left(\frac{p_2}{p_1} \right)^{\left(\gamma - 2 / \gamma \right)} \right]^{\frac{1}{2}} \frac{p_1 A}{1000} (p_1 - p_2)$$ (2)

In equation (2), A is the area of the aperture (in units of cm^2), M is the molecular weight of the gas (g mol^{-1}), T is the temperature of the gas in Kelvin, γ is the ratio of heat capacities at constant pressure and volume and R is the gas constant (ergs mol^{-1} K^{-1}) [6,7].

Under viscous flow conditions, the maximum leak rate occurs when the gas velocity reaches the speed of sound, which occurs for most diatomic gases when $p_2/p_1 = 0.5$. The

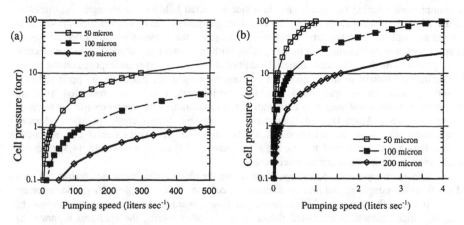

Figure 1. Pressure of hydrogen gas as a function of pumping speed for different diameter apertures (a) p_2 (p_3 in a double aperture limited system) = 10^{-4} torr and (b) p_2 (in a double aperture limited system) = 0.3 torr.

pumping speed needed to achieve this condition is given by the ratio of the leak rate, Q_v, and the pressure in the microscope column outside the cell, p_2, and is proportional to $d^2 p_1/p_2$ where d is the aperture diameter. The proportionality constant depends on the gas.

For a column vacuum of 10^{-4} torr, the pumping speed required to maintain a particular pressure of hydrogen gas in the cell is shown in Figure 1 (a) for different aperture diameters. To construct a cell capable of supporting 10 torr of hydrogen gas, the pumping speed in the vicinity of the aperture would have to be 300 liters per second and the diameter of the apertures used would have to be 50 μm. The same system would be able to support higher pressures of heavier gases. For example, a 50 μm diameter aperture with a 300 liter per second pumping speed on the exit side would support 30 torr of air in the cell. This pumping speed will be difficult to achieve just outside the cell apertures given the volume restrictions in this area. The cell apertures are placed in close proximity to the specimen plane to minimize the electron path length in the gas, which will make practical microscopy difficult (e.g., locating the electron beam after a filament change). To understand this one only has to consider the function of the selected area apertures, their diameter and location in the microscope column.

To achieve higher gas pressures in the environmental cell, a series of pressure zones can be created through the use of multiple apertures above and below the specimen. For example, in a double-aperture limited system, two cell apertures are positioned above and another two below the sample plane. The cell apertures immediately above and below the sample are referred to as the primary cell apertures and the next level as the secondary cell apertures. Using two pairs of cell apertures in this way effectively creates three pressures zones; that inside the cell, p_1, that in the volume between the primary and secondary cell apertures, p_2, and that in the microscope column just outside the secondary cell aperture, p_3. The pressures that can be achieved within the cell can be further increased if additional pumping is available to lower p_3 and p_2. If such pumping is available the type of cell design is referred to as a differentially-pumped, aperture-limited environmental cell.

Assuming that with appropriate differential pumping $p_3 = 10^{-4}$ Pa, the pumping speed outside the secondary cell aperture is 150 liters sec^{-1}, and the diameter of the cell aperture is 200μm, the pressure of hydrogen gas in the volume between the primary and secondary cell apertures, p_2, is according to Figure 1(a) ~ 0.3 torr. If the diameter of the primary cell aperture is 100 μm, a pumping speed in the volume between the primary and secondary cell apertures of 2 liters sec^{-1} is needed to maintain a pressure of 50 torr of hydrogen gas in the specimen volume, see Figure 1 (b). Higher pressures can, of course, be attained if smaller cell apertures are used or higher pumping speeds can be achieved. Using smaller cell apertures will make it more difficult to construct the cell and to use the microscope. The cell apertures are generally fixed in position and must be aligned with each other and with the optic axis of the electron microscope as no movement of the cell apertures is possible once they are mounted in the pole-piece. Attempts to achieve high cell pressures may be pointless as the maximum useful gas pressure, which can be taken as the gas pressure at which structural information is lost due to image degradation because of the interaction of the electron beam with the gas molecules, is generally much less than the maximum gas pressure that can be achieved in the cell. The interaction of the electron beam with the gas molecules also assists in this regard because it causes dissociation and ionization of the gas molecules, which effectively increases the fugacity. Bond et al. [8] have shown that the cell pressure is up to three orders of magnitude higher than the externally monitored pressure.

The scattering of the electron beam by the gas molecules, while increasing the fugacity, decreases the intensity of the electron beam and the image resolution. The decrease in relative beam intensity for 400 and 200 keV electrons is shown as a function of hydrogen gas pressure

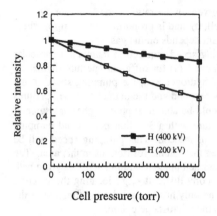

Figure 2. Relative intensity of the electron beam as a function of the pressure of

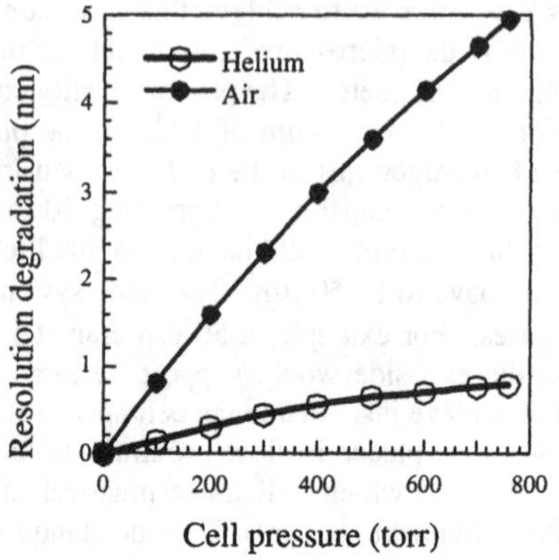

Figure 3. Degradation in resolution due to inelastic scattering as a function of cell pressure. [9]

in Figure 2.   The magnitude of the decrease in the relative intensity of the electron beam would increase with molecular weight of the gas and with the gas pressure, and with decreasing accelerating voltage. The dependence shown in Figure 2 occurs when single scattering events dominate; i.e., at low gas pressures. When multiple scattering events occur, the rate of decrease of the relative beam intensity decreases at a faster rate than shown in Figure 2. Multiple scattering events will also affect image resolution because of an increase in beam convergence angle.

Inelastic energy loss events become increasingly important as the mass thickness of the gas increases. This effect was shown by Lacaze [9], who determined the change in resolution due to inelastic scattering as a function of cell pressure for air and He in a 1 MeV electron microscope with a chromatic aberration coefficient, C_c of 6.5 mm. As shown in Figure 3, the degradation in resolution increases with increasing gas pressure and with gas molecular weight.

From the above discussion it is evident that to optimize the performance of the cell, the primary cell apertures should be close to the specimen plane to minimize the time the electrons spend in a high gas pressure environment, and they should be small to reduce the rate at which gas leaks into the microscope column. From an applications point-of-view, having the primary cell apertures close to the specimen plane restricts tilting capability, and, hence, the image analysis techniques that can be applied, and limits diffraction information, further limiting the usefulness of the environmental cell. The inclusion of the cell inside the objective pole-piece may also prohibit energy dispersive x-ray spectroscopy.   Therefore, in designing an environmental cell it is necessary to consider the requirements of the cell as well as the desired capabilities of the microscope. The two are not necessarily compatible and some compromise in the performance of the cell and of the electron microscope may be necessary.

DESIGN OF THE CONTROLLED ENVIRONMENT TRANSMISSION ELECTRON MICROSCOPE AT THE UNIVERSITY OF ILLINOIS.

The controlled environment transmission electron microscope at the University of Illinois [10,11] is based on a JEOL 4000EX transmission electron microscope. The objective pole-piece has a gap of 15 mm and has several additional ports to accommodate the gas handling device and the pumping systems used to provide the differential pumping, Figure 4. The main component of the cell is T–shaped and is constructed from two 310S stainless steel tubes that are permanently mounted in the pole-piece. The longer tube surrounds the specimen axis and the other the gas handling system; only the tube surrounding the gas handling system is visible in Figure 4b. The cross-section of the primary tube was reduced and flattened in the center to reduce the electron path length in a high-pressure gas environment, and to allow insertion of the objective aperture rod, Figure 4(b). The primary cell apertures (nominally, 100 micrometer in diameter) were machined in the bottom of conical cups made from Inconel 625 and located in the center of the flattened section of the tube and sealed in place. The tips of the cones are visible in Figure 4(c). To completely seal the objective pole-piece from the microscope column it was necessary to establish seals between the microscope column and the ports in the pole-piece for the specimen rod and objective aperture drive; the seals at the specimen ports are indicated in Figure 4b. The secondary cell apertures were placed in the bore of the upper (200-micrometer diameter) and lower (300 micrometer) sections of the objective pole-piece. An additional cell aperture (300 micrometer) was inserted in the bottom of the condenser lens stack to provide further isolation of the electron gun from the cell. Differential pumping between the apertures is achieved by means of magnetically levitated turbo molecular pumps that are attached via vibration isolators to the microscope column.

With this design, pressures of 50 torr of hydrogen gas can be maintained indefinitely and pressures of 200 torr of hydrogen gas can be maintained for several minutes before

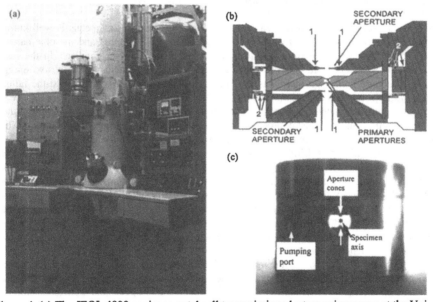

Figure 4. (a) The JEOL 4000 environmental cell transmission electron microscope at the University of Illinois. (b) Schematic cross-section of the objective pole-piece showing the tube that surrounds the specimen stage. The location of the primary and secondary apertures are indicated . The arrows numbered 1 and 2 show the o-ring seals used in the secondary apertures and to isolate the pole-piece from the rest of the microscope column. (c) The objective pole-piece.

deterioration of the gun vacuum is detected. Placing the primary cell apertures at the bottom of conical cups (Figure 4c), minimizes the electron beam path length (~5 mm) in the high gas pressure region, while retaining $\pm$ 30 degrees of tilt about both axes. Diffraction maxima to 12 nm^{-1} are visible. The nature of this design precludes EDS capabilities.

The cell design allows use of standard specimen holders. The following holders are currently available: a single and a double-tilt stage, a single-and a double-tilt straining stage, a double-tilt heating stage, a single-tilt high temperature straining stage, and a single-tilt low temperature straining stage. It is therefore feasible to study gas-solid interactions in the temperature range from 100 K to 700K while also deforming a sample. The dynamic reactions are recorded via a Gatan TV rate camera to super VHS video tape.

The environmental cell design installed in the JEOL 4000 at the University of Illinois is capable of sustaining relatively high gas pressures for long periods of time. The instrument resolution in this configuration is ~0.9nm and no chemical analysis capability is available. Other environmental cells with better resolution and chemical analysis capabilities have been built [12], but these are limited in the maximum operating pressure to about 10 torr of hydrogen gas.

EXAMPLES OF THE USE OF THE ENVIRONMENTAL CELL IN THE STUDY OF HYDROGEN EMBRITTLEMENT MECHANISMS IN METALS.

To illustrate the use of the environmental cell transmission electron microscope, examples from our work on the fundamental mechanisms of hydrogen embrittlement will be described [13-23].

The effect of gaseous hydrogen on the mechanical properties of metals is well known and is characterized by a loss of ductility, a reduction in the toughness and in some cases a change in the fracture mode from ductile transgranular to brittle intergranular. In the early seventies, these effects were generally attributed to hydrogen reducing the cohesive energy (decohesion model) although there was evidence that plasticity often accompanied the failure process [24,25]. Based on the observation of dimples and tear ridges on the fracture surface of hydrogen embrittled materials, Beachem [26] first proposed that hydrogen would enhance rather than diminish the plastic processes that were occurring during deformation. This proposal was basically ignored by the research community until the early eighties when separately Lynch[27], and Birnbaum, Robertson and coworkers [13-23] presented further evidence for hydrogen enhancing plastic processes. Lynch proposed that hydrogen, like liquid metals, caused embrittlement by promoting the injection of dislocations from the surface into the material. Birnbaum, Robertson and coworkers [13-23] provided direct evidence by using the environmental cell TEM that hydrogen enhanced the mobility of isolated edge, screw and partial dislocations, and of dislocation emission from dislocation tangles, from crack tips and from grain boundaries. In this case, the enhancement was not restricted to surface dislocations but occurred throughout the material.

The effect of hydrogen on the mobility of dislocations in iron is shown in the sequence of videotape images presented in Figure 5. The dislocations were produced by first straining the sample in vacuum. When the stage displacement was stopped and held constant, the dislocations stopped moving. On introducing hydrogen gas into the cell and consequently into the sample, the dislocations started to move and their velocity increased with increasing hydrogen pressure. In Figure 5, the dislocations move from left to right and new dislocations appear in the field of view from the left side as the hydrogen gas pressure is increased to 200 torr [13]. On removing the gas from the cell and the sample, the dislocations stopped moving. Reintroducing the gas caused the dislocations to start moving again. Enhancements of factors

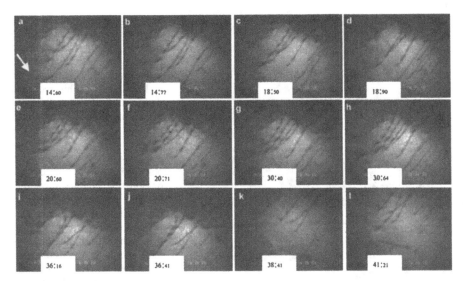

Figure 5. The effect of hydrogen on the mobility of dislocations in iron. The gas pressure was increased to 200 torr during the sequence. The time is indicated in seconds and hundreths of seconds on the images. The arrow indicates the direction of motion [13].

of two in the dislocation velocity were possible in the presence of hydrogen. Such enhancements have been seen in a number of materials with different crystal structures; Fe (bcc) [13], Ni (fcc)[15], Ni$_3$Al (L1$_2$)[19], Ti, (hcp)[18], Ti$_3$Al (DO$_{19}$)[28], Al (fcc)[16,17], Beta Ti-alloy 21S (bcc) [29], FeAl (B2) [30], and In 903 (FCC)[31]. The generality of the observation suggested that the mechanism responsible had to be a universal one. This led to the proposal that the enhancement in the velocity is due to hydrogen shielding the dislocation from interaction with other elastic obstacles [32,33]. One consequence of the elastic shielding model is that the separation distance between a dislocation with a hydrogen atmosphere and another elastic obstacle should be less than that between the dislocation without the hydrogen environment and the obstacle[33].

The change in spacing in a dislocation pileup in 310S stainless steel as hydrogen gas is introduced and the gas pressure is increased is shown in Figure 6[23]. As the gas pressure is increased, the dislocations move closer to the obstacle, a grain boundary in this case, and the distance between the dislocations decreases, Fig 6(a –c). These effects are more obvious in Figure 6 (d), which is a comparison image formed by superimposing a negative image (white dislocations) of the dislocation positions in 90 torr of hydrogen gas (Fig. 6(c)) on a positive image (black dislocations) of the dislocation positions in vacuum (Fig 6(a)). Clearly, all of the dislocations have moved toward the barrier and the distance between them has decreased.

The effect of hydrogen on the dislocation position should be reversible. That is, the dislocations should move in the reverse direction and the distance between them should increase when hydrogen is removed from the sample. This reversibility has been observed for dislocations in high-purity aluminum, see Figure 7[23]. The images are again comparison images with the black dislocations (positive image) showing the initial position and the white dislocations (negative image) the final position. As the gas pressure is increased from 15 to 72 torr of hydrogen gas, the dislocations move from right to left, Fig. 7(a). Decreasing the gas

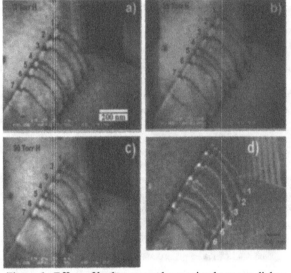

Figure 6. Effect of hydrogen on the spacing between dislocations in 310S stainless steel [23].

pressure results in the dislocations moving from left to right, Fig. 7(b). Increasing the pressure again results in the dislocations moving from right to left, Fig 7c. This sequence demonstrates the reversibility of dislocation motion as hydrogen is added and removed from the sample.

Macroscopic observations of the sides of bulk deformation samples show that the slip planarity increases in the presence of hydrogen. This effect was attributed to hydrogen reducing the stacking-fault energy to such an extent that the separation distance between partial dislocations increased significantly. Consequently, this increases the force needed to form the constriction necessary for cross-slip to occur; hence, making the process more difficult. Altstetter and coworkers [34,35] suggested an alternate mechanism in which cross-slip was restricted because hydrogen promoted edge-character dislocations. While this is correct, they provided no supporting evidence .

The series of images presented in Figure 8 shows that hydrogen can prevent cross-slip by preventing edge dislocation segments from reorienting to the screw orientation necessary for cross-slip[36]. In looking at Figure 8, first consider only the change with time. With time the line direction of the dislocations develops a sharp bend (Figure 8(c)), which corresponds to the dislocations cross-slipping. The change in slip direction is indicated by arrows 1 and 2, which delineate the surface slip traces of the two slip systems. The arrowheads in Figure 8(c) indicate the dislocation segment that has cross slipped. Second, consider the change in

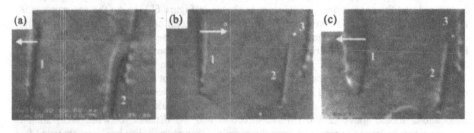

Figure 7. Effect of increasing and decreasing the hydrogen content on the direction of motion of dislocations in high-purity aluminum. (a) hydrogen pressure increases from 15 to 75 torr. (b) hydrogen pressure decreases from 75 to 9 torr and (c) hydrogen pressure increases from 15 to 75 torr [23].

64

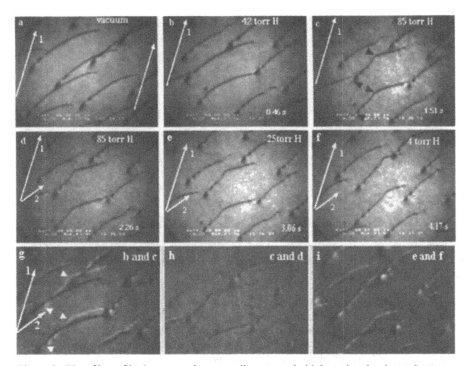

Figure 8. The effect of hydrogen on the cross-slip process in high-purity aluminum. Images (g)-(i) are comparison images formed by superimposing a negative of one image (second letter) on a positive image of another (first letter) [36].

hydrogen pressure with time. The hydrogen pressure is first increased (Figs. 8(a) - 8(c), held constant (Figs. 8(c) and 8(d)), and then decreased (Figs. 8(d) -(f)). With a constant pressure of 85 torr in the environmental cell the dislocations are locked in the cross-slip configuration, with a portion of the dislocation remaining on the primary slip plane and the rest on the cross-slip plane. This locking is perhaps easier to see in the series of comparison images presented in Figure 8 ((g)-(i)); again the black dislocations show the initial position and the white ones the final position. In Figure 8(g), the dislocations have started to cross slip as the hydrogen pressure is increased from 42 to 85 torr. The contrast in Figure 8h, which compares the positions of the dislocation in 85 torr of hydrogen (Figures 8c and 8d), is poor due to the cancellation of features when the positive and negative images are added. As the pressure of hydrogen gas is reduced, the cross-slip process is allowed to continue. In Figure 8(i), which compares the dislocation positions after the gas pressure has been reduced from 25 to 4 torr, most of the dislocation has cross-slipped and only a short segment remains on the original slip plane.

This locking of the dislocation in the process of cross-slipping can be understood as follows. The interaction between hydrogen and an edge dislocation is stronger than that with a screw dislocation. Consequently, a greater atmosphere will form about an edge than a screw dislocation. This atmosphere will have two effects, both of which will inhibit the cross-slip process. Firstly, for the edge dislocation to reorient to the screw configuration so that it can cross-slip the hydrogen atmosphere will have to be dissolved and the hydrogen returned to the lattice. This will increase the total system energy and will not occur readily. Secondly, the line

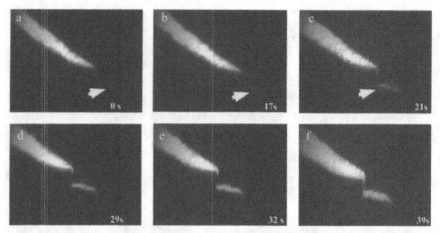

Figure 9. Propagation of a stationary crack due in IN903 to the addition of hydrogen gas to the environmental cell and consequently to the sample. The arrowhead indicates a region ahead of the crack where thinning occurs.

energy of the edge dislocation with the hydrogen atmosphere is lower than that of an edge dislocation although not as low as that of a screw dislocation [36]. Therefore, there will be less of a driving force for the dislocation to reorient to the screw configuration. Of course, these restrictions will be removed when the gas is removed and the cross-slip process can continue, as is observed.

These observations suggest that hydrogen will stabilize edge dislocations, which would prevent cross slip from occurring and can explain the increase in slip planarity that is often reported to accompany hydrogen embrittlement.

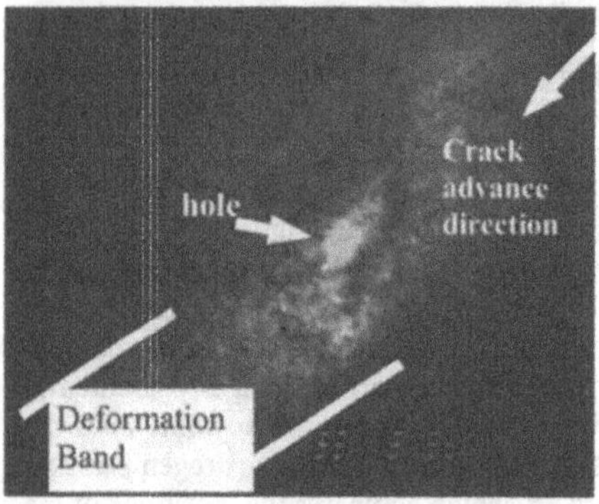

The presence of a hydrogen environment can also cause a crack to propagate at a constant applied load. An example of this is shown in Figure 9, in which a stationary crack in the precipitation hardened alloy IN 903 is caused to propagate by the introduction of up to 200 torr of hydrogen gas to the environmental cell[31]. The first indication that something is happening is the formation of a thin region ahead of the crack; this region is indicated by arrowheads in Figure 9(a)-(c). With time this region grows, the main crack widens and advances a short distance. Large crack advances occur when the main crack links with the cracks that have formed ahead of it. More detailed examination of crack advance in IN903 in hydrogen gas shows that small holes nucleate ahead of the crack, Figure 10, and that large crack advances occur by linking of these cracks.

Figure 10. Nucleation of a hole in a deformation band ahead of a propagating crack in IN903. The hole was produced in hydrogen gas with the stage displacement held constant. The dislocation activity and the nucleation and growth of the hole were caused by hydrogen.

SUMMARY

In the proceeding sections the use of an environmental cell transmission electron microscope to obtain a fundamental understanding of hydrogen embrittlement mechanisms was illustrated. Direct evidence was presented for hydrogen increasing the mobility of dislocations in Fe; thus establishing the viability of the mechanism first proposed by Beachem in 1972. Hydrogen enhancement of the dislocation velocity has been observed for edge, screw, mixed and partial dislocations in materials with different crystal structures. The pervasiveness of the effect suggested that the underlying mechanism had to be a general one, which led to the development of the hydrogen shielding model [32,33]. In this model the hydrogen atmosphere associated with the dislocation effectively shields it from interactions with other elastic obstacles. Consequently, the interaction energy between a dislocation-hydrogen complex and an obstacle is reduced, allowing the dislocation-hydrogen complex to move at a lower applied stress. The microscope has also been used to show by measuring the spacing of dislocation nodes created in vacuum and in a hydrogen environment that hydrogen reduces the stacking-fault energy of 310S stainless steel by up to 20% [37]. If similar reductions occur in other materials, it is insufficient to explain the increase in slip planarity reported to occur when bulk samples are deformed in the presence of hydrogen. Direct evidence for hydrogen stabilizing the edge component of a dislocation, which prevents it from reorienting to the screw configuration necessary for cross-slip, has been found by using the environmental cell TEM. This preference for edge character dislocations will increase slip planarity.

ACKNOWLEDGMENTS

I am indebted to the following people for their assistance in designing, constructing and maintaining the JEOL 4000 environmental cell transmission electron microscope; Drs. T. C. Lee, D. K. Dewald, D. Teter and J. A. Eades, Mr. D. Lillig and Professor H.K. Birnbaum. Without their help and dedication this facility would not have been developed. The controlled environment transmission electron microscope is part of the Center for Microanalysis in the Frederick Seitz Materials Research Laboratory. This work was supported by the Department of Energy through grant DEFG02-91-ER45439.

REFERENCES

1. L. Marton, Acad. r. Belg. Cl. Sci. 21, 553 (1935).
2. E. P. Butler and K. F. Hale, in *Practical Methods in Electron Microscopy*; *Vol. 9*, edited by A. M. Glauert (North-Holland, Amsterdam, 1981), p. 239 .
3. P. Baules, P. Millet, M.J. Casanove, E. Snoeck, and C. Roucau, J. Microsc. Spectrosc. Electron. (France) 14, 305 (1989).
4. F.M. Ross and P.C. Searson, in *Dynamic observation of electrochemical etching in silicon*, Oxford, 1995 (Institute of Physics Conference), p. 511.
5. F.M. Ross and P.C. Searson, in *In situ observation of an electrochemical etching reaction in silicon*, Boston, 1995 (Materials Research Society, Boston), p. 69 .
6. A. Berman, *Vacuum Engineering Calculations, Formulas and Solved Exercises* (Academic Press, San Diego, 1992).
7. *Vacuum* (Leybold Vacuum Products INC, Export, Pa, 1990).
8. G.M. Bond, I.M. Robertson, and H.K. Birnbaum, Scripta Metall. 20, 653 (1986).
9. J.C. Lacaze, Memoire CNAM Central Regional agree de Toulouse, France, Cited in

Butler and Hale (1977).

10. T.C. Lee, D.K. Dewald, J.A. Eades, I.M. Robertson, and H.K. Birnbaum, Review of Scientific Instruments 62, 1438 (1991).

11. D. Teter, P. Ferreira, I.M. Robertson, and H.K. Birnbaum, in *An environmental Cell TEM for studies of gas-solid interactions, in New Techniques for Characterizing Corrosion and Stress Corrosion*, Clevland, Ohio, 1995 (TMS, Warrendale, Pa), p. 53 .

12. I.M. Robertson and D. Teter, Journal Microscopy Research and Technique 42, 260 (1998).

13. T. Tabata and H.K. Birnbaum, Scripta Metall. 17, 947 (1984).

14. H.K. Birnbaum, D.S. Shih, and I.M. Robertson, in *HVEM Environmental Cell Studies of Hydrogen effects in Alpha-Titanium*, Osaka, 1985, p. 53 .

15. I.M. Robertson and H.K. Birnbaum, Acta Metall. 34, 353 (1986).

16. G.M. Bond, I.M. Robertson, and H.K. Birnbaum, Acta Metall. 35, 2289 (1987).

17. G.M. Bond, I.M. Robertson, and H.K. Birnbaum, Acta Metall. 36, 2193 (1988).

18. D.S. Shih, I.M. Robertson, and H.K. Birnbaum, Acta Metall. 36, 111 (1988).

19. G.M. Bond, I.M. Robertson, and H.K. Birnbaum, Acta Metall. 37, 1407 (1989).

20. P. Rozenak, I.M. Robertson, and H.K. Birnbaum, Acta Metall. Mater. 38, 2031 (1990).

21. H.E. Hanninen, T.C. Lee, I.M. Robertson, and H.K. Birnbaum, Journal of Materials Engineering & Performance 2, 807 (1993).

22. P.J. Ferreira, I.M. Robertson and H.K. Birnbaum, Mater. Sci. Forum, 93 (1996).

23. P.J. Ferreira, I.M. Robertson, and H.K. Birnbaum, Acta Mater. 46, 1749 (1998).

24. R.A. Oriani, in *A review of proposed mechanisms for hydrogen-assisted cracking in metals*, Philadelphia, Pa., USA. 29 May - 1 June 1973., 1973.

25. R.A. Oriani, Corrosion 43, 390 (1987).

26. C.D. Beachem, Metall. Trans. A 3, 437 (1972).

27. S.P. Lynch, Scr. Metall. 13, 1051 (1979).

28. H. Xiao, Ph. D. Thesis, University of Illinois, 1993.

29. D. Teter, Ph. D. Thesis, University of Illinois, 1996.

30. D. Lillig, Unpublished work, University of Illinois, 1999.

31. I. M. Robertson, Unpublished work, University of Illinois, 1999.

32. E. Sirois and H.K. Birnbaum, Acta Metall. 40, 1377 (1992).

33. P. Sofronis and H.K. Birnbaum, Fatigue and Fracture of Aerospace Structural Materials American Society of Mechanical Engineers, Aerospace Division 36, 15 (1993).

34. D.G. Ulmer and C.J. Altstetter, Acta Met. et Mat. 39, 1237 (1991).

35. D.P. Abraham and C.J. Altstetter, Metallurgical and Materials Transactions 26A, 2859 (1995).

36. P.J. Ferreira, I.M. Robertson, and H.K. Birnbaum, Acta Mat. 47, 2991 (1999).

37. P.J. Ferreira, I.M. Robertson, and H.K. Birnbaum, Materials Science Forum, 93, 2091, (1996).

ANALYSIS OF THE ATOMIC SCALE DEFECT CHEMISTRY IN OXYGEN DEFICIENT MATERIALS BY STEM

Y. ITO, S. STEMMER, R. F. KLIE, N. D. BROWNING, A. SANE*, T. J. MAZANEC*

Department of Physics, University of Illinois, 845 W. Taylor St., Chicago, IL 60607-7059.
*BP Chemicals Inc, 4440 Warrensville Center Road, Cleveland, OH 44128-2837.

ABSTRACT

The high mobility of anion vacancies in oxygen deficient perovskite type materials makes these ceramics potential candidates for oxygen separation membranes. As a preliminary investigation of the defect chemistry in these oxides, we show here the analysis of $SrCoO_{3-\delta}$ using atomic resolution Z-contrast imaging and electron energy loss spectroscopy in the scanning transmission electron microscope. In particular, after being subjected to oxidation/reduction cycles at high temperatures we find the formation of ordered microdomains with the brownmillerite structure.

INTRODUCTION

Perovskite-type oxides that combine high electronic and ionic conductivity are very promising materials for use as dense ceramic membranes for oxygen separation [1]. As with many of the transition metal perovskite oxides (ABO_3), the physical properties of these materials are controlled primarily by their degree of non-stoichiometry (i.e. the defect chemistry). Non-stoichiometry can be controlled by the temperature, the oxygen partial pressure and by the substitution of cations of different valence. However, by using such parameters to create non-stoichiometry, there may also be the formation of ordered superstructures that may adversely affect the structural, chemical and mechanical stability of the membrane as a whole. To fully understand the parameters leading to the successful implementation of ceramic membranes in practical applications, we therefore need to understand the defect chemistry on the fundamental atomic scale.

The route to characterizing oxide materials on this level is afforded by the combination of Z-contrast imaging [2] and electron energy loss spectroscopy (EELS) [3] in the scanning transmission electron microscope (STEM). These correlated techniques [4] allow direct images of crystal and defect structures to be obtained, the compositions to be quantified and the effect of the structures on the local electronic properties (i.e. oxygen coordination and cation valence) to be assessed [5]. In this paper we discuss the use of these techniques in the JEOL 2010F STEM [6] to analyze the defect chemistry in $SrCoO_{3-\delta}$. This material is chosen to demonstrate the use of the techniques as it is known to exist in a variety of phases with different crystal structures, compositions and valence states of cobalt and can be highly oxygen deficient [7].

EXPERIMENTAL TECHNIQUES

The basis of the STEM techniques is the ability to form a probe of atomic dimensions on the surface of the specimen (Figure 1). In the case of the JEOL 2010F STEM, the size of the probe that can be achieved is 0.14nm [6], which corresponds to the resolution of the Z-contrast image and the energy loss spectrum for a specimen in a zone-axis orientation [4].

Incident Probe

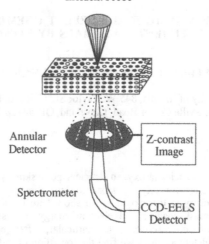

Figure 1: Schematic of the detector arrangement in the JEOL 2010F STEM.

Z-Contrast Imaging

Z-contrast imaging in the STEM uses the high angle scattering from a sample collected on an annular detector (Figure 1). This integrated intensity is then synchronously displayed on a TV screen while the electron probe is scanned across the specimen. The integration over the annular detector averages over the interference effects from adjacent atoms and thus each atom can be considered as an independent scatterer [2,8]. Therefore, the image intensity may be described as an object function, which is sharply peaked at the atomic locations, convolved with the probe intensity profile. For high inner collection angles the cross-section for this scattering is approximately proportional to Z^2, where Z is the atomic number [8]. As a result of this simple image formation mechanism, many of the problems associated with phase contrast microscopy are removed and a simple, direct interpretation of the Z-contrast image is possible [2]. Since the width of the object function is typically of the order of 0.02nm, the spatial resolution of the image is determined by the microscope probe size (0.14nm for the JEOL 2010F). This means that for a specimen aligned with a crystallographic pole along the beam direction, providing the atomic spacing of interest is larger than the probe diameter, atomic columns may be illuminated individually and an atomic resolution image may be obtained.

Electron Energy Loss Spectroscopy

As can be seen from Figure 1, the annular detector used for Z-contrast imaging does not interfere with the low-angle scattering used for EELS. This means that the Z-contrast image can be used to position the electron probe over a particular structural feature for acquisition of a spectrum [4]. The physical principle behind EELS relates to the interaction of the fast electron with the sample to cause either collective excitations of electrons in the conduction band, or discrete transitions between atomic energy levels, e. g. $1s \rightarrow 2p$ transitions. The ability to observe discrete atomic transitions allows compositional analysis to be performed by EELS (the transitions occur at characteristic energy losses for a given element). Furthermore, the transitions to unoccupied states above the Fermi level allows the degree of hybridization between atomic orbitals to be determined, i.e. information on local electronic structure (bonding) changes can be ascertained. In the case of the transition metal elements, the intense features in the $L_{2,3}$-edge spectrum correspond to transitions to the empty Ti 3d states. Hence, the intensity of the "white-line" features can be used to provide some measure of the valence of transition metals, i.e. the number of electrons in the 3d band.

We can investigate the effect of the local atomic structure on the electronic properties further by considering the fine structure of the energy loss spectrum. One means to simulate this is through the multiple scattering (MS) methodology [9]. MS calculations model the density of unoccupied states by considering the scattering of the photoelectron created during the excitation process, from neighboring atoms. The many paths which may be taken by a photoelectron alter the matrix elements for a particular transition due to constructive or destructive interference which occurs between the outgoing and returning photoelectron wave. In effect, the resultant spectrum may be described as a simple absorption edge of hydrogenic form, due to an isolated atom, with intensity modulations due to the atomic structure of the solid. Since this description of the unoccupied density of states is based on a real space cluster of atoms, several unique opportunities are presented in this analysis. A lack of symmetry does not seriously affect the calculation, making it ideal for the study of low-symmetry defects. The effects of dopant atoms may be simply investigated by substituting atom types within the cluster and recalculating the scattering. MS therefore allows spectral changes to be directly interpreted in terms of structural changes(for which we have a starting structural model in the Z-contrast image).

SPECIMEN PREPARATION

Stoichiometric amounts of Sr acetate and Co_3O_4 were ground and calcined at 800°C for 6 hours, followed by grinding and sintering of the compressed powder at 1200°C for 1 hour in air. Electron diffraction showed that after cooling, the sample had the hexagonal $SrCoO_{3-\delta}$ structure. This is the starting structure for our defect chemistry analysis. The sample was then heated in air at a rate of 5K/minute. During heating, a phase transformation to a cubic or pseudo-perovskite structure was observed at ~750°C. After reaching 1014°C, the atmosphere was changed three times from air to nitrogen to air, followed by twice cycling between air and CO_2. After the final CO_2 environment, the sample was cooled in N_2 at a rate of 5K/minute. The weight loss of the sample after this process was 1.68%, which if it is solely attributed to oxygen, gives the final composition as $SrCoO_{2.31+\delta}$. This is the final structure for our analysis.

$[010]_p$ $[110]_p$

Figure 2: Incoherent Z-contrast lattice images of vacancy ordered $SrCoO_{3-\delta}$ acquired along (a) the $[010]_p$ and (b) the $[110]_p$ axis. P denotes the cubic perovskite reference lattice.

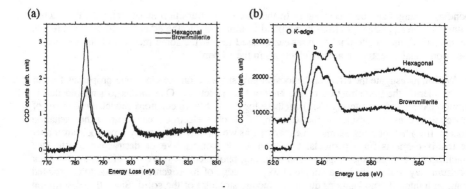

Figure 3: Experimental Co L_{23}-edge and O K-edge energy loss spectra from the as-sintered (hexagonal) and treated (brownmillerite) $SrCoO_{3-\delta}$ samples.

RESULTS

After the oxidation/reduction cycling process, the sample showed two prominent features; cobalt oxide precipitates at grain boundaries and in triple junctions and the presence of ordered micro-domains. Figure 2 shows the Z-contrast images from two projections of the ordered domains. As the image contrast is sensitive to the atomic number, the strontium columns can be distinguished from columns containing cobalt and oxygen atoms. It is clear from the image that in both projections there is a periodical decrease in the intensity of every alternate CoO_2 plane. This superstructure could simply be due to missing atoms in these planes. However, as oxygen is the lightest element in the structure and contributes little to the Z-contrast image, there would have to be cobalt vacancies also. A more likely alternative is that there is a structural reason for the change in contrast (in addition to oxygen vacancies). A shift in the positions of the cobalt atoms to achieve tetrahedral coordination in planes containing oxygen vacancies and a corresponding alternate change of Sr lattice spacing, would be consistent with the formation of the brownmillerite structure [10].

To investigate further the possibility of the formation of the brownmillerite phase, we can acquire energy loss spectra from the micro-domains. Figure 3 shows cobalt L_{23}-edge spectra and oxygen K-edge spectra from the untreated and treated samples. As can be seen from the cobalt spectrum, there is a change in the L_3/L_2 edge intensity ratio. Comparing the values obtained from this experiment with those in the literature [11], implies that there is a change in cobalt valence from 2.9±0.3 in the hexagonal sample to a value of 2.4±0.3 for the treated sample. This is consistent with the presence of an increased number of oxygen vacancies in the treated sample. A similar conclusion can be drawn from the changes in the fine-structure of the oxygen K-edge spectra. In particular, it can be noticed that the first peak decreases considerably in the treated sample. As the oxygen 2p levels are hybridized with the transition metal 3d levels, this decrease is consistent with a filling of empty states in the 3d-band, i.e. a decrease in the valence state.

Figure 4 shows the self-consistent multiple scattering simulations [9] of the oxygen K-edges expected from hexagonal, perovskite and brownmillerite $SrCoO_{3-\delta}$. The simulated brownmillerite and hexagonal spectra contain the sum of contributions from all the oxygen sites in the unit cell, whereas the perovskite structure only has a single unique oxygen environment. As can be seen from the figure, there are significant differences between the spectra obtained from each different structure. The simulated results of hexagonal and brownmillerite structures agree well with experimental EEL spectra (Figure 5). In particular, the behavior of the major peaks (a, b and c) as well as broad extended fine structure peak between 560 and 580 eV are well reproduced. Discrepancies of finer peaks observed in simulated spectra may possibly be due to local structural differences around a particular oxygen site in the experimental spectrum. However, these changes are minor and it is clear from the simulations that the oxidation/reduction treatment results in the formation of micro-domains of the brownmillerite structure.

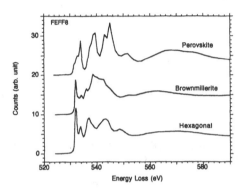

Figure 4: Multiple scattering simulations of the hexagonal, perovskite and brownmillerite structures

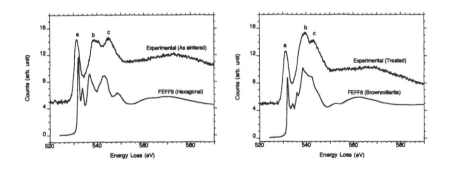

Figure 5: Comparison between the experimental spectra and the multiple scattering simulations for the as-sintered and treated samples.

CONCLUSIONS

The results show that the combination of Z-contrast imaging and EELS in the STEM can be used to perform atomic scale measurements on the defect chemistry in oxygen deficient perovskites. In particular, the Z-contrast technique allows the presence of ordered micro-domains to be imaged directly and the fine-structure of the energy loss spectrum allows the cation valence and the oxygen coordination to be determined. In the future, it should be possible to perform many of the same experiments at elevated temperatures. This should allow the defect chemistry models at or near the actual operating temperatures to also be quantified on the atomic scale, thereby permitting the structural, chemical and mechanical properties of the ceramic membranes to be determined.

ACKNOWLEDGMENTS

The authors would like to acknowledge support from BP chemicals for this work. The JEOL 2010F used to obtain the experimental results was purchased with support from the National Science Foundation under grant number NSF-DMR-9601792, and is operated by the Research Resources Center at the University of Illinois at Chicago.

REFERENCES

[1] H. J. M. Bouwmeester, A. J. Burggraaf, *CRC Handbook of Solid State Electrochemistry* (eds. P. J. Gellings, H. J. M. Bouwmeester), Ch 14. pp. 481-553, CRC Boca Raton, FL (1997).

[2] S. J. Pennycook and D. E. Jesson, *Phys. Rev. Lett* **64**, 938 (1990).

[3] R. F. Egerton, *Electron Energy Loss Spectroscopy in the Electron Microscope* (Plenum, 1996).

[4] N. D. Browning, M. F. Chisholm and S. J. Pennycook, *Nature* **366**, 143 (1993).

[5] E. M. James and N. D. Browning, *Ultramicroscopy* **78**, 125 (1999).

[6] S. Stemmer, A. Sane, N. D. Browning, T. J. Mazanec, in press *Solid State Ionics*

[7] J. Rodriguez, J. M. González-Calbet, J. C. Grenier, J. Pannetier, M. Anne, *Solid State Commun.* **62**, 231 (1987).

[8] P. D. Nellist and S. J. Pennycook, *Ultramicroscopy* **78**, 111 (1999).

[9] A. L. Ankudinov, B. Ravel, J. J. Rehr and S. D. Conradson, *Phys Rev B* **58**, 7565 (1998).

[10] J. Berggren, *Acta Chemica Scandinavica* **25**, 3616 (1971).

[11] Z. L. Wang, J. S. Yin, Philos. Mag. B **77**, 49 (1998).

ENERGY DISPERSIVE X-RAY SPECTROMETRY WITH THE TRANSITION EDGE SENSOR MICROCALORIMETER: A REVOLUTIONARY ADVANCE IN MATERIALS MICROANALYSIS

DALE NEWBURY*, DAVID WOLLMAN**, KENT IRWIN**, GENE HILTON**, and JOHN MARTINIS**
*National Institute of Standards and Technology, Gaithersburg, MD 20899-8371
**National Institute of Standards and Technology, Boulder, CO 80303

ABSTRACT

The NIST microcalorimeter energy dispersive x-ray spectrometer provides important advances in x-ray spectrometry. The high spectral resolution, approaching 2 eV for photon energies below 2 keV, the wide photon energy coverage, 250 eV to 10 keV, and the energy dispersive operation enable a wide range of materials characterization problems to be addressed. This performance is especially critical to high spatial resolution, low-voltage x-ray microanalysis performed with the field-emission gun scanning electron microscope.

INTRODUCTION

Materials microanalysis by electron-excited x-ray spectrometry currently depends upon semiconductor (usually Si) energy dispersive x-ray spectrometry (Si-EDS) and diffraction-based wavelength dispersive spectrometry (WDS). EDS and WDS are highly complementary. Si EDS provides a continuous view of the entire photon energy range of analytical interest (0.1 keV to 15 keV) that is critical for rapid qualitative and quantitative analysis, but the modest spectral resolution (130 eV at $MnK\alpha$) leads to frequent peak interferences and poor detectability. WDS offers high spectral resolution (2 eV - 12 eV) that is critical to overcoming peak interferences and enhancing detectability, but WDS requires mechanical scanning to view the spectrum and its quantum efficiency is poor. Both factors impose a significant time penalty when a wide range of the spectrum must be viewed. The combination of EDS and WDS has proven to be highly effective for conventional microanalysis practice with incident beam energies $E_0 \geq 10$ keV and beam currents (for WDS) in the 10 nA - 500 nA range [1]. Since the electron range decreases sharply as $E_0^{1.7}$, the development of the high performance, low-voltage field-emission scanning electron microscope (FEG-SEM) has led to increased interest in analysis with beam energies below 5 keV. For such beam energies, the lateral and depth dimensions of x-ray excitation lie in the range 10 nm - 100 nm, depending on E_0 and the material composition. Such analytical spatial resolution approaches that previously achievable with analytical transmission electron microscopy, but with the enormous additional benefit that the specimen does not need to be thinned to achieve electron transparency, permitting "as-received" or process stream samples to be directly analyzed with the low-voltage FEG-SEM.

However, low-voltage operation of the FEG-SEM is not without cost. Typical probe currents achieved under low-voltage conditions are in the pA to low nA range for nanometer probe sizes with cold field emission. Thermally-assisted field emission can achieve higher total

75

beam currents but at the expense of substantially larger probe size because of the lower brightness of such sources. Low beam current effectively precludes using conventional WDS because the time penalty becomes too great for practical applications.

A second inevitable consequence of low-voltage operation is the restriction in the atomic binding energies that can be probed. The overvoltage, U, which is the ratio of the incident energy to the binding or critical excitation energy, $U = E_0/E_c$, must exceed unity to begin to ionize a given shell. The ionization efficiency is a complex function of overvoltage, but for a bulk target, the ionization is proportional to the overvoltage to a power n, $I \sim U^n$, where n is in the range 1.3 to 1.7, depending on atomic number. The background of the spectrum is, ideally, due only to the x-ray bremsstrahlung. This continuous radiation has an intensity proportional to an overvoltage factor U, with the bremsstrahlung energy E_v substituted for E_c. The peak-to-background ratio, P/B, which is a critical factor in determining the limit of detection, is thus U^n/U, or U^{n-1}, which gives a value of approximately $U^{0.5}$. The P/B thus drops very sharply as the primary excitation energy decreases. In practice, a value of U of at least 1.25 is required as an effective minimum value to detect a peak from a major constituent in the sample (e.g., present at a level of 0.1 mass fraction or higher). These restrictions frequently force the analyst to use shells that differ from the normal choice, e.g., L rather than K, M rather than L, and N rather than M. Unfortunately for many elements of intermediate and high atomic number, these shells have low fluorescence yields. Finally, many of the L, M, and N lines for these elements lie in the spectral vicinity of carbon and oxygen. C and O are found in many specimens of interest, and their fluorescence yields are relatively high compared to the heavier elements. The spectral interference of C-K and O-K may eliminate access to these elements when they are present as minor (0.01 to 0.1 mass fraction) and trace (≤ 0.01 mass fraction) constituents. The combination of low P/B, low fluorescence yield, and C and O interference effectively eliminates many elements from detection with a low resolution spectrometer such as Si EDS.

NIST MICROCALORIMETER X-RAY SPECTROMETER

The NIST microcalorimeter energy dispersive x-ray spectrometer (µcal EDS) represents a significant leap forward in x-ray spectrometry capabilities, effectively combining the energy dispersive operation of the Si EDS with the resolution of WDS. The structure and operation of the NIST microcalorimeter have been described in detail previously [2], and therefore this paper will concentrate on applications to x-ray microanalysis. Resolution and photon energy range are illustrated in Figure 1 in comparison to the performance of a conventional Si EDS and various WDS diffractors. The first generation µcal EDS gave approximately 8 eV resolution at MnKα, similar to the resolution of the LiF diffractor in WDS. The resolution performance of the latest generation of microcalorimeters specifically adapted to low photon energies below 2.5 keV is shown approaching 2 eV. It is noteworthy that this is much better resolution than that achieved for this energy range by WDS with layered synthetic microstructure (LSM) diffractors. The limiting count rate of the µcal EDS is approximately 1 kHz, which can be compared to a value of 3 kHz for an Si EDS operating at the best resolution limit. Because of its small cross section, typically 0.25 mm² and the specimen-to-detector distance of approximately 40 mm, the µcal EDS has a small solid angle compared to the Si EDS (e.g., 10 mm² at 20 mm specimen-to-detector

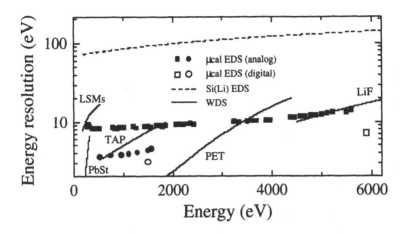

Figure 1 Resolution as a function of photon energy for various spectrometers.

distance). This deficiency has been partly compensated in the current design of the NIST μcal EDS by incorporating a double-focussing polycapillary x-ray optic to couple the emission from the beam-excited location on the specimen to the detector. A gain in solid angle of approximately a factor of 300 that is reasonably constant over a wide photon energy range, e.g., Ti-L (450 eV) to Ti-K (4500 eV), has been realized with this polycapillary x-ray optic. In this paper, results are reported from a variety of specimens examined in a conventional lanthanum hexaboride gun SEM fitted with a μcal EDS/polycapillary optic. From these results, the performance of an analytical system based upon a FEG-SEM can be predicted.

RESULTS
1. Separation of Components in a Complex Target

Figure 2 shows spectra of BaTiO$_3$ obtained under low-voltage analysis conditions with a conventional Si EDS [2(a)] and with the μcal EDS [2(b)]. Oxygen is by far the most prominent peak although oxygen comprises only 0.206 mass fraction of the target. In the conventional Si EDS spectrum, the Ti-L peaks (0.205 mass fraction) can only be detected as a slight distortion on the low energy shoulder of the oxygen peak. The Ba-M family peaks (0.589 mass fraction) are observed, but they do not match the database well. The μcal EDS spectrum shows the peaks for all three elements well separated, and with this spectrum we can understand the limitations imposed by the poor resolution of the Si EDS. Despite the concentration, the Ti-L family peaks occur at a much lower intensity relative to the O-K, due to fluorescence yield effects and detector window transmission. The Ba-M family peaks are also found at low intensity relative to the O-K due to low fluorescence yield. The Ba-M peaks that are actually observed differ substantially from those expected from the database due to incorrect data on relative within-family yields. Indeed, with the restriction to low-voltage excitation conditions, it would be difficult to determine

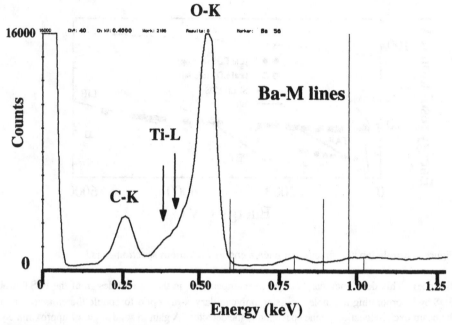

Figure 2(a) Conventional Si EDS spectrum of BaTiO₃; beam energy 3 keV.

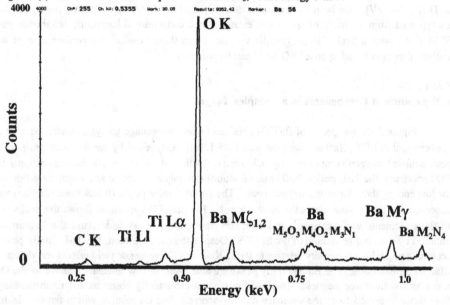

Figure 2(b) NIST μcal EDS spectrum of BaTiO₃.

from the conventional Si EDS spectrum in Figure 2(a) that the target is actually BaTiO$_3$, while there is no such ambiguity with the μcal EDS spectrum.

2. Limits of Detection

Under conventional SEM/semiconductor EDS analysis conditions, analytical sensitivity is sufficient in the absence of peak interferences to simultaneously measure major, minor, and trace constituents. Can the same sensitivity and dynamic range be achieved with low-voltage SEM/microcalorimeter EDS? An interesting case illustrating the limits of detection imposed when using unconventional shells for analysis is provided by the spectrum of barium titanate, shown in Figure 2(b). With beam energies ≤ 5 keV, the conventional choice of Ti-K and Ba-L must be replaced by Ti-L and Ba-M. Figure 2(b) demonstrates that the microcalorimeter EDS can indeed detect these x-ray photons, but this figure also illustrates the low relative yield of Ti-L and Ba-M compared to O-K. Such low yields of these L- and M- peaks inevitably impact detectability negatively. A concentration limit of detection C$_{DL}$ (mass fraction) can be calculated from equation (1) [ref 1]:

$$C_{DL} = \frac{3\sqrt{N_B}}{\sqrt{n}(N_P - N_B)} C_{std} \tag{1}$$

where N$_P$ is the number of peak counts, N$_B$ is the number of background counts, n is the number of repeated measurements, and C$_{std}$ is the concentration of the element in the standard. Using the peak with the best peak-to-background ratio for each element in Figure 2(b), Table 1 gives the limits of detection for a single measurement. For this particular dose and spectrometer efficiency, titanium and barium can only be detected down to the level of minor constituents, while oxygen is detectable well into the trace range below 0.01 mass fraction. Some improvement in the limit of detection for titanium and barium could be obtained by adding all of the family peaks.

Table 1 Calculation of Limit of Detection (weight fraction, single measurement)
(Material: BaTiO$_3$; beam energy 3 keV; prototype low energy NIST μcal EDS)

Element-Line	C$_{STD}$ (mass frac.)	Peak (cts)	Background (cts.)	Detection limit C$_{DL}$
O - K	0.206	17627	1083	0.0012 mass frac.
Ti - L	0.205	1470	663	0.020
Ba - M	0.589	3009	1093	0.030

Further exploration of the nature of L-, M-, and N- shell x-ray spectra from pure element and compound targets measured with the NIST microcalorimeter will be required to fully elucidate the practical analytical situation for low beam energy excitation and unusual shell selection.

3. Measurement of Chemical Shifts in X-ray Spectra

Characteristic photons below 2 keV in energy are created as a result of transitions of outer shell electrons, the energy of which may be modified by chemical bonding effects, leading to shifts in the peak position and in peak shape. Moreover, for the L-family of elements in the first transition series, at least one of the absorption edges occurs among the peaks, so that relative peak heights can change simply as a result of the concentration of the element in the target. Both of these effects can be used to "fingerprint" the "chemical" nature of the measured volume, giving

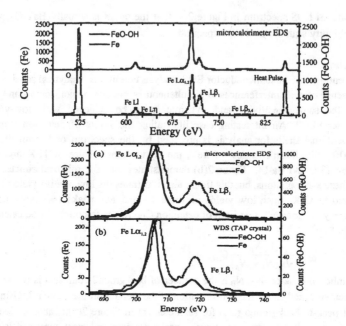

Figure 3 Iron (lower plot) and goethite (FeO-OH; upper plot) measured with (a) the μcal EDS and (b) WDS (TAP diffractor); note that the WDS numerical scale is normalized for dose.

valuable additional information to complement the elemental information which is normally obtained and which is often used to make indirect inferences as to the presence of chemical compounds. This chemical information has previously been accessible only with high resolution WDS measurements. Figure 3 demonstrates that the μcal EDS can achieve similar results with the added benefit that the entire excited spectrum is measured simultaneously. Moreover, the μcal EDS offers substantially better resolution for low energy photons below 600 eV than the synthetic diffractors used with WDS.

CONCLUSIONS

The NIST μcal EDS equipped with a polycapillary x-ray optic to increase the effective solid angle will provide important new capabilities for low-voltage microanalysis performed in the FEG-SEM. Indeed, many materials characterization problems cannot be addressed with existing semiconductor EDS and WDS and will require performance at the level of the μcal EDS.

REFERENCES

1. Goldstein, J. I., Newbury, D. E., Echlin, P., Joy, D.C., Romig, A. D., Jr., Lyman, C. E., Fiori, C., and Lifshin, E., Scanning Electron Microscopy and X-ray Microanalysis, 2nd edition (Plenum Press, New York, 1992).
2. Wollman, D. A., Irwin, K. D., Hilton, G. C., Dulcie, L. L., Newbury, D. E., and Martinis, J. M., "High-resolution, energy-dispersive microcalorimeter spectrometer for X-ray microanalysis", J. Micros., 188, (1997) 196-223.

SURFACE SENSITIVITY EFFECTS WITH LOCAL PROBE SCANNING AUGER-SCANNING ELECTRON MICROSCOPY

D. T. L. VAN AGTERVELD, G. PALASANTZAS, J.TH.M. DE HOSSON
Department of Applied Physics, Materials Science Centre and Netherlands Institute for Metals Research, University of Groningen, Nijenborgh 4, 9747 AG Groningen, The Netherlands, e-mail: hossonj@phys.rug.nl

ABSTRACT

Ultra-high-vacuum segregation studies on in-situ fractured Cu-Sb alloys were performed in terms of nanometer scale scanning Auger/Electron microscopy. S contamination leads to the formation of Cu_2S precipitates which, upon removal due to fracture, expose pits with morphology that depends on the precipitate size and shape. Local variations of S and Sb distributions inside the pits were correlated to local surface orientations as Atomic Force Microscopy analysis revealed.

INTRODUCTION

Auger Electron Spectroscopy (AES) has opened a wide range of surface analysis possibilities for fundamental and applied research. AES yields chemical composition information for surface layers within the range of depth ~0.3-5 nm depending on the (element) peak energy. Application of AES in the fields of microelectronics, catalysis, polymer and metallurgy technology has made it an extremely powerful tool, e.g. for understanding "impurity" induced failure mechanisms, morphology related preferential absorption of impurities, surface degradation mechanisms, etc.

Segregation of impurities to grain boundaries (GBs) in polycrystalline solids produces zones with discrete compositions and properties, which may control the overall performance of the solid. More precisely, segregation can alter the interface energy and may affect the cohesion between GBs, resulting in material failure [1,2]. Solute segregation to the heterophase interfaces is also known to affect the adhesive strength at interfaces [3]. However, in contrast to segregation at grain boundaries, only very few studies have addressed segregation in particular at metal/ceramic interfaces [4]. Especially the dissolution of 4 at.% of Sb in the Ag matrix showed according to HRTEM observations two major effects on the Mn_3O_4 precipitates [5] : (i) a change from a precipitate sharply facetted by solely {111} to a globular shape with sometimes also short {220} and (002) facets and (ii) a partial reduction of Mn_3O_4 to MnO for a part of the precipitates. Macroscopic properties of materials can be better understood if, and only if, the related mechanisms are thoroughly investigated at nanometer and/or subnanometer length scales. If these mechanisms can be controlled, future materials can be fabricated for specific applications. Therefore, we employ nanometer scale (lateral resolution of ~15 nm) UHV- Auger Electron Spectroscopy (AES), combined with Scanning Electron Microscopy (SEM) to study segregation phenomena in Cu-Sb alloys. Since the zone where segregation effects are most pronounced extends only a few monolayers from the grain boundary and the escape depths of Auger electrons are in the nanometer range for all elements, Auger Electron Spectroscopy is a suitable technique for segregation studies.

Polycrystalline Cu-Sb alloys were chosen as a model system for segregation studies at GBs, since Sb mainly segregates at GBs due to size effect (Sb atoms are larger than Cu atoms). At first sight Sb is expected to segregate preferably at GBs consisting of higher index planes because it will be more easily accommodated there than at more densely packed low index planes. On the other hand, the open (excess) volume in the GB or heterophase interface may be the most important factor for segregation sites, and low index planes in GBs tend to be low energy (low excess volume) boundaries. S, which appeared in our samples as contamination, also segregates to GBs but with a different mechanism. The solubility of S in Cu is limited because of the formation of the very stable intermediate phase Cu_2S. A strong interaction between S impurities and vacancies in copper was observed previously, which effectively leads to the formation of highly stable and mobile sulfur-vacancy defect complexes. The S-vacancy and S-S pair interactions favor precipitation of Cu_2S at lattice defects like GBs and dislocations [6]. The presence of Cu-sulphide precipitates at GBs is rather undesirable due to structural transformations that occur with temperature, favouring, in combination with the mechanical properties of Cu-sulphide, crack nucleation and propagation at GBs under creep conditions.

Our aim in the present work is to determine locally the amount of S segregation in connection with the corresponding surface morphology and orientation.

EXPERIMENTAL

Cu-Sb samples were prepared by melting together Cu and Sb in a graphite crucible at 1100° C (for 1 hour), and homogenised afterwards at 750° C for 5 hours. The specimens were cylinders with diameter 3 mm and length 15 mm. The graphite crucible was the source of the S 'contamination' in the specimens, as investigations revealed. Although the presence of S is caused by an uncontrolled process, S induced the same characteristics on all fracture surfaces. The SAM/SEM instrument that is currently used is a modified field emission JEOL JAMP 7800F Microscope, equipped with an impact-bending device for in-situ fracture at a base pressure of $\sim3*10^{-8}$ Pa. The electron probe size is ~15 nm at 10 keV accelerating voltage and $2.4*10^{-9}$ A emission current. Depth profiling is performed in-situ by 1 keV Ar^+-bombardment. Finally, the AFM is an air Nanoscope II from Digital Instruments equipped with a scanner of 9x9 µm maximum scan area.

RESULTS-DISCUSSION

Local Scanning Auger Analysis

Cu-Sb alloys with Sb concentrations between 1 and 2 at.% were fractured in-situ. In intergranularly fractured areas, the exposed surface consists of former GBs. On the exposed GB areas a number of pits with various shapes was found. S contamination from the graphite crucible caused the formation of Cu_2S precipitates. Upon fracture these precipitates are (in some cases partially) removed, leaving behind pits (Fig. 1). Energy Dispersive X-Ray Spectroscopy (EDS) on precipitate leftovers showed the concentrations of Cu and S to be ~68 and ~32 at.% respectively which correspond to Cu_2S. Inside pits, facets can be discerned with shapes and relative orientations characteristic of an fcc structure (Fig. 1, left). In other pits, stepped structures (with ledges and terraces) are formed in between facets, depending on the shape and size of the formerly present precipitate. Below a certain pit size (~4 µm), only facets are observed inside the pits (Fig. 1, right).

During acquisition of Auger maps, the Cu-LMM (916 eV), Sb-MNN (452 eV) and S-LVV (146 eV) direct spectrum peaks were used. The ratio (P-B)/B is displayed as intensity reducing topology effects. P and B are respectively the intensities of Auger electrons at the corresponding peak and background energy at the high energy side of the peak. Auger maps showed that S was localized solely in various amounts on the facets inside pits (Fig. 2-4). As the arrow shows (in Fig. 2) the S distribution on ledges with a size <200 nm is clearly resolved. S influences the fracture behavior of Cu-Sb merely through the formation of weak spots, by precipitation, which may promote fracture. Based on crystallographic arguments, we assume the hexagonally shaped facets, where the amount of S is very low or zero, to be {111} planes in Cu. On ledge-terrace structures, S is almost absent on the terraces that must have the same orientation as the hexagonal facets, since they are parallel.

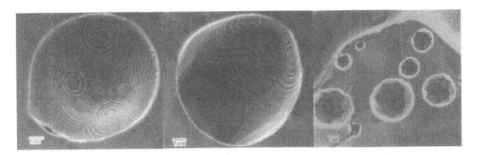

Figure 1 SEM images of pits that were observed on the fracture surface upon in-situ fracture of Cu-Sb alloys.

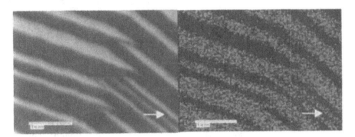

Figure 2 Scanning electron and scanning Auger images of S on in-situ fractured Cu-Sb alloys.

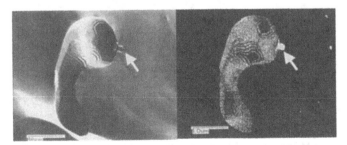

Figure 3 Large scale scanning electron and scanning Auger images of S on in-situ fractured Cu-Sb alloys. The arrow shows a precipitate

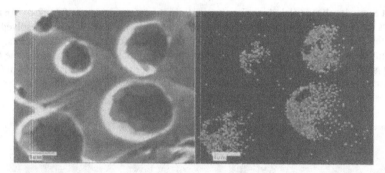

Figure 4 Scanning electron and scanning Auger images of S inside faceted pits.

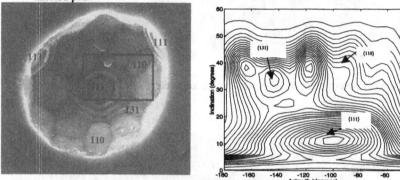

Figure 5 SEM image of the pit used for AFM analysis. The rectangle depicts the AFM scanning area that was used to obtain the radial histogram, shown on the right.

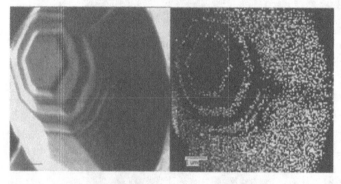

Figure 6 SEM image and Auger map of S inside a faceted pit enabling direct comparison of the S distribution on facets and step-terraces.

Atomic Force Microscopy Analysis

Atomic Force Microscopy (AFM) analysis was used to obtain absolute angles between facets inside pits. The images were taken in air. The technique used to obtain quantitative orientations of facets, from which angles between facets can be calculated, is an adaptation of the radial-histogram transform [7], described in [8]. Using a Savitsky-Golay filter, the AFM image is converted into a histogram showing the distribution of the relative amount of image area as a function of its surface normal, where the surface normal is described by the spherical coordinates θ (inclination angle) and ϕ (azimuth angle). The poles in Fig. 5 (right) depict relatively large areas of the image with the same orientation. The pole at $\phi \sim 100°$ and $\theta \sim 11°$ corresponds to the central facet and the terraces that have the same orientation. The three other poles and the two shoulders correspond to the five facets in the pit examined with AFM, which are under an angle between 26 and 35° with the central facet.

From Fig. 5, it becomes clear that around the hexagonally shaped central facet shown in Fig. 6 are actually 12 ledges, which become clearer when the distance to the hexagonal central facet increases. The intensities of S are different from ledge to ledge. The ledges have the same orientation as the facets that would otherwise be adjacent to the {111} central facet and form, together with the terraces, a 'transition' between facets, which is necessary to accommodate asymmetrically shaped, larger precipitates. Shape and symmetry arguments suggest that of these 12 adjacent facets, the three between the central {111} and other hexagonal facets (also {111} planes) must be {110} and the three between the {111} and square {100} facets might be {131}, as depicted in Fig. 5.

Therefore, regarding the information obtained by AFM, and considering that the six remaining unknown facets must be of the same family of planes, we suggest that these facets and therefore also the accompanying ledges, are {135} planes which make angles of ~29° with {111} planes. The poles and shoulders in Fig. 5 then correspond to, from left to right, {135}, {131}, {135}, {110} and {135} facets and ledges that are in between facet and central facet. Returning to Fig. 6 (where the S distribution is depicted on facets and step-terraces), the ledges and facets with the highest amount of S at the surface are likely to be the highest index planes, where it is easier to accommodate impurities. The small facet on the right, where the amount of S is lower, is likely to be a {110} facet in between two {111}'s.

S and Sb site competition

Sb was found to be distributed rather uniformly on the exposed fracture surface and was also found inside pits. Ar^+-ion depth profiling showed that the depth over which the amount of Sb is elevated (segregation zone), compared to the bulk, is ~2 nm or less. Inside the pits, the amount of Sb was relatively high at those areas where S was almost absent and was lower, but never absent, at the areas where the amount of S was higher (Fig. 7). The fact that Sb is always detected might be due to the higher inelastic mean free path of Sb Auger electrons (~0.9 nm at 452 eV peak) enabling detection of Sb from greater depths. However, for S the inelastic mean free path is rather low (≤ 0.5 nm at the 146 eV peak) thus yielding purely surface sensitive information.

Since the Cu_2S -precipitates have a very high melting point (1130-1131 °C), these are formed first during the cooling process of the Cu-Sb alloy. Later, Sb starts to segregate to grain boundaries and Cu/Cu_2S interfaces. Sb may segregate to {111} planes (see Fig. 7 hexagonal shape) of Cu at the precipitate-matrix interface, where the amount of S is very low. This behavior is different from 'pure' polycrystalline Cu-Sb alloys where the amount of Sb at {111} planes (the most densely packed ones) is expected to be less than other planes.

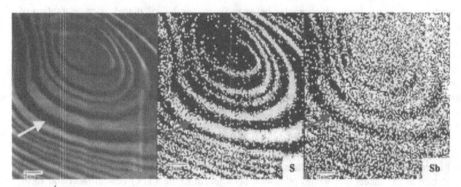

Figure 7. SEM and Auger images of S and Sb distribution (after in-situ fracture) at a step-terrace structure formed on a GB. The arrows shows the presence of S on a ledge and the corresponding positions on the SAM images.

CONCLUSIONS

Templating of Cu_2S precipitates in a Cu-Sb matrix causes, by accommodation of the matrix, the formation of various pits which are exposed after removal of the precipitates by in-situ fracture. Inside the pits facets and step-terrace structures are observed depending on the shape and size of the precipitate. Scanning Auger maps revealed that the intensities of S and Sb are different on different facets inside pits. From Auger maps, AFM measurements, and crystallographic shape arguments, we may conclude that on the hexagonally shaped {111} facets and other low index planes, the S content is lower than on high index planes and vice versa for Sb.

ACKNOWLEDGEMENTS

Financial support from the Netherlands Institute for Metals Research and the 'Stichting voor Fundamenteel Onderzoek der Materie' (FOM-Utrecht) is gratefully acknowledged. We would like to acknowledge assistance help from B. Kooi, H. Hegeman, and P. Balke.

REFRENCES

[1] M.P. Seah in Practical Surface Analysis, Vol. 1 (Ed. D. Briggs, M.P. Seah), John Wiley & Sons Ltd., Chicester, UK, 1990, Chapter 7
[2] R.W. Balluffi in Interfacial Segregation (Ed. W.C. Johnson, J.M. Blakely), American Society for Metals, Metals Park, Ohio, USA, 1979.
[3] J.R. Smith, T. Hong, D.J. Srolovitz, Phys. Rev. Lett. **72** , p.4021 (1994).
[4] D.A. Shashkov, D.N. Seidman, Appl. Surf. Sci. , **94/95**, 416 (1996).
[5] J.Th.M. De Hosson, H.B. Groen, B.J. Kooi, V.Vitek, Acta Mater. , 47, 4077, 1999.
[6] P.A. Korzhavyi, I.A. Abrikosov, B. Johansson, Acta. Mater., 47 (199), 1417.
[7] D. Schleef, D.M. Scheafer, R.P. Andres, R. Reifenberger, Phys. Rev. B **55** (1997) 2532.
[8] J.B.J.W. Hegeman, B.J. Kooi, H.B. Groen, J.Th.M. De Hosson, J. Appl. Phys. **68** (1999), 3661.

X-RAY ELEMENTAL MAPPING OF MULTI-COMPONENT STEELS.

ADAM J. PAPWORTH AND DAVID B. WILLIAMS
Lehigh University, Materials Research Center, 5 East Packer Avenue, Bethlehem, PA 18015.

ABSTRACT

Steels used in steam turbine applications are susceptible to temper embrittlement if operated at temperatures above 400°C. The cause of this embrittlement is the segregation of impurity elements, mainly phosphorus, to the prior austenite grain boundaries. Molybdenum can act as an effective scavenger for phosphorus, but the scavenging effect is lost when the molybdenum is precipitated in carbides during service at these elevated temperatures. Thus, the very slow temper embrittlement is controlled by the rates of alloy carbide formation, rather than by the diffusion of phosphorus. The presence of vanadium apparently retards the embrittlement process even more by interfering with the formation of the molybdenum-rich carbides. Vanadium carbonitrides, are small, only a few nanometers in size, and difficult to see by standard TEM contrast mechanisms. Analysis of segregating elements to the grain boundaries has to be carried out on areas which are devoid of precipitates, especially small V precipitates or other precipitates that could be interpreted as segregation rather than precipitation. One method of detecting precipitates in the matrix is by X-ray mapping areas of interest. This paper describes a new method of X-ray mapping of multi-component steels that have overlapping lines. The method has been found to be effective in removing thickness effects from the maps and discriminating between real and artificial compositional components, which are created by the overlapping lines.

INTRODUCTION

"Clean" 3.5NiCrMo steels containing trace elements (P, Sn, As, Sb) are commonly used in the manufacture of large rotor shafts, which are used in the "low pressure" component of steam turbines. Unfortunately, if these steels are used at temperatures above 400°C, they become temper embrittled with time [1]. Increases in efficiency have raised the temperature at which these steels now operate. In an attempt to overcome the problem of temper embrittlement, "super-clean" steels have been developed, in which the Mn, Si and P content have been drastically reduced [1]. Unfortunately, these super-clean steels still embrittle over time [2], when operating at temperatures above 400°C. The embrittlement of these steels is due to the segregation of P to the prior austenite grain boundary (PAGB) at these elevated temperatures [1,2,3]. Although Mo, in the right proportion, acts as an effective scavenger for P, this effect is lost as Mo carbides slowly precipitate during long-term exposure at elevated temperatures [3]. The addition of V has been found to slow the embrittlement process, but not stop it. It is believed that V, a strong carbide former [4] slows the embrittlement process by competing with Mo in the formation of carbides/carbonitrides, therefore reducing the rate that P is released. Although it is known that V forms very small carbides/carbonitrides a few nanometers in size, the mechanism by which P segregation is retarded is unclear as not all strong carbide formers reduce the rate of temper embrittlement [5]. Studying the effects of small carbides/carbonitrides on temper embrittlement requires very accurate analysis, which can only realistically be performed in a STEM as compositional changes will be small. The main problem with these small precipitates is that they can not be detected in either the TEM or STEM images due to their size and lack of contrast, but as shown here, these small precipitates can be detected by the use of elemental X-ray mapping.

87

This paper also deals with the problems associated with detecting such small precipitates in multi-component steels where there are overlaps between the characteristic X-ray lines from the various elements.

EXPERIMENT

TEM specimens of a 3.5NiCrMoV low-carbon steel were made by manually polishing 3 mm discs to a nominal thickness of 40 μm. The discs were then ion beam thinned at an angle of 4°, using a Gatan precision ion polishing system (PIPS) until a hole appeared. The analysis of the chemical composition was carried out using a VG HB603 FEGSTEM giving a probe size of 1.4 nm (FWTM) with a beam current of 0.5 nA. The STEM uses a windowless Si(Li) X-ray detector, which has a large solid angle of detection (0.3 sr). X-ray acquisition was carried out on an Oxford (Link) exl system, where elemental windows were defined over the K_α lines of the following elements, C, O, Al, Si, P, V, Cr, Mn, Fe, Ni, Cu, Mo and the L_α line of Mo. Two normalizing backgrounds were defined at 3.3 to 3.8 and 10.0 to 12.0 keV energies respectively. Elemental X-ray maps had an acquisition time of 100 ms per pixel with a 128×128 pixels resolution. Digital line-scans had a scan length of 64 nm with a total of 64 spectra being taken along the length of the scan. Each spectrum in the line-scan had an acquisition time of 5 seconds, and was normalized with respect to the background to remove the effects of thickness.

RESULTS AND DISCUSSION

The acquisition of elemental maps by the use of energy windows has been used for many years, and a typical use of this type of map is shown in Figure 1.

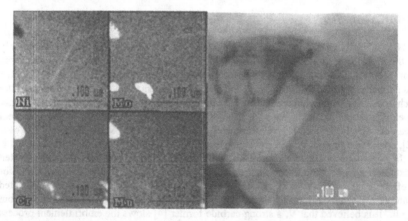

Figure 1. Elemental X-ray maps for Ni, Mo, Cr and Mn, with corresponding BF image. The Ni and Mo maps show that the PAGB is not uniform in composition, which changes as the boundary plane changes.

The four maps in Figure 1 are very effective in showing the precipitates that are not visible in the BF image, and that Mo and Ni have segregated to different parts of the PAGB. There are problems with this type of mapping, and it is these problems that are addressed in this paper. It must be noted at this stage that, modern developments in spectrum imaging, especially "position tagged spectrometry" [6] will in the future overcome most of the problems. Unfortunately such techniques have been developed for use in the SEM and not for the high spatial requirements of STEM EDX microanalysis, in which X-ray counts are limited.

The first problem is that there are no compositional look-up tables to these maps, i.e. the relationship between intensity and composition is not defined. Other work has shown that there is 3.2 ±0.3 wt% Ni in solid solution with the matrix Fe but the simple window map can not give this information. X-ray maps produced by this method can be successfully quantified as shown by Williams et al. [7]. Unfortunately, the method of Williams et al. requires the removal of the background at each pixel and no conflicts between the characteristic X-ray lines.   This highlights the second problem with the X-ray maps shown in Figure 1. Comparing the Cr and Mn maps, the precipitate in the top left hand corner is shown to contain Cr and Mn, and although the Fe map is not shown, Fe would also be present in to the precipitate. A typical energy window is seven channels wide, spanning an energy window of 140 eV. The Cr $K_{\beta 1}$ line is only 48 eV away from the Mn $K_{\alpha 1}$ line; therefore any Cr precipitate will also give a corresponding signal in the Mn map, although the signal in the Mn map will only be a tenth of the signal of the Cr map. In this study of 3.5NiCrMoV steel, we need to quantify precipitates that are a few nanometers in size. There are conflicts between the following characteristic X-ray lines, V $K_{\beta 1}$ and Cr $K_{\alpha 1}$, Cr $K_{\beta 1}$ and Mn $K_{\alpha 1}$, Mn $K_{\beta 1}$ and Fe $K_{\alpha 1}$, Mo L_1 and P $K_{\alpha 1}$, Mo L_{α} and S K_{α}, O K_{α} also conflicts with the Cr $L_{\alpha 1}$.

Quantification of thin film microanalysis using EDX spectrometry is performed by relating the intensities (I) of the characteristic X-rays above the background to the compositions (C) in wt% through the Cliff-Lorimer equation [8], for a binary system A,B where $C_A + C_B = 100\%$:

$$\frac{C_A}{C_B} = k_{AB} \frac{I_A}{I_B} \tag{1}$$

where k_{AB} is the Cliff-Lorimer sensitivity factor, called the k factor. This equation requires k factors for every combination of elements in the sample, which would be extensive for our steel. A variation is to determine the k factors to a reference element, such as Fe in our case. The equation then becomes:

$$\frac{C_A}{C_B} = \frac{I_A k_{FeA}}{I_B k_{FeB}} \tag{2}$$

This equation is fine for compositional analysis of a binary alloy of known uniform thickness and density, but in practice absorption, density and especially thickness changes have to be taken into account. This can also be overcome as shown by Watanabe et al [9] by the use of the ζ-factor method. Conversion of the Cliff-Lorimer equation to a multi-element system requires the denominator to be changed into the total intensity of the system:

$$I_A k_{FeA} + I_B k_{FeB} + I_C k_{FeC} + ... = \text{Total} \tag{3}$$

Then

$$C_A = \frac{I_A k_{FeA}}{\text{Total}} \quad \text{and} \quad C_B = \frac{I_B k_{FeB}}{\text{Total}} \quad C_C = ...\text{etc.} \tag{4}$$

Adding all the concentration up should equal 100%. Unfortunately if an element has been missed out from the energy windows then there could be a larger error associated with each concentration. Other sources of errors are the counting statistics in each peak, and the correct determination of the peak size, i.e. background removal and deconvolution of conflicting peaks.

Elemental X-ray mapping is always a compromise. For example over a period of time the sample will drift. Therefore, to minimize this drift the acquisition time has to be short, typically 100 ms per pixel. The total time for a 128×128 pixel map would be just under an hour, depending on the dead time, which is dependent on specimen thickness.

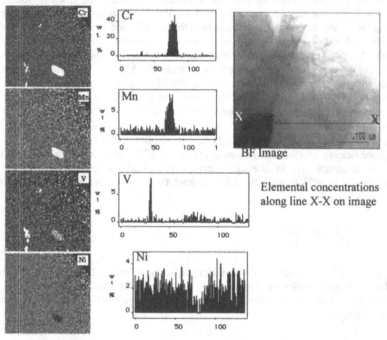

Figure 2. High-resolution X-ray maps that have been compositionally adjusted using equation 4.

High-resolution images of 3.5NiCrMoV steel have poor contrast, making drift correction impossible. Therefore, elemental X-ray maps will have short acquisition times giving poor statistics. This can be an advantage. It is a condition of equation 4 that the peak areas are correct for each element, i.e. all the peak overlaps have been deconvoluted from each other, and that the background is removed. For X-ray mapping this is not necessary; the poor statistics means that the background is negligible at energies above 4.5 keV when using a 100 ms acquisition time. It also means that conflicting peaks are also a minor consideration. For example the V $K_{\beta 1}$ is 12 eV from the Cr $K_{\alpha 1}$ line, but the V $K_{\beta 1}$ is only 11% of the height of the V $K_{\alpha 1}$ peak. As the V precipitate is small, the conflict with the Cr peak is negligible. In contrast the conflict between Cr and Mn does matter, although the $K_{\beta 1}$ line is 48 eV away, the Cr precipitates are large and the strong Cr K_α signal will influence the Mn K_α map.

Therefore to quantify this type of X-ray maps the following procedure can be followed. Assuming accurate k factors have been obtained, then each elemental map is multiplied by its respective k factor, then all the corrected maps are added together to give the total intensity. Each corrected elemental map is then divided by the total and then multiplied by 100 to give a percentage concentration. Figure 2 is an example of this procedure.

From the maps in Figure 2 it can be seen that there is Mn associated with the Cr precipitates. The concentrations possibly show that the Mn concentration is about 7 wt%; in other words we have mapped the Cr $K_{\beta 1}$ peak and therefore the Mn map can be neglected. With the removal of the Mn map all the other maps are then readjusted using equation 4 to give the set of maps given in Figure 3. A fully quantified spectrum with excellent statistics has given the matrix composition for this heat as Ni 2.6 ±0.3 wt% relative error, Cr 0.7 ±0.2 wt% relative error, Mo 0.3 ±0.1 wt% relative error, Mn 0.2 ±0.1 wt% relative error balance Fe. It can be seen that the error is very much greater in these maps, the Ni composition is still shown to be between 2 and 3%. A detection limit of a few percent within the matrix is what would be expected with a 100 ms acquisition time. This method of displaying elemental X-ray maps was developed to remove variation in intensity caused by thickness effects in the specimen. It was also hoped that it would distinguish between real and imaginary compositions where conflicts between different X-ray lines exist. Figures 2 and 3 show that all thickness effects are taken out and that it is reasonably simple to detect real and imaginary compositions. Closer analysis of the V precipitate shows that there is 30% Cr associated with the V. This amount is too much to be just the V $K_{\beta 1}$ peak and, therefore, must be real. Further analysis of V precipitates has shown them to contain ~20 wt% Cr.

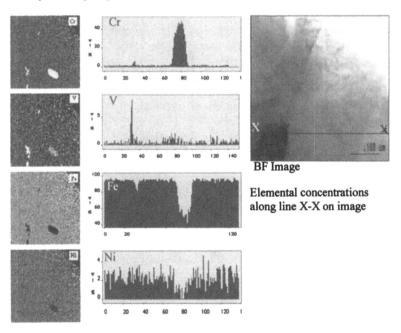

BF Image

Elemental concentrations along line X-X on image

Figure 3. High-resolution X-ray maps that have been compositionally adjusted after the removal of Mn.

CONCLUSIONS

1). Elemental X-ray maps can be quantified by the use of the Cliff-Lorimer equation, without the removal of the background when counting statistics are low.

2). Thickness variations effects are removed from the area mapped, giving only compositional variations.

3). Real and imaginary composition values caused by overlaps between X-ray peaks can be detected and removed.

4). This method gives reasonable estimates of composition in the matrix and precipitates alike, when compared with statistically significant spectra obtained from similar regions.

REFERENCES

1. G. Thauvin, G. Lorang, and C. Leymonie, Met. Trans. A. **23A** August, 2243 (1992).
2. Z. Qu, and C.J. McMahon, Jr, Met. Trans. A. **14A** June, 1101 (1983).
3. J.E. Wittig and A. Joshi, Met. Trans. A. **21A** October, 2817 (1990).
4. J.G. Speer, J.R. Michale and S.S. Hansen, Met. Trans. A. **18A** February, 211 (1987).
5. K.A. Taylor, S.S. Hansen, and R.I. Jaffee, Mechanical Working and Steel Processing Proceedings **147**, (1989)
6. R.B. Mott, J.J. Friel, J. Microsc. **193**, 2 (1999)
7. D.B. Williams, M. Watanabe, D.T. Carpenter, Mikrochim Acta [Suppl] **15** 49 (1998).
8. G. Cliff, and G.W. Lorimer, J. Microsc. **103**, 203 (1975)
9. M. Watanabe, Z. Horita, and M. Nemoto, Ultramicroscopy **65**, 187 (1996).

HIGH SENSITIVITY CONVERGENT BEAM ELECTRON DIFFRACTION FOR THE DETERMINATION OF THE TETRAGONAL DISTORTION OF EPITAXIAL FILMS

C.SCHUER, M.LEICHT[†], T.MAREK, and H.P.STRUNK
Institute for Microcharacterisation, Friedrich-Alexander-University Erlangen-Nuremberg,
Cauerstr. 6, D-91058 Erlangen (Germany), schuer@ww.uni-erlangen.de
[†]now at: Infineon Technologies AG, A-9500 Villach (Austria)

ABSTRACT

We have optimized the sensitivity of convergent beam electron diffraction (CBED) by orienting the specimen such that the central (000) diffraction disc shows a pattern of defect lines that are most sensitive to tetragonal distortion. We compare the position of these lines in the experimentally obtained patterns with results from computer simulations, which need to be based on dynamical diffraction theory. In both experimental and simulated patterns the positions of the defect lines are determined by applying a Hough transformation. As a result of this optimized approach, we can measure the tetragonal distortion of a low temperature grown GaAs layer as low as 0.04%.

INTRODUCTION

We have determined with high spatial resolution and at the same time high sensitivity the tetragonal distortion of epitaxial layers. In our work, GaAs layers serve as exemplary material. These layers were grown at reduced substrate temperatures of about 200°C by molecular beam epitaxy (MBE).

Usually, molecular beam epitaxy of GaAs is performed at substrate temperatures of about 600°C. At this temperature layers grow with high quality and stoichiometric composition. However, if the substrate temperature is lowered to the range 200-300°C (low temperature grown, LT-GaAs), an amount of up to 1.5% of excess As is incorporated into the –still monocrystalline– lattice [1]. This excess As tends to expand the epitaxial LT-GaAs lattice. Thus the pseudomorphic LT-GaAs layer experiences a tetragonal strain, i.e. an elongation of the lattice along the growth direction (c-axis). The degree of tetragonal distortion, $\Delta c/c$, is directly correlated with the excess As content ([As]/([Ga])-1 of the layer [2].

This tetragonal distortion is quite small, in our case 0.04%, as will be shown in the following, and is usually determined by high-resolution x-ray diffraction (HR-XRD) [3]. However, this method has the drawback to average over the entire thickness of the epitaxial layer and to sample a rather large area. As we are interested in a possible change of the As content along the growth direction, we apply convergent beam electron diffraction (CBED) to determine the tetragonal distortion with high spatial resolution, typically 10nm.

In an earlier paper [4] the capability of CBED to resolve the tetragonal distortion of LT-GaAs layers has already been demonstrated. But as it was found in the same work [4] a kinematic approach is not sufficient to achieve consistent quantitative results. In the present work, we determine the tetragonal distortion by quantitatively comparing the experimental CBED patterns with computer simulations based on dynamical diffraction. In order to determine the positions of the defect lines to be evaluated we use Hough transformations of both experimental and simulated patterns, following a suggestion of Krämer and Mayer [5]. The simulations were performed using the ems program package of P.A.Stadelmann [6].

Mat. Res. Soc. Symp. Proc. Vol. 589 © 2001 Materials Research Society

EXPERIMENTAL

We shall discuss results obtained from an LT-GaAs layer grown at a nominal substrate temperature of 180°C onto a semi-insulating GaAs substrate. For CBED experiments, we prepared transmission electron microscopy (TEM) cross-sections of the LT-GaAs layer structure (see Fig. 1). Such a section perpendicular to the growth surface displays the vertical structure of the sample. The substrate, AlAs[1] and LT-GaAs layers are sketched in Fig. 1. The filled circles symbolize the depth distribution of potential positions for diffraction patterns.

In contrast to conventional transmission electron diffraction (TED) with an almost parallel beam, for CBED the electron beam is focused onto the specimen. Thus the diffraction spots expand into diffraction discs. These discs show a structure of dark (defect) and bright (excess) higher order Laue zone (HOLZ) lines. The positions of these lines are very sensitive to changes of the lattice parameters. In the following, we will focus on the HOLZ line pattern of the central (000) diffraction disc, which shows defect lines only (Fig. 2). For our purpose the projection of the [120] zone is favorable [4].

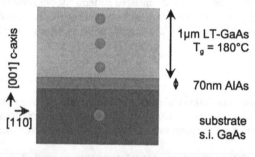

Figure 1 Schematic layer structure of the LT-GaAs sample. The filled circles mark potential positions for diffraction patterns.

Figure 2 Experimental CBED pattern of the GaAs substrate, central (000) diffraction disc, [120] zone axis orientation, accelerating voltage $U_a = 200\text{kV}$.

[1]The AlAs serves as sacrificial layer to facilitate the separation of the LT-GaAs from the substrate e.g. for electro-optical measurements [7] or TEM plan-view examination [8].

To achieve the maximum sensitivity, the positions of the experimental and simulated HOLZ lines were determined by use of a Hough transformation [5]. Starting from the line positions, we calculated the area of characteristic geometrical line arrangements specifically selected for maximum sensitivity. The ratio of two such areas of the same pattern served as a magnification independent measure to compare the experimental and simulated patterns.

The quantitative determination of the tetragonal distortion $\Delta c/c$ was carried out in two steps. First, an experimental pattern from the substrate was taken and fitted with dynamical simulations of undistorted GaAs. The accelerating voltage U_a was the fit parameter, which accounts for small variations of the high voltage calibration of the microscope, as well as for a possible thermal expansion of the specimen due to electron beam heating.

In the second step we simulated a pattern using this determined accelerating voltage value and varying $\Delta c/c$ until the simulated pattern matched an experimental one from the LT-GaAs layer. This second step was repeated to fit several experimental patterns taken at different positions in the LT-GaAs layer along the growth direction (cf. Fig. 1).

The dynamical simulations are very time consuming. To save computing time, we calculated a set of simulated patterns in advance for different values of U_a and $\Delta c/c$. As described before, the line positions were determined. This 'reference set' was later used with an interpolation routine to fit experimental patterns and to evaluate U_a and $\Delta c/c$.

During the TEM examination the specimen was cooled to -174°C in a liquid nitrogen cooling holder to enhance the contrast of the HOLZ lines. (The appropriate lattice constant of GaAs at this temperature was used for simulation.) The temperature stability of the cooling holder was better than ±1°C. Assuming a linear temperature expansion coefficient of GaAs $\alpha = 2 \cdot 10^{-6}$ 1/K [9], a possible influence of the thermal expansion $\Delta a/a = 2 \cdot 10^{-6}$ is negligible compared to the expected tetragonal distortion $\Delta c/c = 4 \cdot 10^{-4}$. We chose a rather large specimen thickness of about 300nm in order to decrease the width of the HOLZ lines as well as to minimize the influence of specimen relaxation due to thinning. All examinations were performed in a Philips CM200 TEM at an accelerating voltage of 200kV.

RESULTS AND DISCUSSION

Details from two (000) diffraction discs are shown in Fig. 3a and b. Figure 3a was taken in the LT-GaAs layer, Fig. 3b in the substrate. In Fig. 3c and d simulated patterns for LT-GaAs layer (Fig. 3c) and substrate (Fig. 3d) are shown for comparison to the experimental patterns. The arrangement of the most sensitive HOLZ lines encircled in the experimental patterns is reproduced very well by the simulations.

Our results of the evaluation of patterns taken from several positions z under the sample surface are summarized in Fig. 4. The accelerating voltage fit yielded $U_a = 198.90$kV. The tetragonal distortion of the LT-GaAs layers was found to be 0.04%. Several patterns were evaluated for the same depth z. The error bars drawn in Fig. 4 represent their statistical error. For the substrate the statistical error was found to be about four times larger than for the LT-GaAs layer. This reflects the fact that the specific geometrical line arrangement used to determine $\Delta c/c$ was selected such that the evaluation of the stressed state is optimized. However, this does not affect the accuracy of the high voltage determination, as for this step another set of lines was used. From the error imposed on the determination of the line positions by the Hough transformation, we estimated an absolute error $\Delta(\Delta c/c) = 0.014\%$.

The measurements show that the tetragonal distortion is constant throughout the LT-GaAs layer. The tetragonal distortion vanishes deep in the substrate. However, about 50nm below the

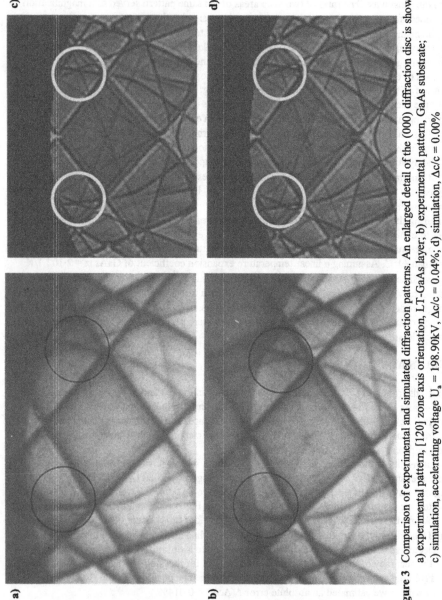

Figure 3 Comparison of experimental and simulated diffraction patterns. An enlarged detail of the (000) diffraction disc is shown. a) experimental pattern, [120] zone axis orientation, LT-GaAs layer; b) experimental pattern, GaAs substrate; c) simulation, accelerating voltage U_a = 198.90kV, $\Delta c/c$ = 0.04%; d) simulation, $\Delta c/c$ = 0.00%

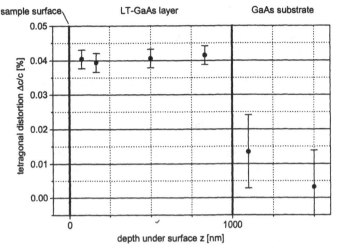

Figure 4 Tetragonal distortion of LT-GaAs layer and substrate in different depth z under the surface.

AlAs layer we find a tetragonal distortion $\Delta c/c = 0.013\%$ of the substrate. Presently, it is not clear whether this is an effect due to the influence of the AlAs layer on the substrate or not.

The value of $\Delta c/c = 0.04\%$ determined by CBED agrees very well with $\Delta c/c = 0.040\%$ measured by HR-XRD for the same sample. It corresponds to a value that is expected for the imaginary 'heteroepitaxy' of GaAs on GaAs with $f = -0.02\%$ misfit.

CONCLUSIONS

By proper choice of the experimental conditions, computer simulations based on the dynamical diffraction theory, and application of digital image processing, we achieved a sensitivity to tetragonal distortion of 0.014%. This demonstrates that the high sensitivity that was reached by Krämer and Mayer for Al can also be realized in GaAs.

We found the tetragonal distortion of the examined LT-GaAs layer to be constant in depth within the margin of error. The quantitative results for the tetragonal distortion were confirmed by HR-XRD.

Concluding, we have demonstrated an electron microscopy technique that is complementary to XRD but in addition offers information on the tetragonal distortion with high spatial resolution.

ACKNOWLEDGMENTS

The authors wish to thank H. Grothe, Institute for General Electrical Engineering and Applied Electronics, Technical University of Munich, for the XRD measurements. The LT-GaAs samples were grown by W. Geißelbrecht and S. Malzer at the Institute for Technical Physics I, Friedrich-Alexander-University Erlangen-Nuremberg.

This work was supported the Deutsche Forschungsgemeinschaft under grant no. DO 356/20.

REFERENCES

1. M.R.Melloch, J.M.Woodall, E.S.Harmon, N.Otsuka, F.H.Pollak, D.D.Nolte, R.M. Feenstra, M.A.Lutz; Annu. Rev. Mater. Sci. **25**, 547 (1995).

2. X.Liu, A.Prasad, J.Nishio, E.R.Weber, Z.Liliental-Weber, W.Walukiewicz; Appl. Phys. Lett. **67**, 279 (1995).

3. M.Fatemi, B.Tadayon, M.E.Twigg, H.B.Dietrich; Phys. Rev. B **48**, 8911 (1993).

4. C.Schür, M.Leicht, T.Marek, H.P.Strunk, S.Tautz, P.Kiesel, W.Geißelbrecht, S.Malzer, G.H.Döhler in *Symposium On Non-Stoichiometric III-V Compounds*, edited by P.Kiesel, S.Malzer, T.Marek (Verlag Lehrstuhl fuer Mikrocharakterisierung, Erlangen, 1998, ISBN: 3–932392–12–4) p.41.

5. S.Krämer, J.Mayer; Journal of Microscopy **194**, 2 (1999).

6. P.A. Stadelmann; Ultramicroscopy **21**, 131 (1987).

7. M.Ruff, D.Streb, S.U.Dankowski, S.Tautz, P.Kiesel, B.Knüpfer, M.Kneissl, N.Linder, G.H.Döhler, U.D.Keil; Appl. Phys. Lett. **68**, 2968 (1996).

8. T.Marek, R.Berta, C.Schür, S.Tautz, P.Kiesel, S.Kunsagi-Mate, H.P.Strunk in *2nd Symposium On Non-Stoichiometric III-V Compounds*, edited by T.Marek, S.Malzer, P.Kiesel (Verlag Lehrstuhl fuer Mikrocharakterisierung, Erlangen, 1999, ISBN: 3–932392–19–1), p.103.

9. O. Madelung (editor), *Semiconductors: Group IV Elements and III-V Compounds (Data in Science and Technology)*, (Springer -Verlag, Berlin, 1991, ISBN: 3–540–53150–5) p.104.

DETERMINATION OF COHERENCY STRAIN FIELDS AROUND COHERENT PARTICLES IN NI-AL ALLOYS BY HREM AND CBED

H. A. CALDERON*, L. CALZADO*, C. KISIELOWSKI**, C. Y. WANG**, R. KILAAS**
*Instituto Politécnico Nacional. Apdo. Postal 75-707, México D.F. 07300, hcalder@esfm.ipn.mx
**NCEM-LBNL, Berkeley CA 94720, USA.

ABSTRACT

Quantitative evaluation of coherency strains around particles in the alloy Ni-12 at.% Al has been performed by means of high resolution electron microscopy (HREM) and convergent bean electron diffraction (CBED). Evaluation of the elastic strain fields around particles can give insight into the elastic interaction between particles during their coarsening mechanism and especially for the late stages where the elastic energy becomes particularly important. Typically, particles form spatial arrangements after long aging times. In the case of fcc-type structures, they align along the <001> soft elastic directions and form groups of many particles. HREM has been used to acquire images including the particle matrix interface. The selected zone axis is [001]. The quantitative analysis to determine the positions of the intensity maxima and the subsequent evaluation of the lattice parameters, is made by using the software *Darip*. Two different directions with respect to the particle matrix interface (parallel and perpendicular) are considered for the evaluation. These measurements are compared to determinations made by means of quantitative evaluation based on calibrated CBED patterns taken from similar regions of the samples.

INTRODUCTION

Processes of microstructural evolution are of importance in the application of materials because they can produce significant changes in their properties. For example Ni-basis superalloys with addition of Al, Ti, Cr, etc., are the most popular materials for the fabrication of turbines. These alloys are formed by a mixture of different phases among which the most important for their excellent mechanical properties at high temperature is the ordered Ni_3Al or γ' phase ($L1_2$ structure). This phase appears in the form of coherent ordered particles homogeneously distributed in a Ni basis solid solution (fcc structure). Use at high temperature of Ni-based superalloys implies a coarsening mechanism of the particle distribution that promotes a degradation of the mechanical properties. Coarsening of particles is controlled by the decrease of the total energy of the system. In the case of solids, two important contributions can be found i.e., the reduction of the interfacial and the elastic strain energies. There is an important reduction of the interfacial energy when larger particles grow at the expense of smaller ones. However the lattice mismatch between particles and matrix induces a strain field that gives rise to elastic interaction between particles. The elastic energy influence depends on the particle volume and thus larger particles imply an increase of the elastic contribution, especially in the late stages of the coarsening mechanism. There is clear evidence of the elastic effects e.g. particle shape changes as a function of size, particle alignment along elastically soft directions, formation of particle arrangements, particle splitting, etc. [1, 2]. There is also a great deal of investigations on the subject based on numerical simulation [see e.g. 3, 4]. According to Johnson et al. [3], it is possible to have inverse coarsening i.e., smaller particles growing at the expense of larger ones, for some elastic conditions. In the same way it has been suggested that growing particles can achieve a critical size beyond which they will split into smaller particles [4]. In this investigation microscopical techniques are employed to evaluate the strain fields around coherent particles in the late stages of the coarsening mechanism i.e., when the influence of the elastic interaction is strong.

99

EXPERIMENTAL PROCEDURE

Monocrystalline samples of the alloy Ni –12 at.% Al were aged at two different temperatures to form a bimodal particle size distribution. The first aging was made at 1133 K and the second one at 923 K. This microstructure was selected in an attempt to create an especially strong elastic interaction between particles. High resolution electron microscopy (HREM) and convergent beam electron diffraction (CBED) are used to evaluate elastic deformations around particles with a high spatial resolution. The quantitative analysis of the HREM images has been carried out by means of the software *Darip*. CBED patterns have been evaluated by means of the software *ems*.

Fig. 1. Dark field image of the alloy Ni-12 at.% Al after aging 1700 h at 923 K and 5h at 1133 K.

RESULTS AND DISCUSSION

The evaluation of strain fields around γ' particles is addressed to the late stages of the coarsening process. The samples are initially constituted by two different particle size distributions. A first aging at a temperature very close to the solvus line produces large particles in a short time. A second aging treatment at a lower temperature produces smaller particles. These particles grow continuously as a function of aging time. After approximately 1000 h of aging, there is no distinction between the formerly different size distributions. All particles can be considered to belong to a single distribution. The initially smaller particles grow faster than the originally larger ones. The details concerning the kinetics and particle size distributions will be given in another publication [5]. Figure 1 shows a typical distribution of particles in the Ni-12 at.% Al alloy after 1700 h at 933 K and 5 h at 1133 K. The spatial distribution of particles is no longer at random but rather groups of particles develop. Some of these groups can be interpreted as a result of splitting of large particles reaching a critical size. They can also be produced by a

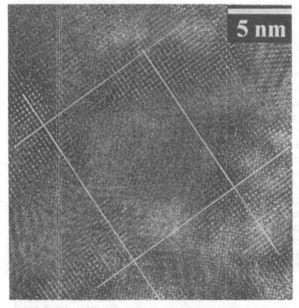

Fig. 2. HREM of a group of four particles. Particles have different order domains as shown by the superimposed lines.

mechanism of coalescence among smaller particles. γ' particles have an ordered L1$_2$ structure which can produce up to four different order domains and three antiphase boundaries. Fig. 2 shows a HREM image of a particle arrangement similar to those in Fig. 1. The contrast shows clearly the order domains in each particle. The superimposed lines help to show that the particles in the array have different order domains and thus they are not a result of a process of splitting of a larger particle. Image simulation has been used to test whether or not the contrast can be affected by variations of defocusing or thickness, but this is not the case and the contrast arises mainly from the different chemical compositions of the {001} planes in the L1$_2$ structure. Thus Fig. 2 clearly shows that migration of particles is driven by elastic interaction and that splitting of particles is unlikely.

Figure 3 shows the interface between a γ' particle and the γ matrix in a particle array. The zone axis is [001]. The original image has been Wiener filtered to eliminate random noise. Quantification of the elastic distortion of the cell is done by determining the positions of the intensity maxima on images similar to the one in Fig. 3. The software *Darip* allows locating the maxima and evaluates average values of their spacing in both parallel and perpendicular directions to the γ-γ' interface. Fig. 3b shows a typical result of the pattern recognition routine to locate the intensity maxima (see the overlapped dots). Fig. 4a shows the evaluation of average lattice spacing in a parallel and perpendicular direction with respect to the γ-γ' interface i.e., a_{par} and a_{perp}, respectively. Each point represents an average value for the corresponding column or line. The value of the standard deviation for each measurement is also included. Fig. 4b shows the variation of the ratio a_{par}/a_{perp} as a function of distance along the transversal section of the interface.

CBED patterns can also be used for determination of the cell dimensions. Experimental patterns from nanosized diffraction volumes are normally compared to simulated ones for evaluation

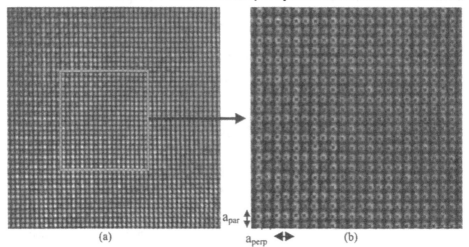

(a)　　　　　　　a_{par}　　　a_{perp}　　　(b)

Fig. 3. (a) Wiener filtered HREM image of γ/γ' interface in Ni-12 at.%Al. (b) Determination of intensity maxima by a pattern recognition routine.

Such a fitting procedure has been done systematically as shown in Fig. 5 for a CBED pattern taken at 200 keV. The simulated and experimental patterns (Figs. 5a,b) are first drawn with fine lines, then the drawings are compared (see Figs. 5c - d). Such a procedure helps to have a higher accuracy since comparing the simulated and experimental patterns can lead to confusion. Contrast variations and extinction contours do not allow direct comparison. Fig. 5b shows the simulated pattern (thickness of 180 nm) that fits the experimental pattern. This is done by means of simulation taking into account dynamic corrections and an accurate determination of the accelerating voltage of the microscope. It is possible to distinguish a variation of 0.0003 nm in lattice dimensions; this represents an accuracy better than 0.1% in the evaluation of strain.

Figure 6 shows CBED patterns taken from a γ' particle, the corresponding γ–γ' interface and the γ matrix. The patterns can be used to evaluate the mean lattice dimensions. The diffraction volume corresponds to a beam size of 1 nm with a zone axis parallel to $[\bar{1}14]$. The right hand side of Fig. 6 shows the drawings of the experimental patterns that were used in the evaluation procedure. The CBED patterns can be fitted to tetragonal structures with the lattice di-

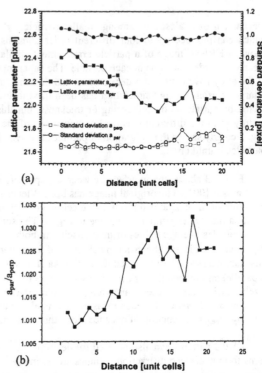

Fig. 4. (a) Average values of lattice spacings both parallel (a_{par}) and perpendicular (a_{perp})to the γ/γ' interface in Fig. 3. (b) a_{par}/a_{perp} as a function of distance from the interface.

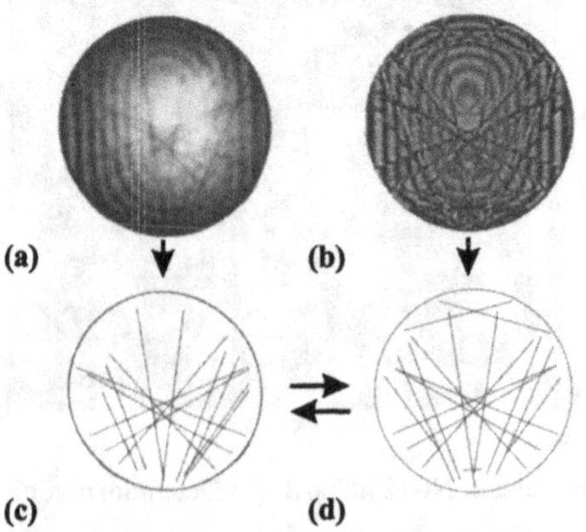

mensions as indicated in Table I. The CBED patterns shown in Fig. 6 have some features that can be related to lattice strains. For instance double (see arrows) and diffuse broader HOLZ lines can be seen in some of the patterns in Fig. 6. Both double and diffuse HOLZ lines are normally close to orientations near the elastically soft <100>. Fig. 6 shows the case of the $(\bar{3}11\bar{3})$ HOLZ line (near (010)). This can be interpreted as a localization of the elastic strain along such directions. Near the γ–γ' interface many diffuse lines are obtained in the experimental CBED pattern which allows determination of only a range of lattice parameters (Fig. 6b). The patterns corresponding to the γ' particles have normally a cubic symmetry (Fig. 6c). Most of the elastic deformation is apparently located in the matrix phase owing to

Fig. 5. Evaluation procedure for CBED patterns. (a) Experimental pattern; (b) simulated CBED, (c) drawing of (a); (d) drawing of (b). B = $[\bar{1}14]$, 200 keV.

the larger elastic modulus of γ' particles.

Figure 7 shows a schematic array of particles and a series of CBED patterns taken from different positions. The evaluation of the lattice parameters is given in Table II. In all cases the CBED patterns that were taken from the matrix can be fitted to tetragonal or orthorombic cells

Table I. Determination of lattice parameters (a and c) by CBED (Fig. 6).

Position	a (nm)	c (nm)
γ' particle	0.3593	0.3593
γ-γ' interface	0.3575-0.3593	0.3593-0.3585
γ matrix	0.3575	0.3585

but the difference in the a and b dimensions of the orthorombic cells are normally minor. In addition, the simulated pattern has to be obtained by assuming that the tetragonal cell is oriented with the c axis parallel to the γ- γ' interface in order to achieve a good match to the experimental pattern. Table II suggests that the elastic strains around particles depend very strongly on the surroundings i.e., most likely other neighboring particles or arrays. For instance positions 1, 2, 5 and 8 can be considered equivalent since they all were taken between two particles and

Table II. Determination of lattice parameters by CBED (Fig. 7).

Position	Lattice parameter a (nm)	Lattice parameter c (nm)
1	0.3577	0.3585
2	0.3580	0.3586
3	0-3583	0.3591
4	0.3586	0.3590
5	0.3584	0.3591
6	0.3595	0.3595
7	0.3588	0.3593
8	0.3589	0.3589
9	0.3588	0.3592
10	0.3574	0.3574

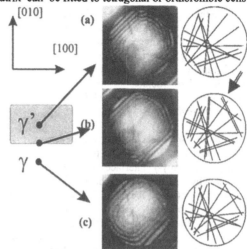

Fig. 6. CBED patterns taken from different locations around γ' particles. (a) γ' particle, (b) γ–γ' interface, (c) γ matrix. B = [Ī14], 200 keV. The arrow indicates (3̄113̄).

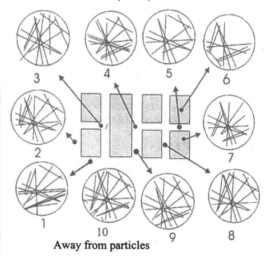

Fig. 7. Schematic array of particles and simulated CBED from experimental CBED patterns taken at different locations indicated by numbers. The determination of cell parameters is given in Table II.

half way along the length of the interface. Nevertheless, the evaluated cell dimensions are different (see Table II) for all these areas, although some similarities can be seen between positions 1 and 2 (outside the arrangement) and positions 4 and 5 (inside the arrangement). In addition Table II suggests that higher strains develop near and outside the particle arrangement. This strongly suggests that the microstructure is not yet in equilibrium as can be expected for the coarsening regime.

Figure 8 shows a comparison between the two techniques used in the present investigation. Two γ-γ' interfaces are shown separated by a narrow matrix channel. Fig. 8 is representative of the HREM observations since in most cases a relatively strong variation of lattice parameters is observed for a_{perp} while smoother tendencies are obtained for a_{par}. Nevertheless typically higher absolute values of a_{par} have been found. On the other hand, Fig. 8 also shows some measurements made by CBED (see (*) and (+)). CBED determinations also show higher lattice parameters parallel to the γ-γ' interface i.e. the lattice parameter c in the tetragonal cells. Both techniques thus produce similar results. However, further investigation is necessary to completely characterize the complex strain fields that develop during the coarsening of coherent particles and to determine experimentally the influence of elastic interaction on this mechanism of microstructural evolution.

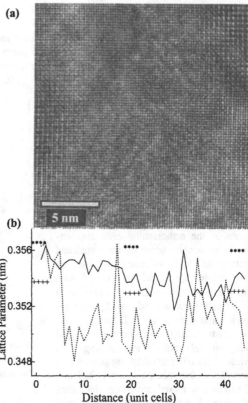

Fig. 8. (a) HREM image of two γ-γ' interfaces, (b) evaluation of HREM image, the solid line represents values of a_{par} and the dotted line is a_{perp}. CBED measurements are indicated by (*) for parameter c and (+) for parameter a.

ACKNOWLEDGEMENTS
HAC wishes to acknowledge financial support from CONACYT (Project 28925 U) and COFAA-IPN. HAC gratefully acknowledges the use of equipment at NCEM (USA), NIRIN (Japan), ININ (Mexico) and IAP-ETH Zürich (Switzerland) for this investigation.

REFERENCES
1. A.J. Ardell, R.B. Nicholson, J.D. Eshelby, Acta Metall. 14, 1295 (1968).
2. M. Doi, T. Miyazaki, T. Wakatsuki, Mater. Sci. Eng. 67A, 247 (1984).
3. W. C. Johnson, T.A. Abinandanan, P. W. Voorhees, Acta Metall. 38, 1349 (1990).
4. Y. Wang, L.Q. Chen, A. Khachaturyan, Acta Metall. Mater. 41, 279 (1993).
5. H.A. Calderon, J. J. Cruz, L. Calzado, T. Mori, C. Kisielowski, C.Y. Wang, PTM 99 Kyoto (Japan). In press.

AN ELECTRON MICROSCOPE STUDY OF DIFFUSION ASSISTED DISLOCATION PROCESSES IN INTERMETALLIC GAMMA TiAl

F. APPEL, U. LORENZ, M. OEHRING
Institute for Materials Research, GKSS-Research Centre, Max-Planck-Strasse,
D-21502 Geesthacht, Germany

ABSTRACT

The paper reports an electron microscope study of diffusion controlled deformation mechanisms in two-phase titanium aluminides which apparently cause the degradation of the strength properties at elevated temperatures. Climb velocities were analyzed in terms the critical vacancy supersaturation necessary for the operation of diffusion assisted dislocation sources. Particular emphasis was paid on structural changes occurring during long-term creep, which are apparently associated with dislocation climb.

INTRODUCTION

Titanium aluminides are one of the promising candidates for the next generation of high-temperature structural materials. The most important prerequisites for advanced applications are probably good creep resistance and structural stability. The service temperature currently considered for titanium aluminides are in the range 650-750 °C, or $(0.53-0.59)T_m$ when referred to the absolute melting point T_m. Conventional disordered metals deform under such conditions by diffusion assisted dislocation climb and this has also been recognized in single-phase [1, 2] and two-phase [3, 4] titanium aluminides. Dislocation climb is known to be a stress driven thermally activated process [5]. Thus, the strength properties of the material become strongly rate dependent and degrade at low strain rates. These features are also characteristic of the high-temperature deformation of the titanium aluminides and certainly harmful for their creep resistance [6]. However, many aspects of dislocation climb in γ-base titanium aluminides are not yet solved. In particular, little information is available about climb assisted dislocation multiplication. Furthermore, TiAl alloys of technical significance are multiphase assemblies with complex microstructures and dense arrangements of interfacial boundaries. At elevated temperatures the misfit dislocations are likely to climb, which under long-term thermal exposure may lead to significant structural changes. Thus, for the high-temperature capability of the materials dislocation climb may be important in several different ways. Investigation of these processes is the subject matter of the present paper.

EXPERIMENTAL INVESTIGATIONS

The investigations were performed on a Ti-48Al-2 Cr alloy (in at.%) with a nearly-lamellar microstructure [4], which might be considered as a model material for engineering two-phase titanium aluminide alloys. Long-term creep studies were performed at T = 973 K and applied stresses σ_a = 80 to 140 MPa [6]. These conditions led to very low creep rates, which are certainly conducive to the initiation of diffusion assisted climb processes.

Electron transparent foils were prepared from deformed samples by mechanical grinding and electrolytic polishing [3, 4]. Conventional and high-resolution electron microscope (TEM) observations were performed in order to characterize deformation induced defect structures. These investigations were supplemented by TEM in situ heating experiments. In these cases the thin foils were prepared from samples which had been deformed at room temperature to strain ε = 3 %. This produced a suitable density of dislocations and certainly a small supersaturation of point defects because jog dragging occurs [4]. Thus, during in situ heating inside the TEM the dislocations are probably subjected to osmotic forces due to excess point defects. The kinematics and dynamics of the observed defect processes might therefore be affected by the applied method. However, in order to establish desired microstructures, processing routes of TiAl alloys often involve high temperature annealing followed by rapid cooling. This certainly leads to a significant point defect supersaturation. The dislocation behaviour occurring under such conditions is therefore not only a scientific matter but also of technical significance. The TEM in situ experiments were carried out at low acceleration voltages of 120 and 200 kV, respectively, in order to avoid radiation damage. For the TEM investigations described here the Philips instruments CM30, CM200 and 400T were used.

CLIMB ASSISTED DISLOCATION MULTIPLICATION

Dislocation climb often occurs at low stresses and strain rates. Under such conditions dislocation multiplication by conventional glide sources is difficult because the applied stress is too low to initiate cross glide and to by-pass the dislocation segments trailed at jogged screw dislocations. In the present study information about climb assisted multiplication has been deduced from in situ heating experiments performed inside the TEM. Climb was preferentially observed on 1/2<110] ordinary dislocations, which is plausible as these dislocations have a compact core structure [7]. During the in situ heating experiments, the dislocations moved under the combined action of thermomechanical stresses and osmotic climb forces due to the point defects generated during the predeformation. The experiments were performed between 800 and 930 K, which is close to the intended service temperature of the material.

Multiplication at elevated temperatures was found to start from different defects incorporating climb, suggesting that different variants of the classical Bardeen-Herring mechanism were operative. Figure 1 demonstrates the growth of climb sources during a long period of about 350 min by using a sequence of micrographs, part of an in situ study performed at 820 K. Accordingly, dislocations were generated by the nucleation and growth of prismatic loops (arrow 1). Since the expansion of each loop is realized by the removal of one atomic plane, the mechanism seems to be exhausted after only one cycle of the source. The expanding loop designated with arrow 2 contains a jog so that climb on different atomic planes occurs. After one cycle of the source a new dipole is generated so that the mechanism is regenerative. The climb processes often lead to the formation of spiral sources, which generate complex configurations of interconnected multiple loops, as has been demonstrated in [4, 8]. The character of the point defects involved in the formation of the climb sources was not identified in the present study. However, there is an increasing awareness that vacancies predominate in thermal equilibrium over interstitials in many intermetallics including γ(TiAl) [9]. Thus, it might be speculated that the defects relevant for the present study are vacancies. Apparently, the critical vacancy concentration c/c_0 required to operate a Bardeen-Herring source is relatively low. The geometrical situation of loop growth observed during the in situ experiments has been described in [5]. For a loop expanding from a source of length L, the critical value is

$$\ln (c/c_0) \geq [\mu \, b \, \Omega \, / \, L \, 2\pi(1 - v) \, kT] \ln (L \, \alpha/1.8b), \qquad (1)$$

where Ω is the atomic volume, $\alpha = 4$ and v is Poissons' ratio. For the present experimental conditions T = 820 K and L = 150 b to 350 b, the values c/c_0 = 3 to 1.7 were obtained. These supersaturations are small in comparison to those produced initially after rapid cooling, which are often in the order $c/c_0 \sim 10^3$ to 10^4 [5]. Thus, Bardeen-Herring sources can probably operate throughout the entire period where annealing out of excess vacancies takes place.

Processing routes of titanium aluminides often involve thermal treatments followed by rapid cooling, which certainly leads to large vacancy supersaturations. Highly non-equilibrium conditions may also occur due to hot- or cold-working of the material. Dislocation motion in large vacancy supersaturations seems therefore to be a common situation for the deformation of γ-base alloys. Climb induced defect structures are mostly formed with slow rates, thus they are of minor interest if fast strain rates have to be realized as this is the case in conventional constant strain rate tests. However, the mechanisms are expected to play an important role for creep deformation, where only small strain rates are imposed.

DEFORMATION INDUCED STRUCTURAL CHANGES

Among the various microstructures, which can be established in two-phase titanium aluminides, the fully-lamellar microstructures exhibit the best high-temperature strength and creep resistance. Thus, particular attention will be given to the stability of this structure under long-term thermal exposure. The lamellar structure consists of thin parallel $\alpha_2(Ti_3Al)$ and $\gamma(TiAl)$ platelets with the orientation relationships $\{111\}_\gamma \parallel (0001)_{\alpha_2}$ and $<110]_\gamma \parallel <11\overline{2}0>_{\alpha_2}$. There is a random occurrence of six γ variants, which can formally be described by rotations of one lamella relative to its neighbour by multiples of 60° about $\{111\}_\gamma$. Thus in addition to the α_2/γ interfaces, different types of γ/γ interfaces occur. Accommodation of misfit which arises because of differences in lattice parameters and crystal structure leads to dense structures of interfacial dislocations and significant coherency stresses [8].

At low temperatures the coherency stresses acting on the mismatch structures of the interfaces are in equilibrium with the high glide resistance impeding dislocation motion. At elevated temperatures this glide resistance is reduced due to thermal activation, thus relaxation of

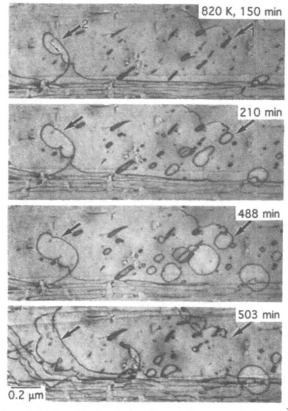

Figure 1: Operation of Bardeen-Herring dislocation climb sources during an in situ heating experiment inside the TEM at T = 820 K. Details: (1) nucleation and growth of prismatic dislocation loops, (2) expansion of a dislocation loop containing two jogs (small arrow). After one cycle of the source, a new dipole is generated so that the mechanism is regenerative. Predeformation at T = 300 K to strain ε = 3%.

the coherency stresses and emission of dislocation loops can occur [8]. These interface-related mechanisms finally lead to significant structural changes as have been recognized after long-term creep at T = 973 K. A characteristic feature is the formation of multiple-height ledges perpendicular to the interfacial plane [6, 8]; an initial stage of the process is demonstrated in Fig. 2. Multiple-height ledges are commonly observed after phase transformation and growth, respectively, and several mechanisms have been proposed to explain the phenomenon [10]. Analogous to these models it is speculated that the large ledges observed in the crept TiAl alloys arise from one-plane ledges, which moved under diffusional control along the interfaces and were piled up at misfit dislocations (arrowed in Fig. 2). Once a sharp pile up is formed, the configuration may rearrange into a tilt configuration with a long-range elastic stress field. This would cause further perfect or Shockley partial dislocations to be incorporated into the ledge. This hypothesis would explain the large height of the ledges (up to 200 nm) and the observation that in all cases they were associated with misfit dislocations. When the slabs grow further it might be energetically favourable to nucleate a new γ grain, a situation which often has been seen in crept material [6, 8].

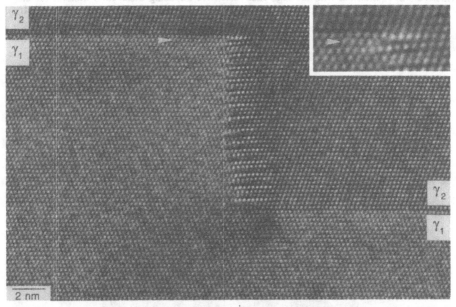

Figure 2: Structural changes of lamellar interfaces observed after long-term creep: T = 700 °C, σ_a = 140 MPa, t = 5,988 hours. Formation of a high ledge, note the misfit dislocation (arrowed) which is manifested by an additional $(111)_\gamma$ plane parallel to the interface.

The driving forces for the interface processes seem in part to arise from changes of the phase equilibria [6]. The alloy investigated was annealed above the eutectoid temperature, where a significantly higher volume fraction of the α phase is expected. An excess concentration of the α_2 phase may therefore exist at the lower temperature of 700 °C, where the creep experiments were performed. Thus, during long-term creep dissolution of the α_2 phase can occur. This view is supported by the TEM observations made on α_2/γ interfaces and shown in Fig. 3. There is clear evidence that the density of structural ledges at α_2/γ interfaces is significantly higher than at γ/γ interfaces. This is probably a consequence of the complex transport processes associated with the $\alpha_2 \rightarrow \gamma$ transformation. Not only must the stacking sequence be changed during migration of the interface, but also the local chemical composition has to be

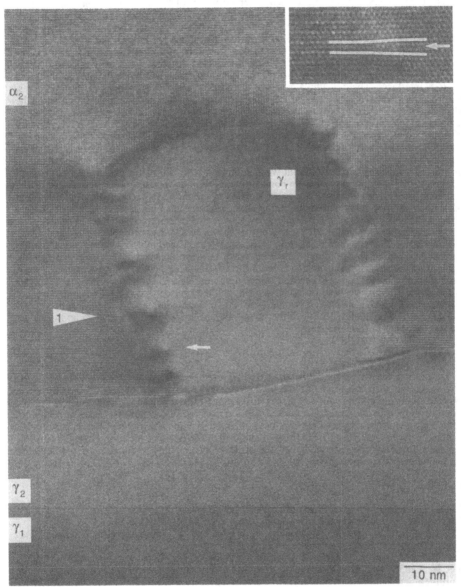

Figure 3: Recrystallization processes in lamellar TiAl observed after long-term creep: T = 700 °C, σ_a = 110 MPa, t = 13,400 hours, ε = 0.46%. A new γ-grain (designated as γ_r) was formed within an α_2 lamella. Note the high density of ledges in the α_2/γ interface and the atomically flat interface γ_1/γ_2. Misfit dislocations situated at the α_2/γ grain boundary can be recognized if the micrograph is viewed under a shallow angle along arrow 1.

adjusted by long-range diffusion. The structure of the ledges exhibits a great variety. Steps up to six $\{111\}_\gamma$ layers in height have been observed which often have an irregular shape. This gives the impression that these large ledges are formed by a diffusion-controlled migration of elementary ledges, which apparently produces the required change in both stacking and composition. The ledges are often associated with misfit dislocations having a Burgers vector component normal to the interface (Fig. 2). During lateral migration of the interfaces these dislocations move along the interface by climb. It is therefore expected that the arrays of interfacial dislocations and ledges play an important role for achieving the phase and point defect equilibrium at the α_2/γ interfaces. There is ample evidence of enhanced self-diffusion along dislocation cores. Likewise interfacial dislocations and ledges are envisaged as regions where deviation from the ideal structures is localized and which may provide paths of easy diffusion. The processes finally result in the formation of new γ grains within α_2 lamellae (Fig. 3). There is a high density of nearly equally spaced misfit dislocations, which give supporting evidence for the mechanism described above. Recrystallization of ordered structures has been investigated in several studies; a detailed review of the data has been given by Cahn et al. [11]. Accordingly, recrystallization of ordered structures is more difficult when compared with disordered metals. This has been referred to the low mobility of grain boundaries and the fact that the ordered state has to be restored. Most if not all new γ grains are situated at the α_2/γ interfaces, which indicates that these interfaces are the preferred nucleation sites. For this nucleation process the misfit dislocations and ledges present at the α_2/γ interfaces play probably an important role. In this respect it is interesting to note that small γ grains as that shown in Fig. 3 are completely ordered giving the impression that the ordered state is immediately established after grain nucleation or that nucleation occurred in the ordered state. This might be a consequence of the heterogeneous nucleation of the grains at the interfacial boundaries. There are certainly crystallographic constraints exerted by the adjacent lamellae, which may control nucleation and subsequent growing of the grains. In view of these observations it might be expected that the high-temperature deformation and work hardening of two-phase titanium aluminides depends significantly on off-stoichiometric deviations. Clearly, further investigations are required in this field and these are the subject of our ongoing research.

CONCLUSIONS

Dislocation/vacancy interactions affect the high temperature deformation of TiAl-base alloys in various different ways. At intended service temperatures of 650 to 750 °C diffusion assisted climb and multiplication of matrix dislocations occur and lead to the degradation of the strength properties at low strain rates. Diffusion assisted dislocation processes are also involved in structural changes which have been observed after long-term creep. For improving the high temperature strength of γ(TiAl)-based alloys in most instances these processes have to be impeded. In this respect, reduction of the diffusibility and implementation of precipitation hardening seem to be most promising.

REFERENCES

1 B.K. Kad and H.L. Fraser, Philos. Mag. A 69 (1994) 689.
2 B. Viguier, J. Bonneville and J.L. Martin, Acta Mater. 44 (1996) 4403.
3 F. Appel and R. Wagner, in: Y.-W. Kim, R. Wagner and M. Yamaguchi (Eds.), Gamma Titanium Aluminides, TMS, Warrendale, 1995, p. 231.
4 F. Appel, U. Lorenz, M. Oehring, U. Sparka, and R. Wagner, Mater. Sci. Eng. A233 (1997) 1.
5 R.W. Balluffi and A.V. Granato, in: F.R.N. Nabarro (Ed.), Dislocations in Solids, Vol. 4, North-Holland Publishing Company, Amsterdam, 1979, p. 1.
6 M. Oehring, F. Appel, P.J. Ennis, and R. Wagner, Intermetallics 7 (1999) 335.
7 K.J. Hemker, B. Viguier and M.J. Mills, Mater. Sci. Eng. A 164 (1993) 391.
8 F. Appel and R. Wagner, Mater. Sci. Eng. R22 (1998) 187.
9 Y. Shirai, M. Yamaguchi, Mater. Sci. Eng. A, 164 (1993) 391.
10 J. van der Merwe and G.J. Shiflet, Acta Metall. Mater. 42 (1994) 1173.
11 R.W. Cahn, M. Takeyama, J.A. Horton, C.T. Liu, J. Mat. Res. 6 (1991) 57.

SPECIMEN CURRENT IMAGING OF DELAMINATION IN CERAMIC FILMS ON METAL SUBSTRATES IN THE SEM

S. RANGARAJAN* and A.H. KING **
*Department of Materials Science and Engineering, SUNY, Stony Brook,NY 11794-2275, USA
**School of Materials Engineering, Purdue University, West Lafayette, IN 47907-1289, USA

ABSTRACT

Plasma spraying of yttria-stabilized zirconia on stainless steel substrates generates rapidly solidified thin film discs called splats. Constraint due to adhesion at their interface generates sufficient stresses during cooling from the splat melting to generate extensive cracking. Stress relief by cracking is always accompanied by delamination of the thin films at the edges of cracked segments. It is shown here that images obtained in the SEM through amplification of the specimen current signal generate the right contrast for delineating regions of edge delamination in the cracked segments of a splat. Image analysis using this contrast provides a quantitative measure of the extent of delamination at the interface. Greater delamination in the splats formed on cooler substrates. Correlation also indicates complimentary nature of stress relief by delamination and cracking. A detailed analysis including data from solidification and cooling and cracking statistics will give a measure of the adhesion strength at the interface. This technique is a quick method for measuring delamination of microscopic degree in thin insulating films on conducting substrates.

INTRODUCTION

Plasma sprayed coatings comprise splats formed by the rapid quenching of molten material obtained from powder passing through a plasma flame and splashing on a cool substrate. On smooth preheated substrates there is a uniform spreading and solidification of the droplets into discs of diameter ranging from 20 - 150 μm and thickness ~ 2 μm depending on the initial particle size. The primary interest of some studies [1] has been the bonding of these coatings to substrates. As a fundamental approach to the formation of these coatings we study the bonding of individual ceramic splats (yttria stabilized zirconia or YSZ) to steel substrates.

The YSZ splats obtained on steel substrates exhibit a dense network of cracks caused by the severe cooling on contact with the substrate. Adhesion to the substrate constrains the shrinkage of the splat material generating tensile stresses in the ceramic film which are relieved by cracking. It has been reasoned for similar thin film cracking that the density of these cracks is a function of the strength of interfacial adhesion[2]. Along with tensile cracking of the ceramic splats there must also occur a delamination due to cracking along the interface as a consequence of the nature of quenching stresses on brittle thin films on ductile substrates [3]. Evidence of delamination has been shown by TEM of splat cross sections [4].

EXPERIMENTAL

The splats for analysis were obtained by single passes of the plasma spray gun operating with typical variables used in the industry (Table. 1), on 2"x1" coupons of stainless steel polished to 0.05 micron grit. Substrates can be preheated to controlled temperatures using a backplate heater. Splats were obtained on substrates heated to 100°C, 200°C, 300°C and on unheated substrates. One polished coupon was also preheated in air to obtain a tint of oxide before spraying at 100°C. Substrate heating has been well known to improve the spreading of molten droplets resulting in disc shaped splats with a continuous rim. Unheated substrates cause the splats to form with a lot of splashing due to break up

111

of the rim during spreading resulting in fingerlike projections. This effect shows a gradual transition with increasing substrate temperature.

Table 1: Plasma Spray Parameters

Gun - Power	PTF4: 650A, 69 V, 45 kW
Spraying gases	Ar 40 slpm[*], H$_2$ 10 slpm
Spraying distance	120 mm
Powder	crushed and fused YSZ (7 wt.% Y$_2$O$_3$)
Powder Size	+10 - 62 μm
Powder feed rate	2g/minute
Carrier gas	Ar/N$_2$ at 2.5 slpm

*standard liter per minute

RESULTS

Scanning electron microscopy has been used extensively in this investigation for characterizing the cracking of quenched ceramic splats. The secondary electron images exhibit regions of bright contrast at the cracked edges (Fig.1). Closer inspection (Fig.1 inset) indicates a gradual increase in brightness over a wider region in the periphery of every cracked piece of the ceramic splat. The effect has been enhanced in the inset by *posterizing* using an image processing operation that assigns singular gray scale values to blocks of gray ranges.

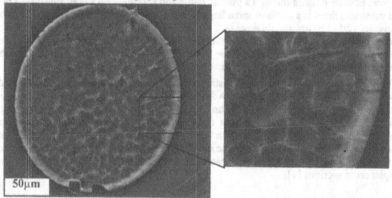

Fig. 1. Secondary electron image of a typical YSZ splat on stainless steel showing charging at cracked edges. Enhanced inset shows the variation of brightness within a piece of cracked film

The ceramic film gets charged in the SEM as it is a good insulator, but because of its small thickness (2 μm) and close proximity of a conducting substrate this does not create distortions in the imaging of the splats at accelerating potentials of 15 and 30 keV for the secondary electron and the backscattered electron imaging modes. However, presence of charge on the surface of insulators is known to affect emission of electrons from the surface used in imaging [5]. The charging of the insulating ceramic in the splat seems to vary from regions in contact with the substrate to delaminated regions. The regions of the ceramic film in contact with the conductive substrate can loose some charge to ground through the specimen stage, whereas the delaminated regions of the thin film that are not in contact with the conductive substrate must hold more of the charge.

Images of the splats obtained using a specimen current amplifier (Fig. 2) capturing current flowing to earth record a contrast that is complementary to the secondary electron images. The central regions of the ceramic splat pieces appear bright as they are in contact with the substrate and conduct the electrons as signal to the specimen current amplifier. The periphery of these thin films appears dark as the charge is not conducted to the substrate. Thus we have a technique to delineate regions of delamination in insulator thin films on conducting substrates.

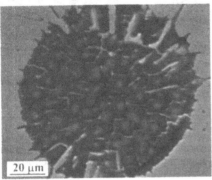

Fig. 2 Specimen current image of YSZ splat on stainless steel substrate showing inverse of secondary electron image contrast.

We have obtained a clear indication of increasing contact area with increasing substrate temperature. Fig. 2 shows a splat obtained on a substrate with a thin oxide film obtained during a preheat, at 100°C. The low area of contact is shown by the bright spots in the center of the splat pieces. Splats on 100°C substrates with no oxide due to preheat showed a larger number of splats sticking but with poorer spreading and more splashing. Fig.1 shows a splat obtained at 200°C on a polished stainless steel surface and there is an improvement in contact, with smaller bright regions at edges indicating less delamination. Specimen current images of a splat obtained on a substrate at 300°C show there is markedly less delamination (e.g. Fig. 5c).

DISCUSSION

Charging Contrast

Images in the SEM are obtained by the projection of signals generated from the interaction of the primary electron beam with the specimen during spatial scanning. The incident electron current is scattered by the atoms in the specimen and dissipated as secondary electrons, backscattered electrons and diffusion current (specimen current) to ground. The variation of the electrostatic field between the specimen and the detector influences the secondary electron intensity and trajectory. The altering potential in the case of insulating specimens is the result of the deposited charge being separated from ground potential by the dielectric specimen (charging).

The scattering yield, σ, is a function of the electron accelerating potential [6]. Charging of specimens in the SEM happens because the initial scattering yield of the material is not equal to unity at the operating voltage V_0. Development of charge on a total insulator specimen alters this potential by V_s until the effective accelerating potential is at a level where the electron yield is 1. For conducting specimens with resistance R_1 the extent of charging and hence the altering potential is limited by current flow to ground. For this case the emission yields can remain less than 1 provided the excess charge flows as specimen current. Change in resistance to current flow is thus related to variations in surface emission yields[6].

Nature of Resistance

The accelerating potential in the SEM also influences the electron scattering range or the average penetration depth of electrons in the specimen and the available population of electrons for specimen current. Monte Carlo simulations of single electron scattering for specified thickness of scattering film and the average atomic weights of the scattering elements have been obtained for accelerating potentials of 5, 15 and 30 kV (Fig. 3 a, b & c).

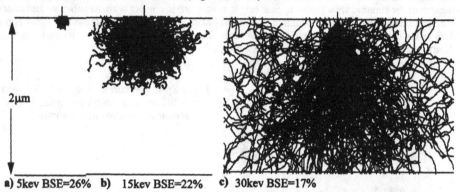

a) 5kev BSE=26%   b) 15kev BSE=22%   c) 30kev BSE=17%

Fig. 3. Monte Carlo simulations of single electron scattering in a 2μm thin YSZ film. Increasing energies of incident electrons result in increasing scattering range and secondary electron population.

At lower accelerating potentials (5 kV) the single electron scattering simulation indicates a small electron range or depth of penetration within the ceramic film. This in turn means a thicker dielectric film for the current conduction, with higher resistance. It results in a general lowering of specimen current magnitudes and the specimen current image of the splat at 5 kV is uniformly dark, indicating no specimen current. The scattering model for 15 kV shows a greater electron range and a greater percentage of secondary electrons indicated by the decrease in the backscattered electron percentage. This means there is more charge within the specimen and also a shorter resistance for the specimen current. The presence of a finite specimen current is essential for observing the effect of variable resistance due to delamination on the image contrast. A higher accelerating voltage of 30 kV indicates that the electron range exceeds the ceramic film thickness (Fig. 3c). A number of electrons exit the dielectric at the interface into the metal by scattering and this leads to a shot noise in the specimen current images. In the delaminated regions this will also result in specimen current due to discharge across the gap to the metal causing loss of contrast.

If we approximate the resistance to current flow in the film to be Ohmic the magnitude of resistance depends on the physical dimensions of the current path as

$$R = \rho \times \frac{L}{A} \qquad (1)$$

where the resistivity ρ is a material property while L is the length and A is the area of cross section. Length

$$L = t - r_b \qquad (2)$$

where t is thickness of film, and r_b is the electron scattering range (Fig. 4), for the region where the film is in contact with the substrate. Area A, is the total area of contact of the film with the substrate at the interface. For the delaminated region, the length L is the length of the shortest path to the point of contact at the interface (r_2 and r_3, Fig.4).

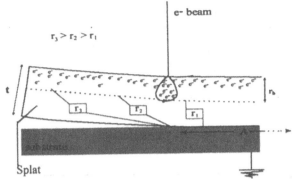

Fig 4. Schematic of electron conduction in the splat in the vicinity of a delaminated edge. r_b is the electron scattering range, a measure of penetration by electrons in the film. r_1, r_2 and r_3 are the effective resistances to specimen current as a function of position in the film.

The conduction path length is variable and maximum at vertical crack boundaries of the splat. The area A, is proportional to the film thickness 't'. The resistivity is also expected to be different for the two regions based on the microstructure of the splat materials. YSZ splats on stainless steel substrates always exhibit a columnar fine grain structure with the columns normal to the substrate[4]. The charge in the regions of contact with the substrate flows along the grain interiors or grain boundaries from the top of the film to the substrate at the bottom. In the delaminated regions the charge must flow across grain boundaries to reach the substrate via the interface. These factors all contribute to a greater resistivity for the charge decay current from the delaminated regions.

The resolution of specimen current images is limited due to beam broadening which increases with depth [7]. The electrons constituting the specimen current originate mostly from the lower regions of the scattered electron range. The capacitance of the dielectric film introduces a time constant for flow of charge constituting the specimen current. Signal to noise ratios can be improved by slower scan rates and longer collection times, and for any particular thickness there is probably an ideal accelerating potential.

Image Analysis

Profile plots of images of cracked pieces in the secondary electron and specimen current modes show the variation of gray scale due to delamination. The region of the film in contact exhibits a uniform intensity. Fig. 5 compares the specimen current image details of two splats obtained on different substrates. Fig 5a is on a substrate with oxide film from preheat at 100°C while Fig 5c is on a substrate at 300°C. Corresponding contrast variations are shown on gray scale profile plots.

The plateau in these graphs indicates the dimension of the thin film in contact with the substrate. The greater delamination in Fig. 5a may be related to the lack of substrate melting and bond development between metal and ceramic at the interface due to presence of oxide. Higher substrate temperatures as in Fig. 5c can lead to substrate melting and interfacial reactions [4]. Hence the greater length of adhered film shown in Fig. 5d. We propose to use this measure to quantify adhesion in this system in future work.

CONCLUSIONS

Along with cracking, delamination acts to relieve stress in the quenched ceramic splats and their effects are complementary. Substrate conditions seem to affect the interfacial adhesion and variations are detected in SEM images.

A novel contrast is observed in the secondary electron and specimen current images of ceramic (YSZ) splats on stainless steel substrates showing regions of delamination. Further refinement and proper quantification of the technique will require experimentation using varying electron accelera-

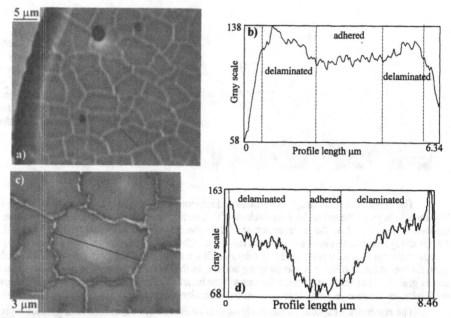

Fig. 5. Intensity profiles across cracked splat elements in gray scale (255=black, 0=white). The profile plots illustrate differences in the dimension of adhered sections.

tion energies and films of controlled thickness. Thus SEM images allow measurement of delamination in the ceramic thin film splats.

ACKNOWLEDGMENT

This work is supported by the National Science Foundation, MRSEC program, Grant No. DMR9632570.

REFERENCES

1. V.V. Sobolev, J.M. Guilemany, J. Nutting, J.R. Miguel, Int. Mater. Rev., 1997, **42**, 3, 117.

2. D.C. Agarwal, R. Raj, Surface and Coatings Technology, 1991, **37**, 4,1265.

3. B.J. Aleck, J. of Applied Mechanics,1949, June, 118.

4. T. Chraska, A. King, Thin Solid Films, to be published.

5. P.R. Thornton, Scanning Electron Microscopy, Chapman And Hall Ltd., London, 1968.

6. L. Reimer, Scanning Electron Microscopy, Springer-Verlag, Berlin, 1985.

7. D.B. Holt, Quantitative conductive mode scanning electron microscopy, in Quantitative Scanning Electron Microscopy, Ed., D.B. Holt, M.D. Muir, P.R. Grant, I.M. Boswarva, Academic Press, London, 1974.

IN-SITU OBSERVATION OF MELTING AND SOLIDIFICATION

H.SAKA*, S.ARAI**, S.MUTO**, H.MIYAI* AND S.TSUKIMOTO*
*) Department of Quantum Engineering, Nagoya University, Nagoya 464-8603,Japan
**) Center for Integrated Research in Science and Engineering, Nagoya University, Nagoya 464-8603,Japan

ABSTRACT

Melting and solidification of metallic materials have been observed directly using an in-situ heating experiment in an electron microscope. In pure Al melting initiates at the surface, while in a eutectic alloy melting initiates preferentially at eutectic interfaces and grain boundaries. The atomic structure of a solid-liquid interface was identified by comparing an experimental image with computed ones.

INTRODUCTION

Melting and solidification are very important problems in materials science. However, only limited numbers of observation of melting and solidification by means of transmission electron microscopy have been carried out [1-5]. Needless to say, this is due to experimental difficulties in preparing thin foil specimens of a liquid phase. However, recently some progresses have been made in observing melting and solidification by means of *in-situ* experiments inside an electron microscope[5-10], which will be described in this paper.

EXPERIMENTAL

The specimens used in this study were pure Al [8], Al(-Si) alloy [9] and a Al-Cu eutectic alloy. The details of specimen preparation and experimental procedures have been described elsewhere [8-10] except for Al-Cu alloy. For the Al-Cu eutectic alloy, an Al foil specimen was mounted onto a Cu mesh and heated inside an electron microscope. When the Al foil melted, the liquid Al reacted with the Cu mesh to form an Al-Cu eutectic alloy. In order to keep the shape of a foil specimen after the specimen melted, some specimens were coated with a hydrocarbon film [11] but some specimens were covered with natural oxide layers.

RESULTS AND DISCUSSION

Fig.1(a) and (b) shows electron micrographs and the corresponding diffraction pattern, respectively, of a pure Al particle before and after melting. After melting thickness contours have disappeared and the diffraction pattern shows only halo rings. The radius of the projected image of the Al particle after melting is larger by 4% than that at room temperature.

Fig.2 shows a sequence of the processes of melting reproduced from a videotape. The area which shows the thickness contours shrinks towards the center of the particle. Since the thickness contours appear only in a crystalline solid, this means that the crystalline solid shrinks. In other words, the liquid phase nucleates at the surface. The solid-liquid interface in Al does not show any anisotropy; it must lie on the isothermal contour surface.

117

Fig.3 shows an interface between solid Si and liquid Al(-Si) alloy. The solid-liquid interface is straight along {111} planes of Si. Fig.4 shows the interface between the solid Si and the liquid Al-Si at higher resolution. The interface is parallel to the (111) plane of the solid Si and atomically straight with a transition layer. Fig.4 (b) shows the intensity profile across the solid-liquid interface. Fig. 5 shows results of image simulations obtained by assuming that the transition layer as a mixture of partially solidified Si on the crystal surface and the liquid state. This layer was modeled by an additional atomic layer on top of the Si-{111} surface, as schematically shown in Fig.5(a). The trial thickness of the transition crystalline layer in the viewing direction was varied from 0.77 to 3.1nm to fit the experimental image, as shown in Fig.5(b), where the total thickness was assumed to be 5nm. The thickness of the crystalline part of the transition layer was estimated to lie between 1.5 to 2.3nm.

Fig.6 shows dynamical behavior of a solid Si/Al(-Si) alloy liquid interface during crystal growth. The solid-liquid interface is travelling from right to left. Fig.7 shows the intensity profiles across the solid-liquid interfaces shown in Fig.6. In (a) the interface is at position 1. In (b), 1/30 sec later, it has advanced to the position 2. The contrast of the atomic columns in the region between 1 and 2 is lower than that of the solid matrix (at the right side of 1) but higher than that of the liquid (at the left side of 2). This suggests that this region between 1 and 2 is a mixture of solid and liquid. In other words, atoms in this region are in a half-molten state. In (c), the solid-liquid interface has advanced to position 3. Again, the lattice images between positions 1 and 3 are fainter than the region at the right side of 1. It is only 5/30sec later that the contrast of this transition region becomes comparable to that of the solid matrix. The velocity of the solid-liquid interface in this particular event is estimated to be approximately 20nm/sec.

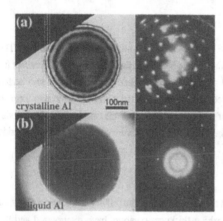

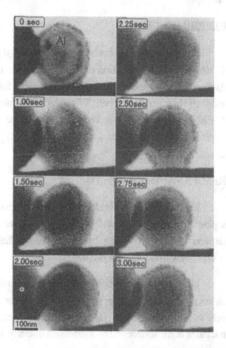

Fig.1 Electron micrographs and the corresponding diffraction patterns, respectively, of pure Al before and after melting.

Fig.2 A sequence of the processes of melting of an Al particle.

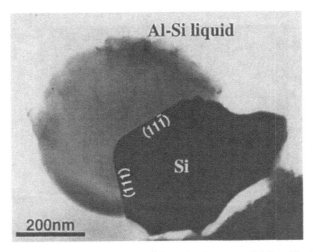

Fig.3　A low-magnification micrograph of an interface between solid Si and liquid Al(-Si) alloy.

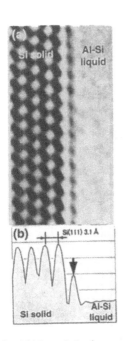

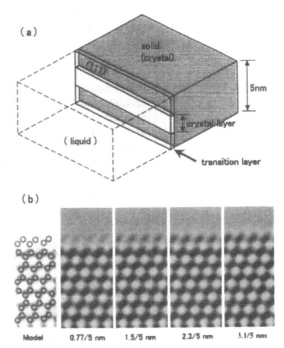

Fig.4　A high resolution image of the interface between solid Si and liquid Al-Si. (b) Intensity profile across the S/L interface.

Fig.5 (a) Model structure used for the image simulation. (b) Projection of the model structure and simulated images.

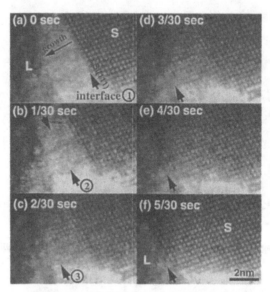

Fig.6　Dynamical behavior of a S/L interface.

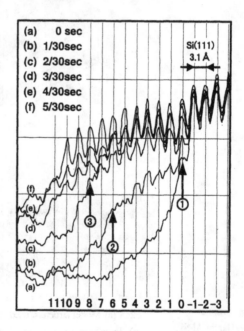

Fig.7　Intensity profiles of the S/L interfaces shown in Fig.6.

Fig.8 shows a series of electron micrographs taken during melting of Al-Cu alloy. Below the melting point the alloy consisted of a typical eutectic structure composed of the Al_2Cu phase and the Al-rich phase (hereinafter denoted by Al). The Al/ Al_2Cu interface was sharp well before the melting took place as shown in Fig.8 (a). On heating near the melting point, the interface became more and more diffuse and the liquid phase appeared at what had been the interface between Al_2Cu and the solid Al. The liquid phase expanded gradually and eventually the whole area melted, as shown in fig.8 (b)-(d). On the other hand, solidification took place very quickly. Fig.9 reproduces from a vide tape two successive frames showing microstructures before and after solidification. During 1/30 sec between these two frames, the solidification has completed.

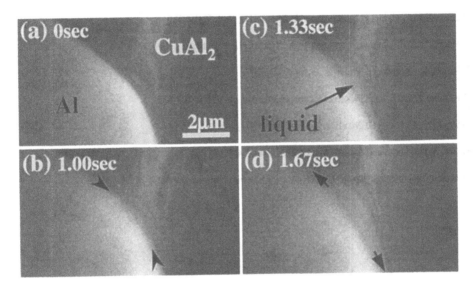

Fig.8   A series of electron micrographs taken during melting of the Al-Cu eutectic alloy.

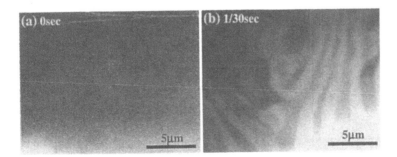

Fig. 9   Microstructure of an Al-Cu eutectic alloy just before and after solidification.

ACKNOWLEDGMENTS

The authors thank Prof. K.Sasaki of Nagoya University for advice in carrying out the experiments and Nippon Laser & Electronics Lab. for coating some of the specimens with hydrocarbon films.

REFERENCES

1) Glicksman,M.E., and Vold,C,D., Acta metall,15(1967),1409.

2) Fujita,H.,Tabata,T., and Aoki,T., 1977, Proc. of the Fifth International Conf. on HVEM, ed. By Hashimoto,H. and Imura,T., pp.439,(Japanese Society of Electron Microscopy).

3) Lemaignan,C.,Camel,D., and Pelissier,J., J. Crystal Growth,52(1981),67.

4) Watanabe,J., Suganuma,S., and Funato,A., Trans,Jpn.Inst.Metals,27 (1986), 939.

5) Sasaki,K., and Saka,H., Phil.Mag.,A,63(1991),1207.

6) Kamino,T. and Saka,H.,Microsc.Microanal.Microstruct.,4(1993)127-135.

7) Sasaki,K., and Saka,H., Mat.Res.Soc.Proc.,466(1997),185.

8) Arai,S.,Tsukimoto and Saka,H.,Microsc. Microanal.,4(1998),264.

9) Arai,S., Tsukimoto,S., Miyai,H. and Saka,H., J.Electron Microsc.,48(1999), 317.

SITE OCCUPANCY DETERMINATION BY ALCHEMI OF Nb AND Cr IN γ-TiAl AND THEIR EFFECTS ON THE α TO γ MASSIVE PHASE TRANSFORMATION

T.M. MILLER*, L. WANG*, W.H. HOFMEISTER**, J.E. WITTIG*, I.M. ANDERSON***
*EECE Dept., **ChE Dept., Vanderbilt University, Nashville, TN 37212
***Metals & Ceramics Division, Oak Ridge National Laboratory, Oak Ridge, TN 37831-6376

ABSTRACT

Atom location by channeling enhanced microanalysis (ALCHEMI) has been used to characterize the site distributions of Nb and Cr alloying additions in the $L1_0$-ordered γ phase of ternary titanium aluminides. Two alloys, $Ti_{50}Al_{48}Cr_2$ and $Ti_{50}Al_{48}Nb_2$, were processed by furnace cooling from 1300°C (within the α-γ two phase field) as well as by rapid solidification using twin-anvil splat quenching of electromagnetically levitated and undercooled samples. ALCHEMI studies of furnace cooled samples yield results generally consistent with those in the published literature. Nb alloying additions are found to partition exclusively to the 'Ti' sublattice, while Cr alloying additions exhibit an 'Al' sublattice preference. However, a higher degree of disorder can be achieved with rapid solidification and high solid state cooling rates (10^5-10^6 K/s). Significant distribution of the ternary elements between the 'Ti' and 'Al' sublattices has been measured in the splat quenched samples, with up to 12% of the Nb atoms occupying the 'Al' sublattice and the fraction of Cr atoms on the 'Ti' sublattice doubling to ~30%. Rapid solidification of TiAl produces an equiaxed hexagonal α phase solidification structure that transforms in a massive fashion to the tetragonal γ phase. Although the amount of massively transformed γ is dependent upon the solid state cooling rate, ternary alloying additions can more strongly influence the transformation kinetics. The Nb-modified alloy exhibits significant amounts of the massively tranformed γ, similar to the $Ti_{52}Al_{48}$ binary alloy, whereas little massively transformed γ is observed in the Cr-modified alloy. These results can be correlated with the relative atomic size, lattice distortion, and sublattice site occupancy of Nb and Cr in the $L1_0$ unit cell.

INTRODUCTION

Titanium aluminidies are attractive candidates for high temperature structural materials because of their low density and high creep resistance. However, poor room temperature ductility has limited their use in actual applications [1]. Current efforts to improve the ductility of two phase (γ-$α_2$) TiAl alloys with 46 to 48 at.% Al have used alloying additions of elements such as Cr, Nb, V or Mo. These alloying additions can influence the lattice parameters and mechanical behavior of the $L1_0$-ordered γ and DO_{19}-ordered $α_2$ phases, as well as the solidification structures, phase stability, and solid state phase transformations in these alloys. The effects of alloying additions in TiAl alloys cannot be fully understood unless their site-substitution behavior in the ordered phases is known. Atom location by channeling enhanced microanalysis (ALCHEMI) provides a method to quantitatively measure the site occupancy of ternary elements in the $L1_0$-ordered structure of TiAl [2-4].

Massive phase transformations, characterized by a 'diffusionless' nucleation and growth process, are well documented in titanium aluminides [5,6]. The rapid solidification process used in this study (electromagnetic levitation, undercooling below the liquidus, and twin anvil splat quenching) provides an ideal method to study this α → γ massive phase transformation. As the amount of undercooling increases, the solidification velocity and resulting splat thickness

123

increases. Samples can be levitated and splat quenched with undercoolings ranging from 0 to 160 K, which results in splat thicknesses from 140 to 400 μm. Since thicker splats experience slower solid state cooling rates, the massive transformation can be observed at various stages of the nucleation and growth process. The purpose of this investigation is to study the effect of ternary alloying additions on the solid-state phase transformations in titanium aluminides to yield a better understanding of $\alpha \rightarrow \gamma$ transformation kinetics.

EXPERIMENT

Arc cast alloys $Ti_{50}Al_{48}Cr_2$ and $Ti_{50}Al_{48}Nb_2$ with < 500 ppm (wt) oxygen were processed by heat treating in vacuum for 4 hr at 1300°C (within the α-γ two phase field) and furnace cooling at ~ 2 K/min. To induce the massive phase transformation, 500 mg spheres were electromagnetically levitated in vacuum, induction melted, and undercooled below the melting point ($\Delta T = T_m - T$) by flowing He gas over the molten sphere. Non-contact optical pyrometry measured the temperature of the undercooled droplet. At the desired ΔT, the electromagnetic field is removed and pneumatically driven twin copper anvils splat quench the falling droplet. Samples for transmission electron microscopy (TEM) were punched from the center of the splat and thinned equally from each side with final thinning by twin-jet electropolishing in 57% methanol, 37% n-butanol and 6% perchloric acid at –25°C and 21 V. The microstructures of the furnace cooled and rapidly quenched alloys were characterized by standard bright-field imaging in a Philips CM20. ALCHEMI was performed by collecting energy-dispersive spectra (EDS) at up to 16 different excitations of the (002) and (220) systematic rows of the $L1_0$-ordered alloy. The characteristic K-series X-ray lines of each element were analyzed. For data reduction, the statistical significance for each measured characteristic intensity is weighted according to Poisson statistics. Delocalization corrected correlation coefficients were extracted from experimental parameters. These experimental methods are described in detail elsewhere [7,8].

RESULTS

Figure 1 shows the relationship between undercooling (ΔT) and splat thickness (t). Increased undercooling (ΔT) results in thicker splats with lower solid state cooling rates. For small undercooling (ΔT~25 K and t ~ 150 μm), finite element heat flow calculations estimate the solid state cooling rate at the splat center to be on the order of 10^6 K/s [9], whereas for large undercooling (ΔT ~ 140 K and t ~ 300 μm), the cooling rate is reduced by approximately an order of magnitude to ~10^5 K/s.

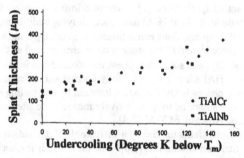

Figure 1 – Splat thickness as a function of undercooling.

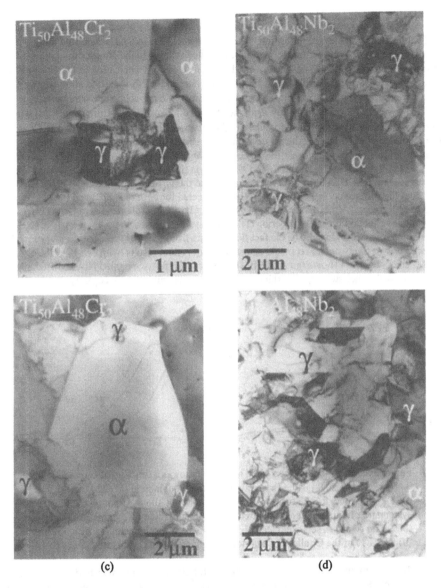

Figure 2 – The effect of ternary additions and solid state cooling rates on the massive $\alpha \rightarrow \gamma$ phase transformation. Bright field images are shown for (a,c) Cr- and (b,d) Nb-modified alloys with undercoolings ΔT of (a,b) 25 K and (c,d) 135 K. The smaller [larger] undercooling results in a splat thickness of 150 [300] μm with a solid state cooling rate at the splat center of $\sim 10^6$ [10^5] K/s. For both undercoolings, the Cr-modified alloy is mostly composed of the α phase, whereas the Nb-modified alloy is mostly composed of the massively transformed γ phase. These results suggest that ternary alloying additions can have a much greater influence than the solid state cooling rate on the $\alpha \rightarrow \gamma$ massive transformation.

Figure 2 shows the amount of the massively transformed γ phase as a function of solid state cooling rate and alloy composition. The γ phase nucleates in the Cr-modified alloy but growth of the massive phase transformation is inhibited relative to the Nb-modified alloy. Solid state cooling rates of ~10^6 K/s produce only the initial stages of γ nucleation in the Cr-modified alloy (figure 1a), whereas similar solid state cooling rates in the Nb-modified alloy (figure 1b) result in extensive γ growth. Larger undercooling produces lower solid state cooling rates, ~ 10^5 K/s. However, the Cr-modified alloy maintains the α solidification structure (figure 1c) whereas the α→γ transformation is almost complete in the Nb-modified alloy (figure 1d). The Cr atoms strongly reduce the kinetics of the massive phase transformation compared to Nb. The Nb-modified alloy exhibits transformation kinetics similar to a binary $Ti_{52}Al_{48}$ alloy [9].

The site occupancies of the Cr and Nb atoms in the $L1_0$-ordered structure were investigated by ALCHEMI for both the quenched and the furnace cooled alloys. Figure 3 shows data from the furnace cooled Cr-modified alloy, collected for both the (220) and (002) systematic rows of reflections. Although the (002) planes have a larger interplanar spacing, the (110) superlattice reflection produces superior 'Ti' / 'Al' site discrimination compared to the (001) superlattice reflection. In spite of the fact that the data from the (001) row have a significantly larger error bar, the two data sets yield statistically equivalent results, with ~85% of the Cr atoms on 'Al' sites. Figure 4 shows the ALCHEMI data from the furnace cooled Nb-modified alloy. Consistent with other studies, Nb exhibits an exclusive 'Ti' sublattice occupancy.

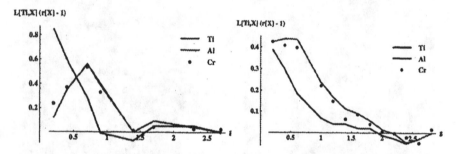

Figure 3 – Delocalization-corrected intensity ratios as a function of orientation g = θ / $θ_B$ [7,8], where $θ_B$ is the Bragg angle for the fundamental (a) (220) and (b) (002) reflections, for the furnace-cooled Cr-modified alloy. Although the (220) systematic orientation exhibits better 'Ti' / 'Al' sublattice discrimination than (002), the two data sets yield statistically equivalent results, with (a) 85 ± 3% and (b) 83 ± 8% of the Cr alloying addition occupying the 'Al' sublattice.

Table I summarizes the site occupancy data for the Cr- and Nb-modified alloys in the furnace cooled (FC) and undercooled rapidly quenched (URQ) conditions. The ALCHEMI data show that Cr exhibits a pronounced (~85%) 'Al' sublattice preference and Nb exhibits an exclusive (~100%) 'Ti' sublattice occupancy in the near-equilibrium furnace cooled alloys. However, with rapid solidification and high solid state cooling rates (10^5-10^6 K/s), a larger degree of disorder is observed, with only ~70% of the Cr occupying the 'Al' sublattice and ~88% of the Nb occupying the 'Ti' sublattice. For both heat treatments, Nb is found to exhibit a stronger site preference than Cr. This behavior can be correlated with the characteristic radii and electronegativity data for these elements, as shown in Table II.

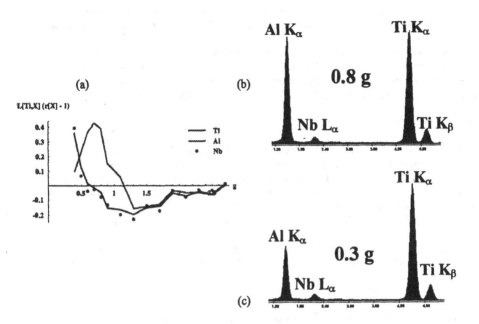

Figure 4 – ALCHEMI data for the furnace cooled Nb-modified alloy. The data in (a) show that the Nb atoms substitute for Ti atoms in the $L1_0$ structure with 100% 'Ti' sublattice occupancy. The size of the Nb L peaks remain proportionate to the Ti K peaks, in contrast to the large relative change of Al K intensity when changing the channeling conditions from (b) 0.8 g to (c) 0.3 g, with $g = \theta / \theta_{220}$.

TABLE I - ALCHEMI Data from $Ti_{50}Al_{48}Cr_2$ and $Ti_{50}Al_{48}Nb_2$

Element	Al Site Occupancy	Ti Site Occupancy
Chromium [FC]	85 ± 3 %	15 ± 3 %
Chromium [URQ]	70 ± 3 %	30 ± 3 %
Niobium [FC]	0 ± 5 %	100 ± 5 %
Niobium [URQ]	12 ± 3 %	88 ± 3 %

TABLE II - Atomic Radii and Electronegativity

Element	Atomic Radius (Å)	Covalent Radius (Å)	Electronegativity
Ti	1.48 (HCP)	1.32	1.3
Al	1.43 (FCC)	1.18	1.5
Nb	1.43 (BCC)	1.34	1.3
Cr	1.25 (BCC)	1.18	1.6

The measured sublattice preferences of the Nb and Cr alloying additions correlate strongly with the covalent rather than the atomic radii in Table II. In particular, whereas the atomic radii of Nb and Al are identical, which might suggest an 'Al' sublattice preference for Nb, there is a much closer correspondence between the covalent radii of Nb and Ti. A similar correspondence exists for the covalent radii of Cr and Al. These correlations are consistent with the strong directional bonding in $L1_0$-ordered γ-TiAl. Electronegativity, an indicator of solid state solubility, also predicts an affinity of Nb for 'Ti' and Cr for 'Al' sites.

CONCLUSIONS

Ternary alloying additions of Cr and Nb exhibit greater distribution between the 'Ti' and 'Al' sublattices in rapidly cooled, splat quenched TiAl alloys relative to corresponding furnace cooled alloys. A significant fraction (~12%) of the Nb alloying additions are found to occupy the 'Al' sublattice in the quenched alloys, whereas Nb has been found to occupy the 'Ti' sublattice exclusively in furnace cooled alloys. The fraction of Cr atoms occupying the 'Ti' sublattice doubles from ~15% in the furnace cooled alloys to ~30% in the quenched alloys.

Alloying additions of Cr significantly reduce the kinetics of the $\alpha \rightarrow \gamma$ massive phase transformation, whereas additions of Nb have little influence on the transformation kinetics.

ACKNOWLEDGEMENTS

Support from the National Science Foundation, Division of Materials Research, DMR9616748, is gratefully acknowledged. Research at the ORNL SHaRE User Facility was supported by the Division of Materials Sciences, U.S. Department of Energy under contract DE-AC05-96OR22464 with Lockheed Martin Energy Research Corp., and through the SHaRE Program under contract DE-AC05-76OR00033 with Oak Ridge Associated Universities.

REFERENCES

1. E.L. Hall, S. Huang, Acta Metall. Mater., 38 (1990) 539-549
2. I.M. Anderson, J. Bentley, Inst. Phys. Conf. Ser. No. 147 (1995) 531-534
3. Y.L. Hao, D.S. Xu, Y.Y. Cui, R. Yang, D. Li, Acta Mater., 47 (1999) 1129-1139
4. C.J. Rossouw, T. Forwood, M.A. Gibson, P.R. Miller, Phil. Mag., 74A (1996) 57-76
5. P. Wang, G.B. Viswanathan, V.K. Vasudevan, Met. Trans., 23A (1992) 690-697
6. J.E. Wittig and W.H. Hofmeister, Microscopy and Microanalysis, 4(S2) (1998) 538-539
7. I.M. Anderson, Acta Mater., 45 (1997) 3897-3909
8. I.M. Anderson, A.J. Duncan, J. Bentley, Intermetallics, 7 (1999) 1017-1024
9. D.E. Sims, Masters Thesis, Vanderbilt University (1996)

ELECTRON CHANNELING X-RAY MICROANALYSIS FOR CATION CONFIGURATION IN IRRADIATED MAGNESIUM ALUMINATE SPINEL

S. MATSUMURA *, T. SOEDA *, N. J. ZALUZEC **, C. KINOSHITA *

*Department of Applied Quantum Physics and Nuclear Engineering, Kyushu University 36, Fukuoka 812-8581, Japan, syo@nucl.kyushu-u.ac.jp
**Materials Science Division, Argonne National Laboratory, Argonne, IL 60439

ABSTRACT

High angular resolution electron channeling x-ray spectroscopy (HARECXS) was examined as a practical tool to locate lattice-ions in spinel crystals. The orientation dependent intensity distribution of emitted x-rays obtained by HARECXS is so sensitive to lattice-ion configuration in the illuminated areas that the occupation probabilities on specific positions in the crystal lattice can be determined accurately through comparison with the theoretical rocking curves. HARECXS measurements have revealed partially disordered cation arrangement in $MgO \cdot nAl_2O_3$ with $n=1.0$ and 2.4. Most Al^{3+} lattice-ions occupy the octahedral (VI) sites with 6-fold coordination, while Mg^{2+} lattice-ions reside on both the tetrahedral (IV) and the octahedral (VI) sites. The structural vacancies are enriched in the IV-sites. Further evacuation of cations from the IV-sites to the VI-sites is recognized in a disordering process induced by irradiation with 1 MeV Ne+ ions up to 8.9 dpa at 870 K.

INTRODUCTION

Magnesium aluminate spinel $MgO \cdot nAl_2O_3$ is expected to be an excellent candidate for insulators in the fusion reactors, since it exhibits strong resistance to formation of defect clusters under irradiation with high energy particles [1,2]. The stability of $MgO \cdot nAl_2O_3$ in radiation fields is believed to come from a number of vacant sites in the spinel structure. In the ideal or "normal" structure of stoichiometric case, Mg^{2+} and Al^{3+} cations occupy 1/8 of the 4-fold coordinated tetrahedral (IV) sites and 1/2 of the 6-fold coordinated octahedral (VI) sites, respectively, in the *fcc* lattice formed with O^{2-} anions. The rest of the sites remain unoccupied as they are empty holes in the lattice. A significant number of structural vacancies are further introduced into the cation sites in the non-stoichiometric compounds with $n>1$. Under irradiation conditions, these empty holes and the structural vacancies are presumed to accommodate displaced lattice-ions and to suppress aggregation of defects into clusters. However, the behavior of displaced lattice-ions and vacancies in $MgO \cdot nAl_2O_3$ under irradiation has not been conclusively confirmed by experimental measurements. Neutron diffraction [3] and nuclear magnetic resonance [4] were employed in a study of lattice-ion displacements in $MgO \cdot nAl_2O_3$ irradiated with neutrons. These techniques disclosed the average overall behavior of structural change, but not detailed information particularly in heterogeneously damaged materials. The present authors applied a combined use of conventional electron channeling x-ray spectroscopy [5] and large angle convergent beam electron diffraction (LACBED) to this problem and succeeded in determining the lattice-ion configuration of local areas in irradiated $MgO \cdot nAl_2O_3$ compounds [6]. This two step method is somewhat difficult and time consuming, because it consists of two separate experiments and several parameters relevant to dynamical electron diffraction which must be determined. In addition, LACBED is not appropriate for use in to heavily damaged areas, since the diffraction pattern is also highly sensitive to local distortion of lattice.

To overcome the difficulties involved in the conventional electron channeling and diffraction technique, the present study examines the potential of electron channeling x-ray spectroscopy for determination of lattice-ion configuration in partially disordered $MgO \cdot nAl_2O_3$

129

spinel compounds. The characteristic x-ray emission is measured precisely as a function of incident-beam direction in the experiment, which we have termed high angular resolution electron channeling x-ray spectroscopy (HARECXS) to differentiate it from conventional electron channeling techniques [5]. The orientation dependent profiles of emitted x-ray intensity are measured and then compared to theoretical calculations [7-9] analyzed to determine the local lattice-ion configurations. In this work, the HARECXS technique is employed to investigate cation configuration in MgO·nAl$_2$O$_3$ with n=1.0 and 2.4 as well as disordering behavior induced by irradiation with 1 MeV Ne$^+$ ions at 870 K.

EXPERIMENTAL PROCEDURE

Disk specimens of MgO·nAl$_2$O$_3$ with n=1.0 (Union Carbide) and 2.4 (Nakazumi Crystal) were dimpled to 30μm thickness and then ion-milled with 5 keV Ar$^+$ ions to electron transparency. These specimens were subsequently annealed at 1470 K for 48 hours to remove lattice defects produced by ion milling, and allowed to furnace cool. Single-crystalline MgO·nAl$_2$O$_3$ bulk crystals were irradiated at 870 K with 1 MeV Ne$^+$ ions to a dose of about 2 dpa (Φ=4.5 x 10^{20} Ne$^+$/m^2). The irradiated bulk specimens were mechanically thinned with a tripod polisher. The wedge shape cross sectional specimens allowed observation along Ne-ion beam direction. Argon plasma processing [10] was performed in order to suppress the hydrocarbon contamination on the specimens during the following HARECXS measurement. Experimental data was acquired using a Philips EM420T analytical electron microscope equipped with an LaB$_6$ electron source and an EDAX 9900 EDX system operating in the TEM mode at 120 kV. Two-dimensional angular resolution measurements of x-ray emission were carried out in a 128 x 100 pixel scan synchronized with incident beam rocking over an angular range of about 100 by 80 mrad between –4g and +4g (g=400) Bragg conditions, using customized computer control. The typical acquisition time for a complete two-dimensional scan was 18-24 hours.

RESULTS AND DISCUSSION

Fig. 1 illustrates calculated HARECXS profiles of stoichiometric MgO·nAl$_2$O$_3$ spinel crystals with various cation arrangements. These calculations were accomplished using a computer program developed by Rossouw *et al.*, which takes account of dynamical scattering of incident electrons as well as the dechanneling and delocalization effects on induced x-ray emission [7,9]. Fifteen reflections in the 400 systematic row were considered in these calculations and the parameter k along the abscissa refers to the intersection of the Ewald sphere with the axis along 400 systematic reflections. In this nomenclature, k/g_{400}=1 corresponds to the exact Bragg condition for 400 reflection. The intensities of Mg-K, Al-K and O-K x-rays are given as a function of incident electron beam orientation, normalized by the values at the 5g (g=400) Bragg condition. The x-ray intensities drastically change with the beam orientation, especially in the central part of strong dynamical excitation of low order reflections. For the normal spinel structure, where the IV- and the VI-sites are purely occupied by Mg^{2+} and Al^{3+} lattice-ions, respectively, the emission from Mg-K is suppressed, while Al-K and O-K are enhanced within the range –1<k/g_{400}<1, as shown in Fig. 1(a). Outside of this range, the intensity relationship is reversed. If the cations are arranged in the inverse form, where all Mg^{2+} lattice-ions are located on the VI-sites and Al^{3+} lattice-ions are distributed equally on both IV- and VI-sites, the Mg emission is enhanced around the symmetric diffraction orientation and the Al signal is enhanced when 1<|k/g_{400}|<2, as seen in (d). For comparison, Fig. 1(c) shows the case when Mg^{2+} and Al^{3+} lattice-ions occupy randomly both IV- and VI-sites. The profiles of Mg-K and Al-K almost coincide with each other. From these figures we see that the HARECXS profiles change sensitively depending on replacement of Mg^{2+} and Al^{3+} lattice-ions between the IV- and the VI-sites. One may conclude from Fig. 1 that HARECXS would be useful for investigating ion configuration in disordered compounds.

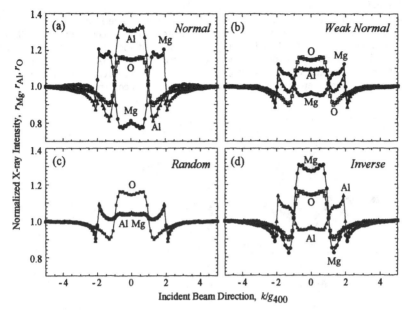

Fig. 1: Simulated HARECXS profiles of MgO·Al$_2$O$_3$ with various cation configuration. (a) P_{Mg}=1, P_{Al}=0 (normal), (b) P_{Mg}=0.5, P_{Al}=0.25, (c) P_{Mg}=0.33, P_{Al}=0.33 (random), (d) P_{Mg}=0, P_{Al}=0.5 (inverse). Accelerating voltage: E=120 kV, Foil thickness: t=150 nm.

Fig. 2 presents our experimentally obtained HARECXS profiles of annealed MgO·nAl$_2$O$_3$ with n=1.0 and 2.4. The profile (2a) for n=1.0 looks analogous to Fig. 1(a), indicating that the stoichiometric compound has a tendency to form the normal structure. In contrast, the profile (2b) for a non-stoichiometric compound with n=2.4 exhibits strikingly different features; for example, Mg-K signal is enhanced in the central part around the symmetric diffraction orientation, and suggests that a considerable amount of Mg^{2+} lattice-ions are located on the VI-sites. Fig. 3 shows the normalized intensities r_{Mg} and r_{Al} plotted against r_O. Here one can observe that there exist approximately linear relationships of r_{Mg} or r_{Al} with r_O around the 400 Bragg condition, as pointed out by Anderson [8]. If the occupation probabilities of Mg^{2+} and Al^{3+} lattice-ions on the IV-sites are defined as P_{Mg} and P_{Al}, then the ratios S (=P_{Al}/P_{Mg}) for both compounds are obtained from the relationships,

$$r_O = \frac{1}{1-S} \cdot \frac{1}{L_{AlO}} \cdot r_{Al} - \frac{S}{1-S} \cdot \frac{1}{L_{MgO}} \cdot r_{Mg}. \tag{1}$$

Here, L is the coefficient of charge density distribution in Al^{3+} or Mg^{2+} lattice-ions relative to O^{2-} lattice-ions. Since the number of unknown parameters in this formulation is now reduced to one when S is evaluated, the value of P_{Mg} is easily determined by comparing the HARECXS profiles directly with the theoretical calculations. Figs. 4(a) and (b) show the calculated results corresponding to Figs. 2(a) and (b), respectively. Quite good agreements are achieved between the theoretical calculations and the experimental results, as shown in Figs. 2 and 4. Table 1 lists the occupation probabilities of cations on the IV- and the VI-sites determined using this procedure. In the stoichiometric compound with n=1.0, most Al^{3+} lattice-ions are located on the VI sites, while Mg^{2+} lattice-ions distribute 60 % and 40 % on the IV- and the VI sites, respectively. The partial disordering tendency of Mg^{2+} lattice-ions has been also revealed by neutron diffraction as $P_{Mg}=0.763 \pm 0.02$ and $P_{Al}=0.118 \pm 0.01$ [3], and has been explained in terms of ionic size and induced lattice strain around the IV-sites. Table 1 indicates that a small amount of vacancies are produced in the IV-sites in the stoichiometric compound. It follows that the same amount of cations should occupy empty holes of the octahedral position. The disordering tendency is more pronounced in non-stoichiometric $MgO \cdot 2.4Al_2O_3$, where 20 % of Al^{3+} lattice-ions are located on the IV-sites and fewer Mg^{2+} ions remain in the IV-sites. Structural vacancies, which appear to be greater than the nominal content, are enriched in the IV-sites.

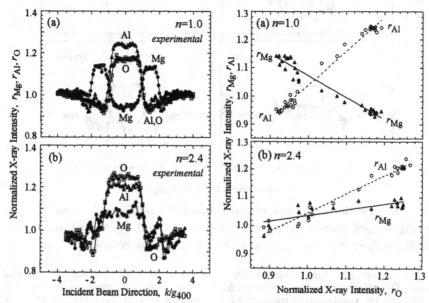

Fig. 2: HARECXS profiles of annealed $MgO \cdot nAl_2O_3$ with $n=1.0$ (a) and 2.4 (b).

Fig.3: Linear relationship between characteristic x-ray intensities around the 400 Bragg positions in the HARECXS profiles of Fig. 2.

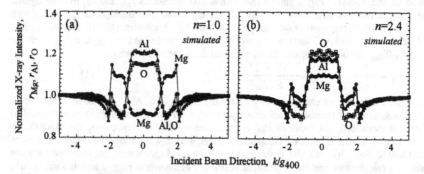

Fig. 4: Simulated HARECXS profiles of $MgO \cdot nAl_2O_3$ with $n=1.0$ (a) and 2.4 (b). Foil thickness: $t=180$ (a), 50 nm (b).

Table 1: Occupation numbers of Mg^{2+}, Al^{3+} and vacancies on the IV-sites and the VI-sites per unit cell in annealed $MgO \cdot nAl_2O_3$. The occupation probabilities on the IV-sites (P_i) are also tabulated.

| | | Number of ions (per unit cell) | | | P_i |
		total	IV-sites	VI-sites	
$n=1.0$	Mg^{2+}	8	4.9 ± 0.1	3.1 ± 0.1	0.61 ± 0.01
	Al^{3+}	16	1.6 ± 0.3	14.4 ± 0.3	0.10 ± 0.02
	Vacancy	-	1.5 ± 0.3	0	-
$n=2.4$	Mg^{2+}	3.9	1.1 ± 0.1	2.8 ± 0.1	0.27 ± 0.01
	Al^{3+}	18.7	3.8 ± 0.3	14.9 ± 0.3	0.20 ± 0.02
	Vacancy	1.3	3.2 ± 0.4	0	-

Fig. 5 demonstrates HARECXS profiles obtained from a pre-peak damaged area at $d=450$ nm and a peak-damaged region at $d=900$ nm in a cross sectional specimen of $MgO \cdot Al_2O_3$ irradiated with 1 MeV Ne^+ ions at 870 K. Here, d is the depth from the irradiated surface. In this sample the calculated dose reaches about 2.0 and 8.9 dpa in the pre-peak and the peak damaged areas, respectively. Well-defined and sharp diffraction spots appeared from both regions, indicating that the spinel structure is still stable even in the peak damaged area. In contrast, the HARECXS profiles have varied significantly from that in Fig. 2 (a) before irradiation. It is apparent that disordering ensues from knocked-on displacements of lattice-ions by irradiation. Table 2 summarizes the occupation numbers of Mg^{2+} and Al^{3+} lattice-ions on the IV- and the VI-sites after 1 MeV Ne^+ ion irradiation at 870 K. The total number of Mg^{2+} lattice-ions decreases, while Al^{3+} lattice-ions increase in the IV-sites in the pre-peak damaged area. This indicates that cations are simply replacing each other on both sites in the pre-peak damaged area. In contrast, the number of cations on the IV-sites decreases in the peak-damaged region. Mainly due to displacement of Al^{3+} lattice-ions from the IV-sites to the VI-positions and as a result, excess vacancies are produced in the IV-sites. Neutron diffraction has suggested that a considerable amount of ions are located on unoccupied sites of the VI-positions in $MgO \cdot Al_2O_3$ after neutron irradiation up to 56 dpa at 678 K [3]. Heavy ion irradiation with 400 keV Xe^{2+} or 1.5 MeV Kr^+ ions at cryogenic temperatures [11] results in extinction of 220 reflections, which indicates removal of lattice-ions from the IV-sites. The experimental results of heavy ion irradiation are qualitatively consistent with the behavior observed in the peak-damaged area. A more extended study on irradiation behavior in $MgO \cdot nAl_2O_3$ is in progress [12,13].

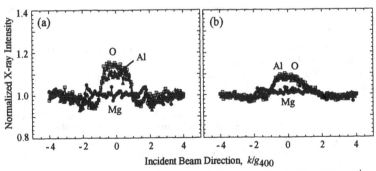

Fig. 5: HARECXS profiles obtained from $MgO \cdot Al_2O_3$ irradiated with 1 MeV Ne^+ ions at 870 K. (a) Pre-peak damaged area ($d=450$ nm), (b) Peak-damaged area ($d=900$ nm).

Table 2: Cation and vacancy configuration on the IV-sites and the VI-sites for n=1 after irradiation with 1 MeV Ne$^+$ ions at 870 K. The probabilities are given in terms of the number of ions per unit cell.

		Number of ions (per unit cell)		
		total	IV-sites	VI-sites
Pre-peak	Mg^{2+}	8	3.0 ± 0.1	5.0 ± 0.1
damaged area	Al^{3+}	16	3.5 ± 0.3	12.5 ± 0.3
(2.0 dpa)	Vacancy	-	1.5 ± 0.3	0
Peak damaged	Mg^{2+}	8	3.0 ± 0.1	5.0 ± 0.1
area	Al^{3+}	16	2.9 ± 0.3	13.1 ± 0.3
(8.9 dpa)	Vacancy	-	2.2 ± 0.3	0

CONCLUSIONS

It has been demonstrated that the high angular resolution electron channeling x-ray spectroscopy (HARECXS) is so sensitive to lattice-ion configuration in the illuminated areas that the occupation probabilities on specific positions in the crystal lattice can be determined with sufficient accuracy. HARECXS has successfully revealed partial disordering behavior in $MgO \cdot nAl_2O_3$ spinel crystals as a function of composition and irradiation with 1 MeV Ne$^+$ ions at 870 K. In contrast to strong preference of Al^{3+} lattice-ions to the octahedral (VI) sites, Mg^{2+} lattice-ions possess relatively weaker tendency to reside in the tetragonal (IV) sites. Ne$^+$ ion irradiation induces not only cation disordering but also slight evacuation of cations from the IV-sites to the VI-positions.

ACKNOWLEDGEMENTS

The authors should like to express their gratitude to Dr. Lynn Rehn for arrangements for this collaborative work of the two groups at KU and ANL. They are also indebted to Dr. C. J. Rossouw at CSIRO for his kind offer of the efficient calculation program and invaluable comments. This research was partly supported by the US Department of Energy under contract BES-MS W-31-109-Eng-38 and by the Grant-in-Aid for scientific research from the Ministry of Education, Science, Culture and Sports, Japan and from the Japan Society of Promotion of Science.

REFERENCES

[1] F. W. Clinard, Jr., G.F. Hurley, R.A. Youngman and L.W. Hobbs, J. Nucl. Mater. **133-134**, 701 (1985).
[2] S. J. Zinkle and C. Kinoshita, J. Nucl. Mater. **251**, 200 (1997).
[3] K.E. Sickafus, A.C. Larson, N. Yu, M. Nastasi, G.W. Hollenberg, F.A. Garner and R.C. Bradt, J. Nucl. Mater. **219**, 128 (1995).
[4] E.A. Cooper, C.D. Hughes, W.L. Earl, K.E. Sickfus, G.W. Hollenberg, F.A. Garner and R.C. Bradt, MRS Symp. Proc. **373**, 413 (1995).
[5] J.C.H. Spence and J. Taftø, J. Microscopy **130**, 147 (1983).
[6] T. Soeda, S. Matsumura, J. Hayata and C. Kinoshita, J. Electron Microsc. **48**, 531 (1999).
[7] L. J. Allen, T.W. Josefsson and C.J. Rossouw, Ultramicrosc. **55**, 258 (1994).
[8] I. M. Anderson, Acta Mater. **45**, 3897 (1997).
[9] C. J. Rossouw, C. T. Forwood, M.A.Gibson and P.R. Miller, Micron **28**, 125 (1997).
[10] N. J. Zaluzec, Proc. Microsc and Microanal. '97, 983 (1997).
[11] N. Yu, R. Devanathan, K.E. Sickafus and M. Nastasi, J. Mater. Res. **12**, 1766 (1997).
[12] T. Soeda, S. Matsumura, C. Kinoshita and N.J. Zaluzec, submitted to J. Nucl. Mater.
[13] T. Soeda, PhD Thesis Research, in progress, Kyushu University.

IN SITU STUDY OF ZIRCONIA STABILIZATION BY ANION EXCHANGE (N FOR O) USING HIGH TEMPERATURE, CONTROLLED ATMOSPHERE ELECTRON DIFFRACTION

RENU SHARMA, EBERHARD SCHWEDA*, DIRK NAEDELE*,
Center for Solid State Science Arizona State University, Tempe AZ 85287;
*Anorganische Chemie, Universität Tübingen Auf Morgenstelle 18, Tübingen D- 72076
Germany

ABSTRACT

Stabilization of zirconia by anion exchange (N for O) is a novel idea. A number of oxy-nitrides with flourite-related (cubic) structure have been reported to form at high temperatures (1100°C). We have used a TEM equipped with environmental cell and Gatan Imaging Filter (GIF) to study the nitridation behavior of zirconia. The *in situ* observations reveal the formation of a cubic structure at ≈800°C when the $Zr(OH)_4*xH_2O$ precursor was heated in ≈2 torr of NH_3. The presence of N in the lattice is confirmed by electron energy-loss spectroscopy.

INTRODUCTION

Zirconium oxide (ZrO_2) forms a monoclinic structure at room temperature, that transforms into a tetragonal structure (>1100°C) and finally into a cubic structure (>1600°C) upon heating. The 4% volume change due to the monoclinic to tetragonal phase transformation results in formation of cracks, thus restricting its use to much lower temperatures [1]. As the metastable tetragonal and cubic ZrO_2 also have improved mechanical and electrical properties, the effort has been concentrated on stabilizing cubic zirconia by cation substitution, e.g. MgO, CaO, Y_2O_3 etc. [2-3]. We have previously reported that a metastable cubic structure is also formed, as nano-crystalline phase (1-2nm particle size), during electron beam heating of amorphous $Zr(OH)_4$ samples [4]. An addition of 5-15% of Mn helps to stabilize the cubic structure for slightly larger crystal size (5-20nm) and up to higher temperatures (850°C) [4].

In the past, the efforts to stabilize cubic zirconia by anion substitution (N for O) have resulted in the formation of fluorite-related structures with ordered oxygen, nitrogen and vacancies in the lattice [5-6]. The ordering of vacancies is reported to have a negative effect on ionic conductivity [7]. Moreover, the reported oxy-nitrides, with a general formula $ZrO_{2-2x}N_{4x/3}$, fall on the ZrO_2-Zr_3N_4 line of the phase diagram instead of the ZrO_2-ZrN line. These oxy-nitrides have been synthesized by heating ZrO_2 or a mixture of ZrO_2 and ZrN in N_2 at 1100°C. The high synthesis temperature encourages ordering of oxygen vacancies. In order to synthesize zirconia with higher ionic conductivity, it is important to stabilize a cubic phase without any ordering of vacancies.

We have used *in situ* time and temperature resolved electron diffraction microscopy in vacuum and in a controlled atmosphere of dry NH_3 gas in order to understand the

135

nitridation reaction mechanism. Electron energy-loss spectroscopy is used for qualitative compositional analysis after the reaction.

EXPERIMENTAL

In order to obtain highly reactive starting material, $Zr(OH)_4*xH_2O$ precursor was prepared by dissolving $ZrOCl_2*8H_2O$ (Aldrich;99.99%) in de-ionized water and precipitating the hydroxide with excess of ammonium hydroxide. The precipitates were then washed in order to remove the traces of NH_4OH till the pH level of the filtrate was ≈7 and dried at 120°C for 48 hrs.

Samples were dry-loaded on 200 mesh Cu or Mo grids for the electron microscopy study. No effect of grid material was observed. A PHILIPS-430 transmission electron microscope, equipped with a differentially pumped environmental cell fitted in between the objective pole pieces, and a post-projector Gatan Imaging Filter (GIF), operated at 300 kV, was used to study gas-solid interactions [8]. The samples were heated using a single-tilt PHILIPS heating holder in vacuum and up to ≈2 torr of dry NH_3. The electron diffraction patterns were recorded at different temperatures during heating. As our starting materials can also be heated by electron beam, the study was restricted to low-dose selected area electron diffraction (SAED) work. Time and temperature resolved SAED patterns were recorded from 8-10 crystals starting from room temperature both in vacuum and in NH_3. Electron energy loss spectra were recorded after the reaction and after stopping the gas flow to detect the presence of nitrogen in the samples using a GIF with an energy resolution of 2.5-3 eV.

RESULTS:

Heating in Vacuum

Fig.1. Selected area electron diffraction (SAED) patterns from $Zr(OH)_4$ x H_2O sample showing (a) an amorphous diffraction pattern at room temperature that crystallizes into (b) a cubic structure at ≈ 300°C. After 1.5 hours, (c) at 450°C the diffraction pattern could be indexed as tetragonal zirconia.

The samples were heated in vacuum to understand the crystallization behavior of the precursor. In order to avoid electron beam heating the samples were only observed using electron diffraction. SAED patterns of starting material consisted of broad diffuse rings due to its amorphous nature (Fig. 1a). No change in the diffraction patterns was observed up to 300°C, confirming the absence of electron beam heating. At 300°C, a few diffraction spots and rings were observed indicating the beginning of the crystallization. The nucleation rate was observed to be very high and in a few seconds well-defined diffraction rings were observed (Fig. 1b). The diffraction pattern could be indexed as cubic zirconia (Fig. 1b). The presence of spots on diffraction rings observed during the present study indicate that the crystal size for the cubic structure is larger than previously reported [5] during electron beam heating. As the temperature of the sample during electron beam heating was completely unknown, the temperature of the sample may be higher, thus transforming the cubic structure in to tetragonal structure as soon as it was formed.

A splitting of $\{220\}_c$ reflections, observed as the temperature was increased to 350°C confirmed the advent of the transformation into the tetragonal phase (Fig.1c). The appearance of new reflections upon further heating, at ≈450°C indicated the beginning of the transformation into the monoclinic structure. Although some of the observed d-spacings can be indexed for all three known structures within the error margins for electron diffraction data (Table.1), the following trends clearly confirm the phase transformation:

(a) The first observed d-spacing (from 2.92 Å) increases slowly, as expected from reported x-ray diffraction data.
(b) Extra reflections appeared as the sample was heated, that do not belong to the other structure.

From our observations it was not possible to rule out the possibility of two or more phases to be present in any of the observed diffraction patterns. For the present study, it was not important to determine the rate of transformation but the mechanism. Except for increased sharpness of diffraction spots, no other changes were observed in the SAED patterns during 30 minutes of isothermal either at 300°C or at 350° heating. The monoclinic structure was found to be stable up to 750°C and 1.5 hr of heating time.

Table.1 Some observed d-spacing as measured from SAED patterns for the sample heated in vacuum and their indices.

d_{obs}(Å) 300°C	hkl cubic	d_{obs}(Å) 350°C	hkl tetragonal	d_{obs}(Å) 500°C	hkl monoclinic
2.92	111	3.03	101	3.11	111
2.6	200	2.67	002	2.80	111
1.82	220	2.59	110	2.61	002
1.51	311	1.97	200	2.23	-211
1.44	222	1.60	211	1.91	211
		1.55	211	1.61	031
		1.49	202	1.56	222

Heating in Ammonia

  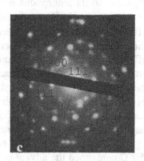

Fig.2. A sequence of SAED patterns recorded during the heating of Zr(OH)₄.xH₂O sample in ≈2 torr of NH₃ showing (a) crystallization of a cubic structure at 450°C which (b) transformed into a tetragonal structure at 475°C. The extra reflections belonging to the tetragonal structure are marked with arrows. (c) The tetragonal structure transformed into a cubic structure upon further heating due to N substitution in the lattice.

Table.2. Some observed d-spacing as measured from SAED patterns for the sample heated in NH₃ and their indices.

d_{obs}(Å) cubic	hkl	d_{obs}(Å) tetragonal	hkl	d_{obs}(Å) *cubic	hkl
2.92	111	2.96	101	2.92	111
2.5	200	2.63	200	2.67	200
1.78	220	2.48	110	1.76	220
1.52	311	2.06	102	1.55	311
1.24	400	1.78	112	1.47	222
1.15	311	1.51	103		
		1.41	202		
		1.27	220		
		1.23	213		

Another part of the precursor sample was then heated in ≈ 2 torr of dry NH₃ gas using the environmental cell. The SAED patterns were recorded after every 100°C up to 300°C, after every 50°C up to 450°C and every 25°C up to 820°C starting from room temperature. In the beginning, a broad diffuse ring pattern similar to the one shown in Fig.1a was observed.

The crystallization was observed to start at 450°C and the first phase observed was indexed as cubic (Fig. 2a). The higher temperature of crystallization (450°C in NH₃ instead of 300°C in vacuum) may be due the difference in gas pressure. At 475°C, extra reflections were observed due to the cubic to tetragonal phase transformation (marked by arrows in Fig.2b). Upon further heating, spots on the ring pattern began to become sharper, indicating an increased level of crystallinity. At 820°C, the extra reflections from tetragonal structure disappeared again and a cubic pattern was observed which was found to be stable upon cooling. No monoclinic phase was observed during present experiment. The reflections observed in Fig.2c are clearly sharper than in Fig. 2a. Moreover, the diffraction rings are also less diffuse indicating larger grain size at higher temperature as expected. But the presence of a cubic structure at 820°C for large particle size is a clear indication of nitridation reaction.

The measured d-spacings were observed to have similar trends as observed for the sample heated in vacuum (Table.2). The first observed d-spacing changed from 2.92 Å to 2.96 Å and then back to 2.92 Å. The extra reflections belonging to the tetragonal structure disappeared, confirming the formation of a cubic structure. As a cubic phase should not be stable at these temperatures, it must be stabilized by N substitution in the lattice.

In order to confirm the nitridation reaction, the electron energy-loss spectrum was recorded after stopping the gas flow in order to rule out the presence of N-edge from NH₃ (Fig.3). A small hump indicating the presence of N was observed along with C, Zr and O. The carbon edge is from the carbon film. Further studies to quantify the amount of nitrogen are in progress and will be published elsewhere.

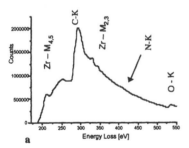 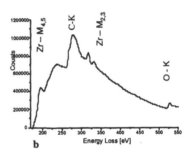

Fig.3. Electron energy-loss spectra (EELS) recorded (a) after the reaction at room temperature and in vacuum (b) from pure ZrO₂ sample.

CONCLUSIONS

The *in situ* study of nitridation of zirconia precursor indicates that a cubic structure can be stabilized by N substitution in the lattice at ≈800°C. No super-lattice reflections indicating the ordering of O and N in the lattice were observed during these experiments. Our studies indicate that a low temperature route to synthesize N-stabilized zirconia will be advantageous to avoid anion/vacancy ordering. The amount of N observed by EELS appeared to be quite low. Quantitative measurements of N content are needed to determine the composition limit for the formation of this phase.

ACKNOWLEDGEMENTS:

Research support from NSF (DMR-9806000), Deutsche Forschungsgemeinschaft. and the use of Center for High Resolution Electron Microscopy are gratefully acknowledged.

REFERENCES:

[1] C.J. Howard, and R.J. Hill, J. Mat. Sci., **26** (1991) 127.
[2] R.C. Garvie, J. Phys. Chem., **82**(2) (1978) 218.
[3] C.F. Grain, J. Am. Ceram. Soc., 50(6) (1967) 288.
[4] R. Sharma, M. McKelvy, and M. Lajavardi, Electron Microscopy, **2** (1992)463.
[5] J.C. Gilles,. Rev. Hautes Temp. Refract., **2** (1965)237.
[6] Martin Lerch, J. Am. Ceram. Soc., **79**(10) (1996) 2641.
[7] H.J. Rossel, Adv. Ceramics, 3 (1981), 47.
[8] Renu Sharma and Karl Weiss, Microscopy Research and Techniques, **42** (1998) 270-280.

ENVIRONMENTAL SCANNING ELECTRON MICROSCOPY AS TOOL TO STUDY SHRINKAGE MICROCRACKS IN CEMENT-BASED MATERIALS

J. BISSCHOP, J.G.M. VAN MIER
Faculty of Civil Engineering and Geo-Sciences, Delft University of Technology,
P.O. Box 5048, 2600 GA Delft, The Netherlands, J.Bisschop@ct.tudelft.nl

ABSTRACT

In this paper a method is described to observe shrinkage microcracks on 'wet' specimen cross-sections of cement-based materials with Environmental Scanning Electron Microscopy (ESEM). A sample cooling device which can be used in the ESEM chamber was built to control the relative humidity above a microscope sample. The accuracy of measuring relative humidity is determined to be 5% at a sample temperature of 3°C. A microscope sample preparation method and a pump-down sequence of the ESEM-chamber, both without any drying of the sample, are described. Preliminary results show that in the studied mortar the visibility of shrinkage microcracks on a 'wet' specimen cross-section is low due to closure of microcracks by swelling of the cement paste.

INTRODUCTION

Cement-based materials like concrete and mortar contain a certain amount of water in the capillary pores of the cement fraction. When this water is removed from the capillary pores by drying or hydration processes the cement paste shrinks due to an increase in capillary stress [1]. Microcracking due to this shrinkage may influence the durability of the concrete in aggressive environments by increasing permeability and thus facilitating ingress of fluids and gases [2]. Besides, severe shrinkage microcracking influences the mechanical behavior of concrete.

To study shrinkage microcracking a reliable microcrack detection technique is needed. The destructive techniques (fluorescence light microscopy and conventional SEM) that have been used in the past to detect microcracks on a *specimen* cross-section in concrete have one major drawback: the concrete is dried during microscope *sample* preparation or in the vacuum of the SEM [3,4,5]. This drying causes formation of additional microcracks on the sample surface. Therefore, studying original microcracks is difficult, because microcracks generated by sample preparation or in the SEM obscure the original crack-pattern.

Unlike conventional SEM, Environmental Scanning Electron Microscopy is a promising visual tool for the study of shrinkage or mechanical microcracking in cement-based materials. In ESEM water vapor is used as a signal-amplifying gas to create an image. The presence of water vapor in the ESEM can be used to control the humidity in ESEM. The phase diagram of water (Fig. 1) shows how the relative humidity (RH) of a system depends on the water vapor pressure and the temperature. By controlling both the water vapor pressure and the temperature of the sample in the ESEM-chamber, it is possible to control the relative humidity above the sample. Thus, by creating a relative humidity of 90-100% above the sample it is possible to prevent the sample from drying during examination in the ESEM.

In this paper the principles and equipment to control relative humidity by regulating both the pressure and the temperature are described. A 'wet' sample preparation method is described and an ESEM pumpdown-sequence without drying of the sample will be given. Preliminary results are presented which illustrate the visibility of shrinkage microcracks on a 'wet' specimen cross-section. For detailed principles and functions of ESEM reference is made to Uwins [6].

141

Mat. Res. Soc. Symp. Proc. Vol. 589 © 2001 Materials Research Society

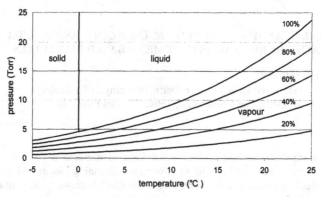

Figure 1: Phase diagram of water at low pressures showing lines of relative humidity of the water vapour. 1 Torr = 133.3 Pa.

EXPERIMENT

Pressure Control

A Phillips XL30 ESEM (Tungsten filament) is used in the present microcrack studies. The water vapor pressure in the sample chamber is controlled by the vacuum system of the microscope. The water vapor pressure of the chamber can be adjusted at any value between 0.1 and 20 Torr with an accuracy of 0.1 Torr. To obtain the required pressure the system will be either in the pump or flood mode. The equilibration mode is maintained by a needle valve which is controlled by an automatic control system (the pressure servo).

The water vapor pressure is an important factor influencing the image clarity. At a working distance of 10 mm the optimum GSE-detector gain is at a vapor pressure of about 3.5 Torr. The practical maximum water vapor pressure at which still satisfactory images can be acquired is about 6 Torr at a 10 mm working distance. This means that at a temperature of 22°C in the ESEM-chamber a maximum practical relative humidity of 30% can be achieved (Fig. 1). To obtain higher humidities, while keeping the pressure below 6 Torr, the sample needs to be cooled.

Temperature Control

In the microcrack studies only control of the relative humidity immediately above the sample viewing surface is needed, because this determines the rate and extent of drying of the upper most part of the sample. By setting the humidity at 90-100% no or negligible drying of the upper most part of the sample will take place, and this enables observations to be made at a 'wet' surface. At lower humidities drying of the sample surface occurs, which will result in shrinkage and microcrack formation on the sample surface. So, at low humidity it is also possible to perform in-situ drying experiments, in which the growth of shrinkage microcracks on the sample surface is followed (see [7,8]).

Assuming that the measured pressure indicated in the microscope software is the same as the actual pressure just above the sample, then it is possible, by regulating the sample surface temperature (or rather the temperature just above the sample), to control the humidity just above the sample. It is assumed that the localized heating by the electron beam has a negligible influence on the overall relative humidity above the sample.

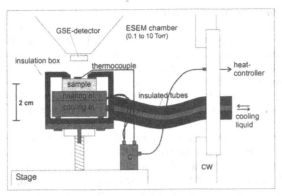

Figure 2: Schematic illustration of set-up to control sample surface temperature in ESEM.
GSE = gaseous secondary electron, c = connector, cw = chamber wall.

In figure 2 the set-up to regulate the temperature of the sample surface (by cooling and/or heating the sample bottom) is illustrated schematically. The set-up exists of a cooling and a heating element. The cooling is provided by a cooling liquid with a temperature of –10°C. The set-up is capable of cooling a 7 mm thick sample below freezing temperature. The sample surface temperature is measured with a thermocouple connected with the heating element to an automatic heat-controller .

Experiments have been performed outside the ESEM chamber to determine the accuracy of the measurement of the temperature just above the sample with the thermocouple used in the set-up. In these experiments it has been determined that the thermocouple in the set-up measures this temperature with a systematic error of +0.7°C and an accuracy of 0.6°C. In the microscope software the measured pressure is given with a precision of 0.1 Torr. The accuracy for temperature and pressure result in an (absolute) accuracy for the measurement of the relative humidity of 5% at a temperature of 3°C (Fig. 1).

During cooling a temperature gradient will develop across the sample. When at a sample surface temperature of 3°C this temperature gradient is more than 3°C, then the bottom of the sample is in freezing conditions. It should be noted that the freezing point of water in capillary pores depends on the radius of the pore [9]. First freezing of water in a concrete mortar has been observed to start at temperatures lower than -7°C [9]. No effects on the sample surface of possible freezing of water in the sample bottom have been observed.

If too much condensation or ice formation takes place on the cooling set-up in the ESEM-chamber, it becomes difficult or impossible to obtain or maintain the required pressure. Therefore, the set-up is insulated in such a way that the sample surface temperature is always colder than all other parts of the set-up.

Sample Preparation

The mortar specimens (prisms) which are used in the drying experiments have dimensions of 40x40x160 mm^3 and a hardening age of one week [8]. After the hardening of one week, the specimen is dried to introduce drying shrinkage microcracks. After the drying experiment, the sample preparation starts with cutting a sample from the centre of the specimen with a 1 mm thick diamond saw. The sample top surface is orientated perpendicular to the long surfaces of the specimen. Sample dimensions of 30x30x7 mm^3 were found to be practical.

The best way to study microcracks in cement-based materials is to look at a flat (ground) surface. This type of sample also allows the use of stereological methods to obtain (3-d) data concerning the density and distribution of microcracks in a dried specimen, which is the main purpose of the microcrack studies. After cutting, the sample is ground by hand with moderate pressure on a moderate-speed lap wheel with sand papers of p230, p500, p800 and p1200, sequentially. It was found that further polishing of the sample with diamond paste did not increase the flatness of the cement paste fraction of the sample surface. The samples are kept wet constantly during sample preparation and are examined as soon as possible after sample preparation.

It is very important to notice that during cutting and grinding the dried material is rewetted by the cooling water. The cement paste will absorb the water and this will result in swelling of the cement paste and corresponding closure of microcracks on/near the sample surface [7] (Fig 3a-c). Unfortunately, dry cutting or grinding is not an option, because the heat developed by the cutting would probably cause evaporation of water from the sample.

Pumpdown Sequence

The conditions that are needed in the chamber of the ESEM before examination can start, are a water vapor concentration of 100% and a pressure of 5.5 Torr. To achieve these conditions a pumpdown sequence has to be applied in which the air in the chamber (at 760 Torr) is totally replaced by water vapor. The optimum pumpdown sequence to achieve this with minimum sample evaporation is calculated by Cameron and Donald [10]. Applying this pumpdown sequence still involves some drying, especially during the first pump-down. In the pumpdown sequence used in the microcrack studies (Table 1), which is similar to the one calculated by Cameron and Donald [10], evaporation of water from the sample is avoided by supplying a little excess water to the sample surface.

Condensation and evaporation of water in and on the sample is accompanied by heat release and consumption, respectively. Because the thermocouple is lying on the surface of the sample, the measured temperature is strongly influenced by condensation and evaporation (see also [11]). During flooding the measured temperature will increase if condensation of water at the surface takes place. This happens when at a surface temperature of 3°C the partial water vapor pressure exceeds 5.5 Torr (Fig. 1). During flooding the measured temperature can increase as much as 7°C . During pumping the measured temperature will instantaneously decrease again with pressure due to the endothermic evaporation of the condesation water on the sample surface.

Table I: Pumpdown sequence used in the microcrack studies.

pumping	pressure drop (Torr)	R.H. at 5.5 Torr and 3°C	flooding	pressure raise (Torr)	water-%
pump 1	760 → 5.5	~ 0%	flood 1	5.5 → 14	61%
pump 2	14 → 5.5	60%	flood 2	5.5 → 13	84%
pump 3	13 → 5.5	80%	flood 3	5.5 → 12	92%
pump 4	12 → 5.5	90%	flood 4	5.5 → 11	96%
pump 5	11 → 5.5	94%	flood 5	5.5 → 10	98%
pump 6	10 → 5.5	96%	flood 6	5.5 → 9	99%
pump 7	9 → 5.5	98%			

Starting conditions: $T_{sample} = 3°C$, $T_{source bottle water} = 22°C$, and excess water supplied to sample surface.

RESULTS

Figure 3a shows a microcrack introduced in ESEM by drying at 10% RH above the sample surface. The micrographs in Figures 3b and 3c show the closing of the microcrack upon rewetting (= increasing RH) and resultant swelling of the cement paste. The microcrack does not close entirely (Fig. 3c) even after submerging the sample surface with condensation water. This can be explained by the drying/rewetting hysteresis of the cement paste [12], or due to local rotations near the crack face. It should be noted that the uppermost part of the sample reaches shrinkage equilibrium in correspondence with the RH above the sample within a few minutes. Thereafter, there is no visible change in crack width anymore.

Figure 3d shows a micrograph of a 'wet' cross-section of a dried specimen of young mortar. The sample was rewetted by the cooling water during cutting and grinding. A closed microcrack can be seen extending perpendicularly from the specimen surface to a sand grain just below the surface. The visibility of the closed microcrack is increased by the abrasion of the edges of the crack due to the grinding.

Shrinkage microcracks introduced by drying in the (E)SEM make right angles with the sample surface, because self-restraining of the sample due to drying shrinkage cause tensile strains to be directed parallel to the sample surface. Because these microcrack formed after grinding, they have sharp edges (Fig. 3a). On the other hand microcracks introduced in the drying experiment can be expected to make any angle with respect to the cross-section plane. The edges of these original microcracks are abraded by grinding and thus are more rounded.

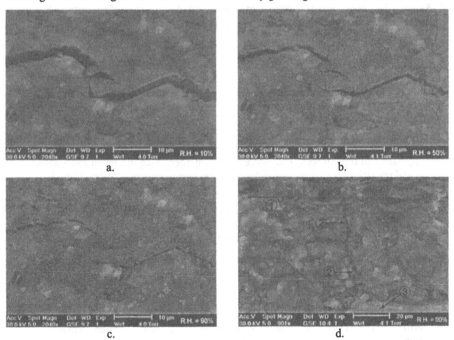

Figure 3. Visibility of shrinkage microcracks in cement-based materials as function of relative humidity above the sample. a. RH is 10% ±5%, b. RH is 50% ±5%. c. RH is 90% ±5%. d. Original shrinkage microcrack on 'wet' specimen cross-section at RH of 90% ±5%. 1 = specimen surface, 2 = closed shrinkage microcrack affected by grinding, 3 = sand grain.

CONCLUSIONS

1. By using the equipment described in this paper it is possible to control humidity immediately above a sample in ESEM with an accuracy of 5% at a temperature of 3°C.

2. The visibility of shrinkage microcracks on a 'wet' specimen cross-section in the studied mortar is limited because microcracks close largely due to the swelling of the cement paste upon water absorption.

ACKNOWLEDGMENTS

The assistance of Mr. A.S. Elgersma with the construction of the ESEM cooling device is gratefully acknowledged. Financial support was obtained from Delft Interfaculty Research Program 'Micromechanics for macroscopic lifetime prediction' (DIOC-10).

REFERENCES

1. C.L. Hwang and J.F. Young. Drying shrinkage of Portland cement pastes. I. Microcracking during drying. Cem. Con. Res. **14**, 585 (1984).

2. A.D. Jensen and S. Chatterji. State of the art report on microcracking and lifetime of concrete – part 1. Mat. Struct. **29**, 3 (1996).

3. E. Ringot, J.P. Ollivier, and J.C. Maso. Characterisation of initial state of concrete with regard to microcracking. Cem. Con. Res. **17**, 411 (1987).

4. A. Bascoul, J.P. Ollivier, and A. Turatsinze. Discussion of the paper 'Fracture zone presence and behaviour in mortar specimens' by N. Krstulovic-Opara. ACI Mat. J. **91** (5), 531 (1994).

5. H.C. Gran. Fluorescent liquid replacement technique. A means of crack detection and water:binder ratio determination in high strength concretes. Cem. Con. Res. **25** (2), 1063 (1995).

6. P.J.R. Uwins, Environmental scanning electron microscopy. Mat. Forum. **18**, 51 (1994).

7. K.O. Kjellsen and H.M. Jennings. Observations of microcracking in cement paste upon drying and rewetting by environmental scanning electron microscopy. Advn. Cem. Bas. Mat. **3**, 14 (1996).

8. J. Bisschop and J.G.M.van Mier. Quantification of shrinkage microcracking in young mortar with fluorescence light microscopy and ESEM. Proc. 7th euroseminar on Microscopy Applied to Building Materials, June 1999, Delft, The Netherlands.

9. V. Penttala. Freezing-induced strains and pressures in wet porous materials and especially in concrete mortars. Adv. Cem. Bas. Mat. **7**, 8 (1998).

10. R.E. Cameron and A.M. Donald. Minimizing sample evaporation in the environmental scanning electron microscope. J. Micr. **173** (3), 227 (1994).

11. B.J. Griffin. Hydrated specimen stability in the electroscan ESEM: specimen surface temperature variations and the limitations of specimen cooling. Proc. 50th annual meeting EMSA. San Francisco, 1306 (1992).

12. H.W. Reinhardt. Beton als constructiemateriaal – eigenschappen en duurzaamheid. Delft University Press. 152 (1985).

MICROCHARACTERIZATION OF HETEROGENEOUS SPECIMENS CONTAINING TIRE DUST

Marina Camatini[★], Gaia M Corbetta[★], Giovanni F Crosta[★★],
Tigran Dolukhanyan[§], Giampaolo Giuliani[¶], Changmo Sung[§]

[★] Dept. of Environmental Sciences, University of Milano – Bicocca, I – 20126 Milano, IT
[¶] Pirelli Pneumatici, I – 20126 Milano, IT
[§] Center for Advanced Materials, Dept. of Chemical Engineering, University of Massachusetts, Lowell, MA 01854
[★] Center for Electromagnetic Materials and Optical Systems, Dept. of Electrical and Computer Engineering, University of Massachusetts, Lowell, MA 01854

ABSTRACT

This work is focused on dust or debris produced by the wear of tire tread. Two problems are addressed, which are solved by analytical electron microscopy (AEM): characterization of tire debris and identification of tire debris particles in a heterogeneous specimen. The characteristic morphology, microstructure and elemental composition of tire debris can all be determined by AEM. The scanning electron microscope (SEM) shows that the surface of a tire debris particle has a typical, warped structure with pores. The characteristic elements of tire rubber are S and Zn, which are detected by energy dispersive X ray (EDX) spectroscopy. The identification of rubber particles in heterogeneous debris containing talc and produced by a laboratory abrader is possible by the analytical SEM. Transmission electron microscope images, EDX spectra and selected area electron diffraction patterns characterize tire debris at the sub–micron scale, where the material can no longer be treated as homogeneous.

INTRODUCTION AND MOTIVATION

Particulate matter produced by motor vehicles mainly originates from the combustion of hydrocarbons and from the wear of tires, brakes and the road surface.

This work is focused on the particles from the wear of tires, also known as tire debris or tire dust.

Detailed knowledge about the size, shape, microstructure and constituents of tire debris serves a dual purpose:

- to model and estimate the impact of particles on human health and on the environment;
- to diagnose and control the tire manufacturing process.

This work provides preliminary characterization and identification results about tire debris particles.

Materials come from laboratory wear tests; specimens are suitably prepared and are examined by analytical electron microscopes (AEMs).

The characteristic morphology, microstructure and elemental composition of tire debris are determined and described.

The identification of tire debris by AEM alone is shown to be feasible at least in some types of specimens.

Tire Rubber and Debris

The amount of debris lost by (passenger and truck) vehicle tires per year has been known for a long time. An estimate for the USA provided by the EPA [1] was $6 \cdot 10^8$ kg in the 1970s. More recent (1990s) figures are $53 \cdot 10^6$ kg for Great Britain [2] and at least $47 \cdot 10^6$ kg for Italy (assumed to be $1/7$ of the total weight of worn tires) [3].

The main ingredients of tread rubber are: elastomers, fillers and additives. The approximate ranges of values are shown in TABLE I.

TABLE I

Typical ingredients of passenger car tire tread and tire weight loss due to wear.

The amounts of all substances, except elastomers, are given in parts per hundred parts of rubber (phr).
HVO = high viscosity aromatic oil   NA = naphtylamines   NR = natural rubber
PB = polybutadiene   PW = petroleum wax   SBR = styrene butadiene rubber

Elastomer masterbatch	Reinforcing Fillers (phr)	Vulca-nizers (phr)	Accel-erators (phr)	Activators (phr)	Anti-oxidants, Protectors (phr)	Softening Fillers (phr)	Weight loss (lbs)
SBR 75% PB 10% NR 15%	carbon black 35 SiO_2 35	2	0.5÷2	ZnO 2	NA 1 PW <10	HVO 20	2 to 5

Previous Work on Rubber Blends and Tire Debris

Transmission electron microscope (TEM) techniques have been most notably exploited by HESS et al. [4], MEDALIA et al. [5] and MERCER et al. [6] to analyze and characterize carbon black aggregates, which are of independent interest. The application of stereoscopic TEM to the imaging of carbon black – elastomer blends is described in [7]. The characterization of dispersions of fillers in elastomers is extensively covered by [8].

Tire debris was investigated in the early 70's [9, 10, 11]. More recently, interest towards tire debris has revived because of its suspected role in respiratory diseases [12] and, more generally, on the environment. The papers by KINDRATENKO, et al. [13] deal with shape analysis and report scanning electron microscope (SEM) images with energy dispersive X ray (EDX) spectra of interest. FAUSER, et al. [14] describe sampling, characterization, identification, chemical analysis and the environmental effects of tire dust.

MATERIALS AND METHODS

Material Selection

The materials, which have been analyzed so far come from laboratory wear tests. Large (sub–millimeter) particles as well as particles in the respirable range are included.

Laboratory debris is produced by two types of abrasion equipment: a) steel brush and b) steel blade abrader, respectively. The former produces tire debris particles (tdp's) alone. The latter uses talc microcrystals as an anti–smear agent. As a consequence, rubber particles are dispersed in talc. The materials will be referred to as type–a and type–b debris.

SEM Specimen Preparation

Double sided adhesive carbon tape laid on top of a standard *Al* SEM stud carries the particles. Up to three different specimens can be mounted on one stud of 1/2" diameter. The type−*a* debris appeared not to need any conductive coating. All other specimens do need *Au* coating.

TEM Specimen Preparation

Type−*a* and −*b* debris particles are sieved and poured into cylindrical molds of 1/4" diameter, which end with a conical tip. Two component *Allied Chemical 145 − 10025* and *145 − 10030* epoxy resin is cast into the molds and cured for 24h at room temperature.

An *LKB Ultratome III* type *8801A* microtome with diamond knife at room temperature is used, as described by [7]. Acceptable performance is obtained by a glass knife, provided the specimen is cooled with LN_2. Slices, which are less than 100 nm thick, are transferred to *Formvar* coated *Cu* grids. No coating is applied to the specimen.

Instrumentation

The following analytical instruments at the Center for Advanced Materials, University of Massachusetts − Lowell, have been used:
* *Amray 1400* SEM operated at 20 kV, equipped with a *Tracor Northern TN 3205 1000V* EDX spectrometer and image processing facilities,
* *Philips EM 400 T* TEM operated at 120 kV, equipped with a *TN 5500* EDX spectrometer.

RESULTS AND THEIR INTERPRETATION

Secondary Electron Imaging (SEI) and EDX Spectroscopy (EDXS)

Type−a Debris
The most remarkable morphological features of type−*a* debris are a warped surface and porosity. The SE image of FIGURE 1.*a* suggests a fractal structure. However, this property needs further investigation.

A typical result from EDXS is shown in FIGURE 1.*b*. X rays from lighter elements (e.g., *C* and *O*) are not shown. *Al* may come from the specimen holder or be an impurity of the *tdp*; *Si* is due to the reinforcing filler; *S* and *Zn* originate from vulcanized rubber, hence *tdp*'s; *Fe* comes from the steel brush.

The material appears homogeneous on this scale. In other words the analyzed volume includes one or more representative elementary volumes (REVs).

Type−b Debris
Particles from abrader *b*) form easily recognizable clusters, where rubber is surrounded by talc microcrystals, as shown on the left side of FIGURE 2.*a*. The EDX spectrum obtained from a talc particle contains the *Si* and *Mg* peaks, which characterize talc, and sometimes *Fe*, from the abrader tool. The cluster (C) of FIGURE 2.*a* yields the spectrum of FIGURE 2.*b*: in addition to *Si* and *Mg* it contains *S* and *Zn*, which characterize *tdp*'s. The *S* and *Zn* peaks are weaker than those from FIGURE 1.*b*, because of higher primary electron absorption and secondary X−ray attenuation by the talc cladding. Once again, *Fe* comes from the abrader.

The identification of rubber in type−*b* debris, regarded as a heterogeneous specimen, is thus possible by SEM combined with EDXS.

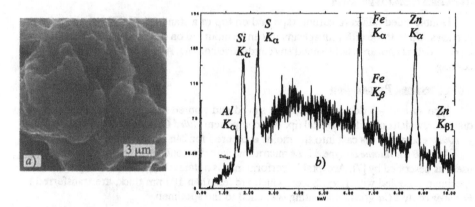

FIGURE 1.*a* : SE image of a particle of type−*a* debris.
FIGURE 1.*b*. The EDX spectrum of the *tdp*, where *S* and *Zn* are typical of vulcanized rubber.
Si is due to the reinforcing filler. *Al* comes from the holder, *Fe* from the abrader.
The physical and chemical properties of the material are almost uniform on this scale,
hence the analyzed volume includes one or more representative elementary volumes .

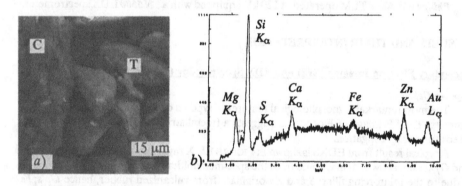

FIGURE 2.*a*: SEI of type−*b* debris. Left: cluster (C) of a rubber particle surrounded by
fractured talc crystals. Center and right: talc (T) crystals. The material is heterogeneous
already on this scale.
FIGURE 2.*b*: EDX spectrum of C. The *tdp*'s it contains can be identified from *S* and *Zn*.

TEM Imaging, Selected Area (SA) Diffraction and EDX Spectroscopy

Type−a Debris

So far, only type−*a* debris has been analyzed. Transmission imaging at a higher
magnification enhances the differences in electron absorption and scattering cross sections
of SiO_2 vs. SBR. The inhomogeneities of the material become visible. FIGURE 3.*a* suggests a
REV of $\simeq 10^{-1}$ μm³. The typical SA electron diffraction pattern (DP) of a *tdp* may at most
exhibit halos as in FIGURE 3.*b*. However, one can easily find microcrystals of foreign
substances e.g., alkali halides, $CaCO_3$, dolomite, silicates of *Mg,Al,K* and *Ca*, alloys of *Fe* and

Mn, coming both from the abrader and the environment. In other words, on the sub−REV scale, *tdp*'s have to be identified.

The available EDX detector is suitable for light elements, hence *C* and *O* can be seen.

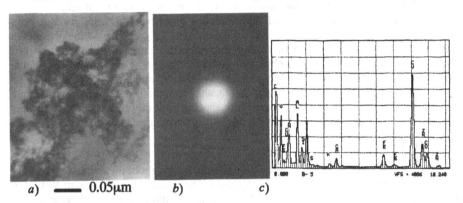

a) ▬▬ 0.05μm   *b)*   *c)*

FIGURE 3.*a*: TEM image of type−*a* debris, which suggests a REV of $\simeq 10^{-1}$ μm^3.
FIGURE 3.*b*: typical SADP of a *tdp* with halos; no microcrystalline structure is expected.
FIGURE 3.*c*: typical EDX spectrum, where the *Zn K$_\alpha$* and *Zn L* peaks are visible; elements other than *C, O, Si* are impurities due to the abrader and to the environment.

The preliminary procedure, necessary to identify a *tdp* can be summarized as follows:

P1) if the TEM image is similar to that of FIGURE 3.*a*, then it corresponds to a *tdp*; otherwise one must obtain a SADP;

P2) if the SADP reveals a microcrystal structure, then it cannot come from a *tdp*, otherwise go to the next step;

P3) carry out EDXS: if *S* and *Zn* are found, then a SBR particle has been hit, otherwise the selected area contains an amorphous cluster (e.g., precipitated *SiO$_2$*, carbon black, ...).

It is always advisable to obtain the DP before performing EDXS, because the latter often irreversibly damages the specimen by local electron beam heating.

CONCLUSION

Preliminary results about the characterization and identification of tire debris have been presented.

SEI allows morphological characterization: the distinct feature of type−*a* debris is the particle surface, which remains irregular and exhibits pores even at 10k × by a SEM; type−*b* debris forms easily recognizable rubber − talc clusters.

The most abundant elements in type−*a* debris are *C, O, Si, S, Fe, Zn*. Moreover, *S* and *Zn* may play the role of markers for *tdp*'s. Type−*b* debris additionally contains *Mg*. As a consequence, the identification of rubber in type−*b* debris, regarded as a heterogeneous specimen, is possible by SEI combined with EDXS.

Within the experimental limits, by SEM, the material properties appear uniform over the explored volume, whereas analysis by the TEM yields information on the sub−REV scale.

As a consequence even type−*a* debris seen by the analytical TEM is equivalent to a heterogeneous material. A preliminary identification procedure has been proposed and implemented.

The material of type–b is closer to the debris produced by normal tire wear. Therefore, the above results, although preliminary, can be already passed on to the models, which estimate the impact of debris on the environment and on human health. These models deal with e.g., the transport of particulate matter and chemicals in air, water and soil.

Another application is the feedback of information to the manufacturing process. Both the quality control of the product and the assessment of its life cycle rely on the observable properties of debris particles i.e., morphology, microstructure and microanalysis.

The TEM analysis of type–b debris and the identification of tdp's in environmental samples by SEM + TEM are needed to complete the present scheme.

In order to validate the whole approach, one shall eventually determine whether and how the above described methods can deal with specimens of different origin and history, where the morphology and chemistry of rubber particles may have been altered.

ACKNOWLEDGEMENTS

The financial support of *Pirelli Pneumatici*, Milano, Italy is gratefully acknowledged: this work is carried out in respect of contract obligations. The authors thank Mr. AMIT KAULGUD, Lowell, MA, for his cooperation in TEM sample preparation. Advice and constructive criticism by Dr. JIM BENTLEY and a Referee about the preliminary version of this manuscript have been of great help.

REFERENCES

[1] ENVIRONMENTAL PROTECTION AGENCY, *Summary of Nationwide Emission Estimates of Air Pollutants, 1969* (EPA – Office of Air Programs, Div of Appl Technol, May 1971).

[2] ENVIRONMENT AGENCY, *Tyres (Tires) in the Environment* (Environment Agency: Bristol, UK, 1998), ISBN 1 873 16075 5.

[3] COMMISSIONE DELLE COMUNITÀ EUROPEE (EC COMMISSION), *Documento informativo sugli pneumatici usati (Information Report on Scrap Tires)*, (Direzione Generale Ambiente, Sicurezza Nucleare e Protezione Civile   delle Comunità Europee: Brussels, B, 1991).

[4] W M HESS, C R HERD and E B SEBOK, Kautsch Gummi Kunstst **5**, 328 (1994).

[5] A I MEDALIA and G J HORNIK, Pattern Recognition **4**, 55 (1972).

[6] H N MERCER, A H BOYER and M L DEVINEY, Rubber Chem Technol **52**, 377–86 (1979).

[7] V E HANCHETT and R H GEISS, IBM J Res & Dev **27**, 348 – 55 (1983).

[8] W M HESS, Rubb Chem Techn **64**, 386 – 449 (1991)

[9] J A CARDINA, Rubber Chem and Techn **47**, 1005 – 10 (1974).

[10] M L DANNIS, Rubber Chem & Technol **47**, 1011–37 (1974).

[11] W R PIERSON and W W BRACHACZEK, Rubber Chem and Technol **47**, 1275 – 99 (1974).

[12] A G MIGUEL, G R CASS, J WEISS and M M GLOVSKY, Env Health Persp **104** #11, 1180–86 (1996).

[13] V KINDRATENKO, P VAN ESPEN, B TREIGER and R VAN GRIEKEN, Env Scie & Techn **28**, 2197 – 2202 (1994); Microchimica Acta Suppl **13**, 355 – 61 (1996).

[14] P FAUSER, *Particulate Air Pollution with Emphasis on Traffic Generated Aerosols*, PhD thesis, Technical University of Denmark (DTU); Report Risø–R–1053(EN) (Risø National Laboratory: Roskilde, Denmark, 1999), ISBN 87–550–2532–3 (Internet).

Microelectronic Materials

EXPERIMENTAL METHODS AND DATA ANALYSIS
FOR FLUCTUATION MICROSCOPY

P. M. VOYLES,*† M. M. J. TREACY,† J. M. GIBSON,‡ H-C. JIN,** and J. R. ABELSON**
*Dept. of Physics, University of Illinois, Urbana, IL 61801, pvoyles@uiuc.edu
†NEC Research Institute, Princeton, NJ 08540
‡Argonne National Laboratory, Argonne, IL 60439
**Dept. of Materials Science and Engineering and Coordinated Science Laboratory, University of Illinois, Urbana, IL 61801

ABSTRACT

We have developed a new electron microscopy technique called fluctuation microscopy which is sensitive to medium-range order in disordered materials. The technique relies on quantitative statistical analysis of low-resolution dark-field electron micrographs. Extracting useful information from such micrographs involves correcting for the effects of the imaging system, incoherent image contrast caused by large scale structure in the sample, and the effects of the foil thickness.

INTRODUCTION TO FLUCTUATION MICROSCOPY

It has long been recognized that the pair distribution function $g_2(r)$ yielded by kinematic diffraction is an insufficient characterization of the structure of disordered materials such as amorphous semiconductors. Several computer simulations of disordered materials have been produced which have identical $g_2(r)$, but different structure at medium range [1, 2]. Moving beyond the pair correlation function has proved difficult, however. Experimental techniques such as Raman scattering, near-edge x-ray absorption fine structure, nuclear magnetic resonance, and electrical noise measurements have offered tantalizing glimpses of structural differences at medium range, but remain difficult to interpret in terms of purely structural models since such signals depend in a complicated way on the details of the interatomic potential. Fluctuation microscopy uses high-energy electrons as a probe, so it does not suffer from those limitations.

Fluctuation microscopy depends on the statistical analysis of hollow-cone dark-field (HCDF) micrographs taken at deliberately low resolution, ~1 nm. In this limit, we can ignore the effects of microscope aberrations and view the image as a map of the diffracted intensity from mesoscopic volumes of the sample, the size of which is set by the resolution, which is in turn controlled by the size of the objective aperture. There are then two parameters which characterize the imaging conditions: the magnitude of the dark-field scattering vector k and the objective aperture diameter Q. We compute the variance of the "diffraction map" image as a function of k and Q

$$V(k,Q) = \frac{\langle I^2(k,Q,\mathbf{r})\rangle}{\langle I(k,Q,\mathbf{r})\rangle^2} - 1 \tag{1}$$

where $\langle\ \rangle$ denotes averaging over the image position coordinate $\mathbf{r}$.

To understand in a qualitative way why the variance is sensitive to medium-range order (MRO), consider two samples: one a homogeneous random assortment of atoms with no

MRO and the other a heterogeneous material with small, randomly-oriented ordered clusters. Set Q so the mesoscopic volume is approximately the same size as those clusters. All of the mesoscopic volumes of the random sample will be statistically the same, so the diffracted intensity from each volume will be the same, and the image will have a low variance. In the heterogeneous sample, some of the clusters will be oriented near a Bragg condition and diffract strongly and others will not. These differences in diffracted intensity lead to an image with large variance. In general, a large image variance indicates some form of MRO.

By varying the imaging conditions k and Q we can obtain information about the character of any MRO present. Varying k at constant Q is called variable coherence microscopy and gives information about the structure and degree of ordering inside any ordered regions. Varying Q at constant k is called variable resolution microscopy and gives information about the size of the ordered regions. Only variable coherence microscopy has been experimentally implemented so far.

The quantitative information fluctuation microscopy provides is complicated and still incompletely understood, but we know that it depends on the four-body pair-pair correlation function $g_4(r_1, r_2, r, \theta)$, where r_1 and r_2 define one pair lengths, r is the distance between pairs, and θ is their relative angle [3]. It has been shown that the pair-pair correlation function contains more information about MRO than $g_2(r)$ [4].

Fluctuation microscopy has been used to show that thin films of amorphous semiconductors contain more MRO than can be explained by the continuous random network, and that that MRO is reduced on thermal annealing [1] (and for hydrogenated amorphous silicon by exposure to light [5]). This has lead to the development of the paracrystalline theory of the structure of as-deposited amorphous thin films [6]. In this paper we focus on recent advances in the variable coherence experimental method and data analysis.

RECOVERING THE TRUE ELECTRON STATISTICS

The advent of charge-coupled device (CCD) cameras and other linear devices for recording electron images has made statistical techniques such as fluctuation microscopy possible. However, care must still be taken when analyzing such images. For fluctuation microscopy, we must carefully correct for the modulation transfer function (MTF) of our CCD camera.

The MTF is the reciprocal-space function which connects the measured electron intensity with the intensity incident on the device:

$$I_{\text{measured}}(\kappa) = MTF(\kappa)I_{\text{incident}}(\kappa) \tag{2}$$

where κ is the Fourier transform coordinate of the image, not a scattering vector. Since Parseval's theorem connects the variance and the power spectrum $P(\kappa)$ of the image, it is simple for us to correct for the MTF by

$$V = \int 2\pi\kappa^2 \frac{P(\kappa)}{|MTF(\kappa)|^2} d\kappa. \tag{3}$$

We have measured the MTF of our Gatan MSC Model 794 CCD at an accelerating voltage of 200 kV using the stochastic image method as given by Zuo [7]. Daberkow et. al. have noted that this technique is sensitive to aliasing in the calculation of the stochastic image power spectrum [8]. We have investigated this by the simple expedient of resampling the acquired images, expanding each original pixel into 4 pixels, then 8, then 16 so that the discrete Fourier transform approaches the ideal continuous transform.

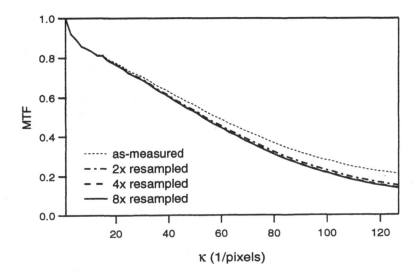

K (1/pixels)

Figure 1: The MTF of the Gatan MSC 794 camera as measured by analyzing stochastic images as measured, then with varying resampling before computation of the Fourier transforms.

Figure 1 shows the results of this analysis. The effects of the resampling are most dramatic at high frequency. At the Nyquist frequency of 128 1/pixels, the MTF without resampling is 0.209, which is similar to previous reports for other cameras [7, 8, 9]. However, the transfer is reduced to 0.15 at 2× resampling, and 0.139 and 0.136 at 4× and 8× respectively. The small reduction between 4× and 8× makes it unlikely that the MTF would be further suppressed by additional resampling. We conclude that 4× resampling is sufficient and necessary for accurate measurement of the MTF by the stochastic image method.

REMOVING INCOHERENT VARIANCE

So far we have assumed that all of the sources of variance are due to the microstructure of the material. This is clearly not the case, as variance (image contrast) can be caused by a variety of other mechanisms in real samples. If there are macroscopic features in the sample, such as thickness variations, bends, ridges, or holes, these will contribute mass-thickness variance. Non-uniform illumination will also contribute variance. What these have in common is that the contrast mechanism is incoherent. Incoherent variance contributes a k-independent systematic error to $V(k)$ which is different for every area of the sample.

We remove incoherent variance from the analysis in two ways. First, we Fourier filter the images [6]. This is easily implemented by adjusting the limits of integration in Equation 3. This is particularly good at removing the effects of non-uniform illumination, which are low in spatial frequency. Macroscopic features in the image are more difficult to remove in this way, as they can contribute power over a range of frequencies.

We correct for variance due to inhomogeneities by direct measurement. At large enough scattering vector ($k = 14.0$ nm^{-1} for Si), HCDF images become incoherent [10]. (This is the optical conjugate of Z-contrast microscopy using a high-angle annular dark field detector in a STEM.) In this limit, the type of variance described in the first section due to differences in coherent diffraction is suppressed, leaving only variance due to mass-thickness. We acquire several such high-angle images for each area of the film investigated, and subtract the average of their variances from the measured variance at lower k. This effectively removes the k-independent systematic error, although it only works if there is negligible specimen drift during the acquisition of the entire series of images. The ability to measure the incoherent variance is the primary advantage of working in HCDF instead of tilted dark field.

THICKNESS EFFECTS

Foil thickness also plays a role in the experimental variance. In order to compare the variance of different samples, rather than the same sample before and after some treatment as we have done previously, we need to understand and correct for the effects of thickness.

We have measured the properties of a set of six films of varying thickness in order to investigate these effects. The samples are a-Si:H thin films deposited by reactive magnetron sputtering of a Si target in an Ar plasma at a substrate temperature of 230 °C at a rate of ~0.29 nm/sec. With H_2 in the discharge, this experimental system routinely produces a-Si:H films whose properties are identical to those grown by optimized plasma-CVD [11]. The films are deposited on top of the commercial Al foil and Corning 7059 glass simultaneously. The thickness of the films was determined by fitting the optical transmittance and reflectance of the films on glass measured with a Cary 5000 dual-beam spectrophotometer using the Tauc-Lorenz model established by Jellison and Modin [12]. TEM samples were prepared by etching off the Al substrate in 20% hydrochloric acid solution then catching the free-standing film on a support grid. Variable coherence measurements were performed in a Hitachi H9000 TEM with a computer-controlled hollow cone unit and automated image acquisition system operated at 200 kV with an objective aperture which yields an image resolution of ~ 1.5 nm. Each $V(k)$ trace is the average of 10–12 areas on the film, requiring ~250 images. Error bars are one standard deviation of the mean.

First, we need a convenient method to measure film thickness in the TEM. If we assume a Poisson distribution for the probability of n scattering events and the average BF intensity $\langle I_{BF} \rangle / \langle I_0 \rangle$ is due to unscattered electrons, we predict that $\langle I_{BF} \rangle / \langle I_0 \rangle$ decays exponentially with increasing film thickness [13], $\langle I_{BF} \rangle / \langle I_0 \rangle = \exp(-t/\Lambda_{BF})$. I_0 is the intensity of the beam, and $\langle \ \rangle$ indicates averaging over the image. Figure 2(a) shows $\langle I_{BF} \rangle / \langle I_0 \rangle$ for our samples, from which we determine that for a-Si:H $\Lambda_{BF} = 106 \pm 3$ nm under the imaging conditions described above. This allows us to measure in situ the thickness of such films so we can correct the experimental variance.

Next, we examine the thickness scaling of the denominator of Equation 1, the average DF intensity. If the average DF intensity $\langle I_{DF} \rangle / \langle I_0 \rangle$ is due to electrons scattered once, $\langle I_{DF} \rangle / \langle I_0 \rangle = C_1 (t/\Lambda_{DF}(k)) \exp(-t/\Lambda_{DF}(k))$, where C_1 is a scaling constant. Figure 2(b) shows $\langle I_{DF} \rangle / \langle I_0 \rangle$ and fits to this functional form for our samples. $\Lambda_{DF}(3.0 \text{ nm}^{-1}) = 125 \pm 9$ nm, $\Lambda_{DF}(4.5 \text{ nm}^{-1}) = 140 \pm 15$ nm, and $\Lambda_{DF}(5.75 \text{ nm}^{-1}) = 164 \pm 5$ nm.

The last piece of the thickness scaling of V is the second intensity moment $\langle I^2 \rangle$. If the sample consists of ordered clusters, the number of clusters increases linearly with thickness. If the clusters are uncorrelated, the variance of their scattered intensity increases linearly with the number of clusters, so $\langle I^2 \rangle = C_2(k)(t - t_0)$. t_0 is ~ 5 nm, which is consistent with the presence of a thermal oxide layer on the Si surfaces. $C_2(k)$ will be determined by the size

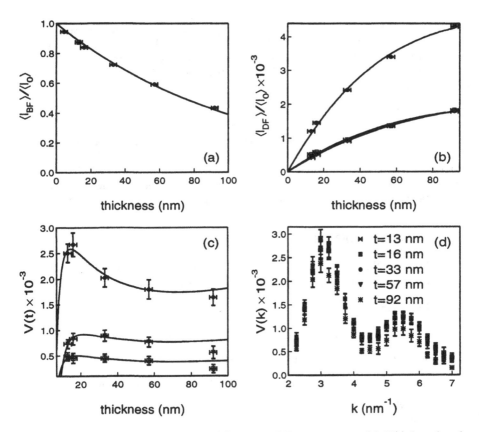

Figure 2: (a) Average BF transmission, (b) average DF transmission, (c) $V(t)$ for select k, and (d) $V(k)$ for films with various thickness, scaled to $t = 20$ nm. In (b) and (c), ● is $k = 3.0$ nm^{-1}, □ is $k = 4.5$ nm^{-1}, and ▼ is $k = 5.75$ nm^{-1}.

of the ordered clusters, which we hope to measure directly by variable resolution microscopy, and the structure factor of the clusters.

Substituting into Equation 1, we predict that V as a function of t should be

$$V(t) = \frac{C_2(t - t_0)}{(\frac{C_1 t}{\Lambda_{DF}})^2 e^{-2t/\Lambda_{DF}}} \qquad (4)$$

with two adjustable parameters C_2 and t_0, and Λ_{DF} and C_1 measured by fitting $\langle I_{DF}\rangle/\langle I_0\rangle$. Figure 2(c) shows $V(t)$ data for a variety of k and fits to Equation 4.

The fit clearly fails for $t > \Lambda_{DF}/2$. Since $\langle I_{DF}\rangle$ is fit well over the full range considered, the failure must be due to a sublinear dependence of $\langle I^2\rangle$ on t at large t. This may be due to small-angle diffuse scattering. Such scattering would tend to smooth the image intensity toward the mean, reducing $\langle I^2\rangle$, but would still be accepted by the objective aperture and not cause deviation in $\langle I_{DF}\rangle$.

Figure 2(d) shows $V(k)$ for films of varying thickness all scaled to a thickness of $t = 20$ nm using Equation 4 and parameters extracted by fitting. With the exception of the thickest film data, all the curves are the same within experimental error, demonstrating that we can accurately correct for the effects of thickness on $V(k)$. It is necessary to repeat the measurement of $V(k)$ for various thicknesses to extract the relevant parameters if one wishes to apply this correction method to a k outside the range measured here or to a material other than a-Si.

CONCLUSIONS

Fluctuation microscopy is a powerful new tool for the investigation of MRO in disordered materials. We have detailed the experimental methods and data analysis necessary to reliably compare samples prepared by different means to different thicknesses. This involves deconvolving the MTF of the imaging system, Fourier filtering the resulting images, correcting for the effects of incoherent variance by high-angle HCDF imaging, and correcting for foil thickness.

ACKNOWLEDGEMENTS

We thank the National Science Foundation (P. M. V., DMR 97-03906) and the National Renewable Energy Laboratory (J. R. A. and H-C. J., ADD-9-18668) for support.

REFERENCES

1. J. M. Gibson and M. M. J. Treacy, Phys. Rev. Letts. **78**, 1074 (1997).

2. N. Mousseau and L. J. Lewis, Phys. Rev. Letts. **78**, 1484 (1997).

3. J. M. Gibson and M. M. J. Treacy, to be published in Ultramicroscopy.

4. P. M. Voyles, J. M. Gibson, and M. M. J. Treacy, to be published in J. Electron Microscopy.

5. J. M. Gibson, M. M. J. Treacy, P. M. Voyles, H-C. Jin, and J. R. Abelson, Appl. Phys. Letts. **73**, 3093 (1998).

6. M. M. J. Treacy, J. M. Gibson, and P. J. Keblinski, J. Non-Cryst. Solids **231**, 99 (1998).

7. J. M. Zuo, Ultramicroscopy **66**, 21 (1996).

8. I. Daberkow, K.-H. Hermann, L. Liu, W. D. Rau, and H. Tietz, Ultramicroscopy **64**, 35 (1996).

9. O. L. Krivanek and P. E. Mooney, Ultramicroscopy **49**, 95 (1993).

10. M. M. J. Treacy and J. M. Gibson, Ultramicroscopy **52**, 31 (1993).

11. M. Pinarbasi, N. Maley, A. M. Myers, and J. R. Abelson, Thin Solid Films **171**, 217 (1989); M. Pinarbasi, M. J. Kushner, and J. R. Abelson, J. Appl. Phys. **68**, 2255 (1990).

12. G. E. Jellison, Jr. and F. A. Modin, Appl. Phys. Lett. **69**, 371 (1996).

13. L. A. Freeman, A. Howie, A. B. Mistry, and P. H. Gaskell in *The Structure of Non-Cryst. Mats.* edited by P. H. Gaskell (Taylor & Francis, London, 1976) p. 245.

Microdiffraction, EDS, and HREM investigation for phase identification with the electron microscope.

P. Ruterana[1], A. Redjaïmia[*]
LERMAT, ESA-CNRS 6004, Institut des Sciences de la Matière et du Rayonnement, 6 Bd Maréchal Juin, 14050 Caen Cedex, France
*Laboratoire de Science et Génie des Surfaces, CNRS, Ecole des Mines, Parc de Saurupt, 54042 Nancy Cedex, France

[1]Author for correspondence: email: ruterana@lermat8.ismra.fr, tel: 33 2 31 45 26 53, fax: 33 2 31 45 26 60

Abstract: For micro or nanophase identification, the ideal method is electron microscopy which combines experimental technique and theoretical analysis in order to solve the problem which is typical in non homogeneous materials. In this work, two heterogeneous materials have been investigated: a Fe- 22Cr - 5Ni –0.3Mo – 0.03C (wt%) duplex stainless steel and a 6H-SiC reinforced Al-Cu alloy fabricated by the Ospray technique followed by extrusion and aging. This aging often gives rise to the precipitation of intermetallic compounds some of which can be new phases. Using microdiffraction, EDS and HREM, it was possible to identify two new phases: the orhorhombic τ phase in the heat treated duplex steel and the tetragonal $Al_7Cu_3Zr_2$ in the reinforced aluminium copper alloy

1. Introduction

The properties of heterogeneous materials like alloys and composites depend critically on their structure and composition[1]. During the aging heat treatment, the properties are optimized by the tight control of the type, size and composition of the intermetallic precipitates which form. Therefore the formation of undesirable precipitates may be detrimental to the particular properties[2]. These precipitates are often of nanometric size and one of the most adequate method for their analysis is electron microscopy.

In this work, two new phases have been identified: the τ phase which is closely related to the δ phase inside the duplex steel; and in the reinforced aluminium alloy, a Cu-rich precipitate in epitaxial relationship with the well known Al_3Zr dispersoid.

2. Experimental

The duplex austenite-ferrite ($\delta+\gamma$) specimens were solution treated at 1350°C for 20 min to achieve homogeneity, they were subsequently quenched in order to retain a supersaturated and fully δ microstructure[3]. Specimens were then annealed in a temperature range between 550° and

650°C for various times up to 355 hours followed by water cooling. The heat treatments were carried out in electric muffle furnaces under vacuum in order to minimize oxidation.

The Al-SiC composite whose composition was Al – 6%Cu – 2% Mn - 0.3% Ag - 0.3 % Mg – 1% (Cr, Ti, V, Zr), was fabricated by the Ospray preform which consisted in mixing the atomized molted alloy with 13% SiC particles of diameter smaller than 15 μm and quenching onto a mandrel. The composite was then extruded at 490°C and solution treated at 520°C for 20 min, followed by water quenching (T4 temper). Specimens were aged at 190°C (T6 temper). The details of the fabrication can be found elsewhere[4]. We analysed samples after extrusion, T4 and T6 tempers.

The TEM samples were prepared by the conventional methods of electropolishing for the duplex steel and by mechanical grinding followed by ion milling for the Al composite. Microdiffraction analysis was carried out in a Philips CM12 electron microscope equipped with a ± 60° double tilt stage and the patterns were obtained using a nearly parallel beam which was focussed on a very small area of the thin foil (few tens of nm). HREM was carried out on microscopes with a point to point resolution better than 2 nm.

3. Results

3.1 The τ phase

The isothermal heat treatment of the δ-ferrite between 550 and 650°C leads to the formation of intergranular precipitation of fine and heavily faulted particles which exhibited a needle-like morphology (fig. 1). Close examination involving various orientations of the δ matrix grains established that the habit planes and growth direction of the particles are $\{110\}_\delta$ and $<110>_\delta$, respectively. Therefore, the number of variants for the particles in a given δ matrix grain is equal to 6 (n=6).

Fig. 1: Morphology of the τ phase which is embedded inside δ matrix.

The determination of the crystal structure of the τ phase was carried out by diffraction along common parallel axes of τ and the δ matrix (fig. 2) in conjunction with crystallographic symmetry investigation. If we consider G^I

and G^{II}, the point groups of δ and τ, the intersection group H is represented by their common symmetry when the precipitates adopt an orientation relationship with the matrix[5]. H is one of the 32 crystallographic point symmetry groups. The point group G^I is 4/m $\bar{3}$2/m and the fact that the new phase forms six needle-like variants with equivalent symmetries in each grain of the matrix is indicative that the order of the group H is equal to eight. Therefore the possible point group H(8) can be one of the following:

a) 4/m, 422, $\bar{4}$2m (tetragonal)

b) mmm, 2/m2/m2/m (orthorhombic)

Examination of the diffraction patterns recorded from the τ phase along [uvw] zone axis parallel to $<001>_\delta$ does not exhibit any four-fold symmetry (fig. 2).

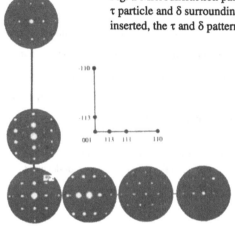

Fig. 2 Microdiffraction patterns from near parallel beam focussed on τ particle and δ surrounding matrix. Zone axes of the δ phase are inserted, the τ and δ patterns are superimposed

Therefore, H is the common group between the cubic δ and the orthorhombic group: 2/m2/m2/m. the point group G^{II} which fulfills this requirement has to contain 3 two-fold axes normal to m mirrors, so only the following point groups can be retained:

1. 4/m mm (tetragonal)
2. mmm (orthorhombic)

3. 6/m mm (hexagonal)   4. m$\bar{3}$m (cubic)   5. m$\bar{3}$ (cubic)

Examination of the microdiffraction patterns along the <uvw> zone axes parallel to $<001>_\delta$, $<110>_\delta$ does not show any four-fold or six-fold symmetry. Therefore, the diffraction behavior of the τ precipitate strongly suggests that its point group is G: 2/m2/m2/m in the orthorhombic system. In order to determine the space group of the precipitate, it was necessary to construct its reciprocal space by first tilting along two orthogonal $<110>_\delta$ zone axes and recording the superimposed patterns. In this case, the incident beams along the $[001]_\delta$, $[113]_\delta$, $[111]_\delta$, $[110]_\delta$ and $[\bar{1}10]_\delta$ directions were enough to cover the two phases reciprocal spaces, as shown in figure 3. By considering the specific shifts and the absence of the periodicity difference between the zeroth and the first layer reflections along the $[001]_\delta$ direction, it can be deduced that the diffraction symbol is mmmF---. This is in agreement with only the F2/m2/m2/m space group[6].

The mutual orientations between the τ and δ phases are directly deduced from the diffration patterns:

$(200)_\tau//(\bar{1}10)_\delta$ $\qquad\qquad\qquad (020)\tau//(110)_\delta$ $\qquad\qquad\qquad (002)\tau//(002)_\delta$

The mismatch along $[010]//[110]$ is nearly 15% whereas in the $(200)_\tau$ and $(002)_\tau$ planes it is close to 0%. This explains why the precipitates form platelets in the $[100]//[\bar{1}10]$ direction along which the misfit is zero percent.

The following parameters are deduced from the diffraction patterns for the τ phase:

a: 0.405 nm,   b: 0.344 nm   c: 0.287 nm

EDS analysis, using standard quantitative methods, shows that the new phase is enriched in Mo, in fact the (Fe+Mo) concentration is constant between δ and τ. The atomic Fe concentration of 66.4% in δ drops to 43.8% in τ, whereas that of Mo which is 2.1 jumps to 22.8%, the other constituents remain unchanged. This means that Mo promotes the formation of the τ phase and the possible mechanism for its formation is a substitutional diffusion.

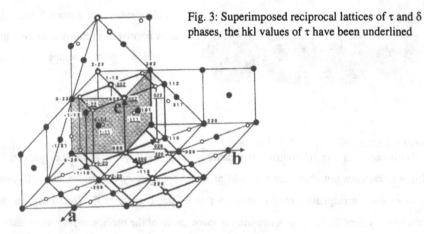

Fig. 3: Superimposed reciprocal lattices of τ and δ phases, the hkl values of τ have been underlined

3.2 The copper rich Zr precipitate in 6H-SiC/Al alloy composite

The new phase was found in epitaxy on the more common Al_3Zr reinforcing precipitates (fig.4). EDS gave a composition of Al 60.5%, Zr 14% and Cu 25.5% atomic concentration which is nearly $Al_7Cu_3Zr_2$. Along the $[110]_{Al3Zr}$ zone axis the two diffraction patterns are similar[7].

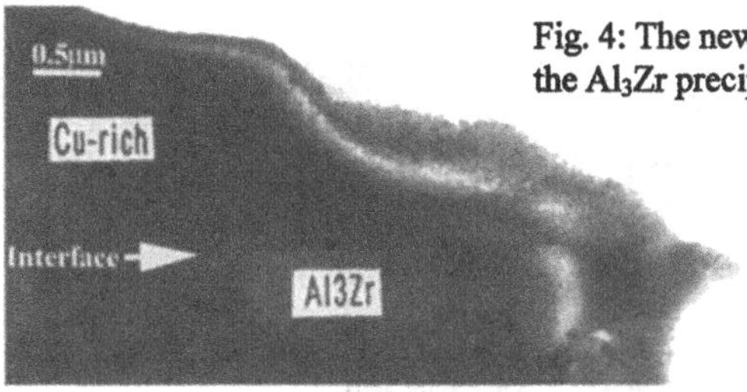

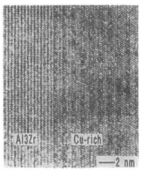

Fig. 4: The new copper Zr phase is in epitaxy with the Al_3Zr precipitates

HREM analysis of the interface area shows that this phase grows epitaxially without any misfit dislocations. This suggests that in the (001) plane, the two lattices are matched. From the diffraction patterns, the new phase has a smaller c parameter of 1.16 instead of 1.73 nm for Al_3Zr. We therefore have a new teragonal compound. The HREM images show typical contrast as seen in fig. 5, there is a 6 lattice fringes superperiodicity whose first half is similar to that of Al_3Zr. The second half exhibits a similar structure but with lower intensity. Superimposing the Al_3Zr image and that of the new phase shows a good match in the high intesity part, which probably means that the lattice spacing in this part is .02165 nm, like in Al_3Zr and then smaller in the second part. In order to determine the structure of the new phase, due to the similarity of the two images, we assume that the new phase is also a stacking of face centred Au_3Cu cubes along the c- direction. This time, there are only three cubes with 12 atoms per unit cell as deduced from EDS. Then, if the first half of the cell is identical to Al_3Zr, with $(2 \times Zr + 2 \times Al)$ in B type planes and one A type atomic plane with 2 Al (fig. 6)

Fig. 5: A HREM image of the two phases

In order to take into account the four-fold symmetry, there are only two possibilities, P1 and P2, for positionning the atoms in the remaining three atomic planes.

P1. 3 planes with (2A and 1B type) with 0.5 Al and 0.5 Cu occupancy

P2. one A type plane with Cu atoms, one B plane with Al and Cu atoms, and one A type plane with Al atoms.

Models at the two experimental images defocus for thicknesses ranging from 2 to 14 nm were carried out and in the case of the P2 model there is no match with a disagreement which increases with the specimen thickness.

Model P1 agrees quite well with the experimental images (fig. 6). It exhibits a space group of P 4/m mm

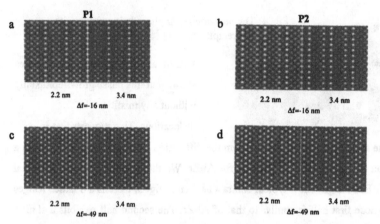

Fig. 6: Simulated images of models P1 and P2, good agreement with the experimental images is obtained for model P1

4. Summary

By using a combination of elecron microscopy methods, it has been possible to identify two new phases. The combination of microdiffraction and group theory analysis was successively applied in order to determined the F2/m2/m2/m symmetry group of the τ phase in the duplex steel, and HREM, image simulation and EDS allowed to analyse the $Al_3Cu_4Zr_2$ precipitate (P/4mmm) in the Al-SiC composite

Aknowledgements Dr. P. Vermaut and J.P. Morniroli are gratefully acknowledged for their most valuable contribution to this work.

References

1. R. Lagneborg Tans. Am. Soc. Metals **60**, 67(1967)
2. E.L. Brown, G. Gross, Metall. Trans. **14**, 791(1983)
3. A. Redjaimia, P. Ruterana, G. Metauer, and M. Gantois, Philos. Mag. **67**, 1277(1993)
4. P.S. Jensen and W. Khal, Proc. XXII Riso Int. Symp. on Mater. Science Metal Matrix Composites Processing, Microstructure and Properties, D.J. Jensen editor, 1991, p.182
5. J.W.Cahn and G. Kalonji, Proc. Int. Conf. on Solid State Phase Transformations, H.I. Aaronson at al., editors, 1982, p. 3
6. J.P. Morniroli in Inst. Phys. Conf. Ser. **98**, 87(1989)
7. P. Vermaut, P. Ruterana, Acta mater. **44**, 2445(1996)

ENERGY-LOSS FILTERED IMAGING OF SEGREGATION-INDUCED INTERFACE BROADENING IN SiGe/Si P-CHANNEL MOSFET DEVICE STRUCTURES

D.J. NORRIS *, A.G. CULLIS *, T.J. GRASBY **, E.H.C. PARKER **
* Department of Electronic and Electrical Engineering, University of Sheffield, Sheffield, S1 3JD, UK, d.j.norris@sheffield.ac.uk.
** Department of Physics, University of Warwick, Coventry, CV4 7AL, UK.

ABSTRACT

SiGe/Si p-channel heteroepitaxial MOSFET test structures have been fabricated using solid-source molecular beam epitaxy. High-resolution transmission electron microscopy and energy-loss filtered imaging have been used to quantitatively determine the nanoscale Ge distributions across the SiGe alloy channel. The Ge profile at the edges of the alloy channel were found to be asymmetrical due to the effect of Ge segregation, with an exponential-like distribution directed toward the surface. The results agree well with the predictions of segregation theory and indicate that the concentration of Ge in the extended distribution lay in the range 10%-1% over a distance of several nanometers from the body of the channel. Secondary ion mass spectrometry measurements upon the same samples were insensitive to this short range extended Ge distribution.

INTRODUCTION

At present, there is a considerable effort directed towards the development of an advanced SiGe-based device technology for the fabrication of high-speed analogue and digital electronics. One particular device structure that is currently under investigation is the strained p-channel field-effect transistor (FET), a component of which consists of a pseudomorphic layer of SiGe grown by molecular beam epitaxy (MBE) on Si. A major requirement for good electrical performance is the need for interfaces to be sharp on either side of the SiGe layer since small scale interface roughness leads to unwanted hole scattering.[1] Interface sharpness can be difficult to control and, indeed, under certain conditions the surface of the strained layer can exhibit wave-like undulations caused by thermodynamically driven growth instability.[2] It is possible to produce more uniform alloy layers by performing growth at a relatively low-temperature, capping with Si, and then annealing at high-temperature to restore the required electrical properties.[3] However, even after implemention, interfaces may still not be optimally sharp due, for example, to the effect of Ge segregation.[4] Hence, it is crucial that the composition profile across such interfaces can be characterised with nanometer-scale resolution.

Interfaces of this type have been analysed using secondary ion mass spectroscopy [3,5] (SIMS); however, this method tends to sample large areas, to give averaged results, and consequently it is difficult to obtain nm-scale resolution in the high-concentration regime. Small electron-probe methods [6,7] involving electron energy-loss spectroscopy (EELS) and energy dispersive-X-ray spectroscopy (EDS) have been used to profile the Ge distribution. Moreover, imaging techniques include energy-filtered TEM (EFTEM) or high-angle annular dark-field (HAADF): previous EFTEM work on SiGe layer structure includes, Si-K deficit profiling of

Mat. Res. Soc. Symp. Proc. Vol. 589 © 2001 Materials Research Society

relatively thin multiple channel layers.[8] Whereas, HAADF imaging was performed on morphologically distorted SiGe layers[9] to analyse the effects of inter-diffusion and segregation.

In this work, we present a study using combined HREM and EFTEM imaging for the analysis of the Ge distribution within a ~9nm SiGe channel heterostructure grown by MBE. For EFTEM, images were formed using the Ge-L energy-loss electrons from which the Ge profile across the layer could be determined with nm-scale resolution.

EXPERIMENTAL DETAILS

Layers were grown on RCA-cleaned, n-Si(001) substrates in a VG V 90S MBE system. After an *in-situ* clean, layer deposition employed a growth rate of 0.1nm/s. First, 250nm of Si was deposited while ramping the heater down to 450°C. A 9nm layer of SiGe (Ge~30%) was next deposited, followed by a 25nm Si capping layer, both at 450°C (this temperature being low enough to suppress SiGe undulation formation). The temperature was then ramped up to 700°C for 30 min for a post-growth *in-situ* anneal, before being reduced to 600°C for the growth of a final layer of boron doped Si.

Cross-sectional TEM specimens, with [110] surface normals, were prepared by mechanical polishing and argon ion milling to electron transparency. These were examined using a JEOL 2010F TEM with a field emission gun; this operated at 200kV and was equipped with a Gatan imaging filter, which provided electron energy loss spectroscopy (EELS) and imaging.

RESULTS

Conventional TEM performed on cross-sectional samples, and a typical region of the channel layers is shown in figure 1. The 200 bright-field (Fig. 1a) and HREM (Fig. 1b) images show no large scale roughening or defects within the SiGe channel region. However, there is a clear difference in the relative sharpness of the buffer/channel and channel/cap interfaces. An intensity profile, taken from Fig. 1a along the growth direction (A to B), is plotted in Fig. 1c and shows that the buffer/channel interface is significantly sharper than the channel/cap interface. However, the presence of strain contrast in such images makes a quantitative comparison of the Ge gradation difficult; therefore, we have used energy filtered TEM.

In figure 2a, an EELS spectrum is shown, obtained using a 3nm electron probe focused into the central region of the SiGe channel. Both the Ge-L and Si-K energy loss edges are shown in the range 1-2keV of the spectrum. For EFTEM imaging, the delayed Ge-L signal was used with the post-edge window placed at 1300eV and pre-edge windows at 1125eV and 1175eV. A series of filtered images were acquired at each of these windows using a 30eV slit width and exposure times of 40s: the specimen drift was no more than 0.3nm during this exposure period. The three-window background subtraction method was used to generate Ge maps of the channel area. In figure 2b, an example of a Ge map is shown derived using the Ge-L edge, from which the profile distribution of Ge across the channel could be determined. The latter, along the growth direction A to B, is shown in Fig. 2(c) and exhibits a central plateau region with asymmetrically sloping sides. The Ge elemental distribution across the channel was

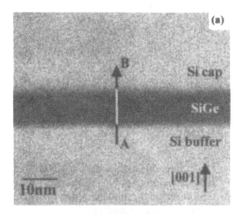
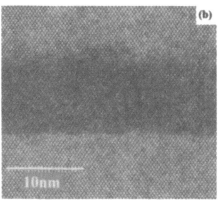

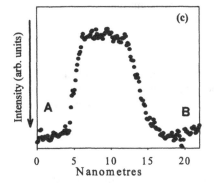

Fig. 1 (a) 002 bright-field and (b) high resolution [110] images of SiGe channel in Si matrix, and (c) inverted intensity profile across the region outlined in (a).

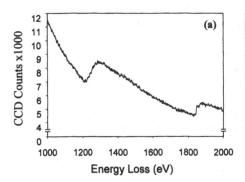
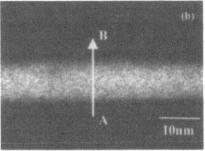

Fig. 2 (a) electron energy-loss spectrum from within the SiGe channel; (b) Ge elemental map acquired using the Ge-L edge (1217eV) shown in (a).

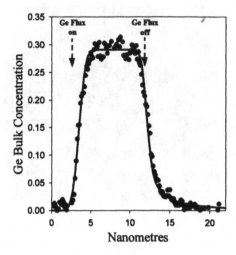

Fig. 3 Experimental Ge profile (dots) taken from image of Fig. 2(b) and theoretical curve (solid line) from segregation theory modified for resolution limitation.

independently calibrated by determination of the Ge peak concentration using the spectrum shown in Fig. 2a. The 3nm electron probe was incident at the relatively flat central region of the distribution and, from the relative Ge and Si edge amplitudes, a Ge concentration of 29±4% was obtained.

DISCUSSION

Although the theoretically expected spatial resolution of the Ge distribution is still unclear,[10,11] for the present work, the spatial resolution has been estimated in two ways. Firstly, by comparing the measured Ge profile with that (Fig. 1c) across an elementally sensitive 200-type image of the channel, a match to significantly better than 1nm is found in both full-width-half-maximum and overall shape. Secondly, segregation theory was found to correlate closely with the experimentally derived profile (see below). Both of these methods give significant confidence that actual resolution is achieved on the 1nm scale.

From the EFTEM Ge-L profile in Fig 3, it is clear that the Ge distribution exhibits a 6-7nm relatively flat central region, across which the measure Ge concentration is close to the 30% value intended during growth. However, the channel interfaces are clearly asymmetrical. The buffer/channel interface is relatively sharp, with a width of ~1.5nm (from 10% to 90% of maximum Ge concentration). In comparison, the channel/cap interface is significantly broader, having an exponential-like tail, in the Ge concentration range 10%-1%, which extends several nanometers from the body of the channel. These main features of the channel shape were a general characteristic of other regions of the SiGe alloy channel investigated. Ge profiles obtained using SIMS, on similar layers, have been found to be insensitive to the extended Ge-distribution in the high-concentration regime.[3]

The predicted Ge profile derived from segregation theory has been fitted directly to the asymmetrical Ge profile and this is shown by the continuous curve in Fig. 3. For this, we have used the kinetic, two-state exchange segregation model, with self-limitation,[10,11] to predict the expected Ge distribution for our layer growth temperature. For the present calculations, the rate equation (1) and mass conservation relation (2) for the bulk (B) and surface (S) concentration were:

$$dS(t)/dt = \Phi_{Ge} + P_1 B(t)(1 - S(t)) - P_2 S(t)(1 - B(t)), \qquad (1)$$

$$S(t) + B(t) = S(0) + B(0) + \Phi_{Ge} t \qquad (2)$$

where $P_1 = v_1 \exp(-E_a/kT)$ and $P_2 = v_2 \exp(-(E_a + E_s)/kT)$.

Parameters for fitting included a lattice vibration frequency ($v_1 = v_2$) of 10^{13}/s; an incident Ge flux concentration (Φ) of 0.29 (as determine from EELS); an activation energy (E_a) of 1.8eV; and a segregation energy (E_s) of 0.2eV (based on previous work with adjustment to fit the long range, low concentration Ge tail measured in our samples by SIMS): k is the Boltzmann constant and T the temperature in Kelvin.

Solving these equations numerically allowed both the segregated Ge concentration in the surface monolayer and the incorporated bulk Ge concentration in the sub-monolayer to be deduced. In addition, to account for the estimated resolution limitation, a Gaussian convolution function of 1nm FWHM was imposed upon the resultant bulk profile. It is apparent that the theoretical profile in Fig. 3 exhibits the characteristic shape expected for the segregation-induced interface broadening and that there is an excellent correlation with the experimental data at the 1nm resolution level.

CONCLUSIONS

The work reported here has demonstrated that electron energy-loss imaging on the EFTEM, employing the Ge-L edge, can quantitatively determine Ge distributions across very narrow SiGe MBE layers with approximately nanometre resolution. For the present layer structure, it is shown that the surface-directed tail of the Ge distribution extended several nanometres in a Ge concentration range greater than ~1%. This has significance for the performance of HMOS devices fabricated using such layers.

ACKNOWLEDGEMENTS

The authors would like to thank Profs M.G. Dowsett and T.E. Whall for helpful discussions, Drs C.P. Parry and P.J. Phillips for the growth of SiGe/Si layer structures and the EPSRC for financial provision.

REFERENCES

1.   T. E. Whall and E. H. C. Parker, J. Phys. D: Appl. Phys. **31**, p. 1397 (1998)
2.   A. G. Cullis, D. J. Robbins, A. J. Pidduck and P. W. Smith, J. Crystal Growth **123**, p. 333 (1992)
3.   T. J. Grasby, R. Hammond, C. P. Parry, P. J. Phillips, B. M. McGreggor, R. J. H. Morris, G. Braithwaite, T. E. Whall, E. H. C. Parker, A. Knights and P. G. Colman, Appl. Phys. Lett. **74**, p. 1848 (1999)
4.   O. Dehaese, X. Wallart and F. Mollot, Appl. Phys. Lett. **66**, p. 52 (1995)

5.   S. Fukatsu, K. Fujita, H. Yaguchi, Y. Shiraki and R. Ito, Appl Phys Lett **59**, p. 2103 (1991)

6.   D. J. Norris, A. G. Cullis, T. J. Grasby and E. H. C. Parker, J. Appl. Phys. **86** (12), p. 7183 (1999)

7.   D. J. Norris, A. G. Cullis, M. G. Dowsett and E. H. C. Parker, (to be published)

8.   T. Walther, C. J. Humphreys, A. G. Cullis and D. J. Robbins, in *Microscopy of Semiconducting Materials 1997*, edited by A. G. Cullis and J. L. Hutchinson (Institute of Physics Publishing, Bristol, 1997) p. 47-54.

9.   T. Walther, C. J. Humphreys and A. G. Cullis, Appl. Phys. Lett. **71**, p. 809 (1997)

10. T. Walther and C. J. Humphreys, J. Crystal Growth **197**, p. 113 (1999)

11. W. Jager and J. Mayer, Ultramicroscopy **59**, p. 33 (1995)

ATOMIC-SCALE IMAGING OF DOPANT ATOM DISTRIBUTIONS WITHIN SILICON δ-DOPED LAYERS

R. VANFLEET **,i, D.A. MULLER *, H.J. GOSSMANN *, P.H. CITRIN *, J. SILCOX **
* Bell Labs, Lucent Technologies, Murray Hill, NJ
** School of Applied and Engineering Physics, Cornell University, Ithaca, NY

ABSTRACT

We report measurements of the distribution of Sb atoms in δ-doped Si, over a wide 2-D concentration range. Both annular dark-field imaging and electron energy loss spectroscopy proved sufficiently sensitive to locate Sb atoms at the atomic scale. Improvements in both detector sensitivities and specimen preparation were necessary to achieve these results, which offer a surprising explanation for the dramatic difference in electrical activity between 2-D and 3-D dopant distributions at the same effective volume concentrations. The prospects for the general identification of individual dopant atoms will be discussed.

INTRODUCTION

Two dimensional dopant sheets, known as δ-doping, are integral parts of many novel semiconductor device concepts [1]. Confinement of one dimension of the doping layer allows exploitation of quantum properties. Practical application of these ideas requires not only the tools to create the thin dopant layer but the ability to measure the resulting structure. The techniques to produce thin doping layers during substrate growth are now currently available [1]. However, traditional compositional profiling techniques such as Secondary Ion Mass Spectrometry (SIMS) and Rutherford Back Scattering (RBS) only have depth resolutions that are comparable to the expected profile widths (several nanometers). Thus, SIMS and RBS results only give upper limits on δ-layer widths. We will discuss here the use of Electron Energy Loss Spectroscopy (EELS) and Annular Dark Field (ADF) imaging in a Scanning Transmission Electron Microscope (STEM) to give atomic level profiling of antimony δ-doped layers in silicon. This data has been used recently to show that these antimony in silicon δ-doped layers have significantly higher dopant activation than the equivalent bulk doped layers [2].

EXPERIMENT

STEM

The specimens were observed in a VG HB501 Scanning Transmission Electron Microscope (STEM) with an Annular Dark Field (ADF) detector and Parallel Electron Energy Loss Spectrometer (EELS). The electron source (cold field emission) is focused to a very small (~2Å) probe of electrons at the sample. A signal, derived from the interaction of the probe with the sample is displayed according to the probe position to form the image. The Annular Dark Field (ADF) detector is a large area detector with a hole in the center. Electrons are collected

---

iCurrent affiliation: Advanced Materials Processing and Analysis Center (AMPAC) and Department of Physics, University of Central Florida, Orlando, FL

Mat. Res. Soc. Symp. Proc. Vol. 589 © 2001 Materials Research Society

that have scattered from a minimum out to a maximum angle from the original beam direction. The electrons that scatter less than the inner angle of the ADF detector are used either for EELS or bright field imaging. Because the EELS/bright field electrons and ADF electrons are spatially separated, they can be collected in parallel, creating two data sets with the same probe position. STEM bright field imaging is equivalent to traditional TEM imaging by reciprocity (although the image is created differently) and will not be discussed here. EELS will be covered later. While the inner and outer dimensions of ADF detectors are physically small (on the order of mm), they correspond to large angles for electron scattering. Electron scattering factors at large angles are mainly due to the nucleus and related to the atomic number (Z). It is because of this atomic number relation that the ADF image is commonly referred to as "Z-contrast". While this atomic number contrast is clearly evident, not all contrast effects seen in ADF images are related to atomic number differences. The effects of crystal orientation, specimen thickness, lattice strain, and local atomic structure can be pronounced. Therefore, the interpretation of an ADF image as having atomic number contrast must be supplemented by further information about the specimen, imaging conditions, and image simulation. The contrast reversals with defocus and thickness variations inherent in conventional TEM imaging do not occur in ADF images. Thus, direct measurements of the scattered intensity can be interpreted more readily with the ADF mode, subject to the complications mentioned above. Images shown in this manuscript are taken under conditions that give the most compact electron probe with the minimum of tails for the Cornell VG STEM. The electron energy was 100 KeV and convergence angle ~10.3 mrad. There are no post specimen lenses, so the ADF detector angles are fixed at ~35 mrad and greater than 200 mrad as the inner and outer angles. The advantage of this method over traditional Transmission Electron Microscopy (TEM) arises from the smallness of the electron probe. The small electron probe allows direct atomic level ADF imaging and compositional analysis by EELS, from regions much smaller than traditional techniques

EELS

The Electron Energy Loss Spectrometer spatially separates the various electron energies in the beam. This allows probing of the inelastic scattering processes or energy loss processes of electron interaction with the sample. Ejection of a bound core electron by the fast imaging electron to the conduction band results is an element specific energy loss in the EELS spectrum. This allows elemental identification and quantification with near the resolution of the electron probe. The exact energy loss and structure in the edge spectra contains information about the local bonding and electronic structure. The integrated intensity in the edge (after background subtraction) contains the information needed to quantify the local composition. The integrated intensity is proportional to the number of atoms of the particular element under the beam (thickness as well as compositional effects), the beam intensity, and the cross section for scattering into the energy window. The feasibility of identifying dopant concentrations by Electron Energy Loss Spectroscopy depends upon the energy loss cross-section, and microscope parameters. The number of counts (I, units of counts) above background for a given edge is given by the scattering cross-section for the selected energy window ($\sigma(\Delta E)$, units of area per atom), the number of that type of atom (N), the electron flux (J, units of electrons per area per time), the detector efficiency (ε, units of counts per electron), and the time for which the data is collected (t).($ I = N\sigma(\Delta E)J\varepsilon t $) Thus, a sequence of points taken under the same detector conditions (same ε and t), close together in time (same J) and processed in a consistent way (same energy window (ΔE) and background subtraction process) will have the number of counts proportional to the number of atoms, interacting with the beam, of the element observed. If further care is taken to have all points in regions of uniform thickness, the counts are then proportional to the atomic density of the element observed.

Because bright field imaging and EELS use the same electrons, conventional TEM requires changes in the microscope conditions to switch between these modes. This introduces uncertainty in the exact positioning of the probe for EELS data collection. One of the strengths of STEM is the parallel collection of ADF and EELS data. The ADF signal can be used to pinpoint the position from which the EELS data originates and can do so at the atomic level. Additionally, picking the focus that optimizes the contrast in the ADF image gives the minimum spatial spread of the electron probe at the specimen. This is the condition for best spatial resolution in the EELS data. Thus, ADF imaging and EELS are natural compliments.

Detectors

To achieve our observed resolution, improvements in the detectors were required. The Cornell VG STEM's ADF detector was redesigned to include a single crystal YAP scintillator that is optically coupled to the PMT by a solid quartz rod. This increases the detector efficiency by matching the scintillator emission spectrum to the sensitivity of the PMT and reducing the signal (or photon) loss between the YAP and the PMT [3]. The PEELS system is a McMullan style spectrometer with a YAP single crystal scintillator, imaged by a cooled CCD camera.

Sample Preparation

Samples were prepared by Tripod polishing to a low angle wedge (~1°) [4]. The final polish was colloidal silica followed by several passes over a wet but clean polishing cloth to remove the residual silica particles from the finished surface. The polishing process is followed up by an HF dip just prior to loading into the microscope. The HF removes most of the, at times significant, surface oxide. The presence or absence of the surface oxide is evident in the ADF images with variations that mask the lowest doped antimony layers. Ion milling results in a significantly thicker damage layer that would also effect the antimony visibility. Within the microscope, under vacuum, the sample is baked to >100°C to remove not only water vapor but most importantly the loosely bound hydrocarbons. With out the bake, contamination sufficient to rapidly degrade the imaging of the layers and the EELS results is common.

RESULTS

Figure 1 shows four ADF images of antimony delta doped layers. The as specified sheet concentrations are: 3e13 Sb/cm² (A), 1e14 Sb/cm² (B), 5e14 Sb/cm² (C), 1e15 Sb/cm² (D). The brighter regions are the antimony layers with the growth direction to the right or upper right of the images. At the higher dopant levels, the asymmetry with tailing in the growth direction is clearly evident. A simple model can be used to quantify ADF images [5]. In this model, each atom contributes to the signal according to its atomic number to the 1.7 power ($Z^{1.7}$). This model, and other versions of Z-contrast, can only apply to sufficiently thin specimens. The extra scattering due to dopants cannot increase unbounded as the silicon signal becomes a significant fraction of the beam. Thus, the contrast should decrease with thickness. While the exact thickness dependence is not yet fully understood this quantification process looks promising. For a ~40 nm thick region of the 1e14 Sb/cm² specimen the quantification process gave a sheet

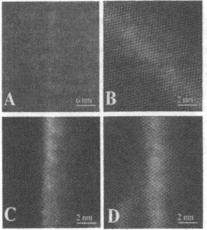

Figure 1. Four delta doped layers with sheet concentrations increasing from 3e13 Sb/cm² in A to 1e15 in D. Silicon <110> lattice can be seen in each with the antimony layer being seen as a vertical strip (diagonal in B) of extra brightness (or scattering).

concentration of 1.17e14 Sb/cm² which compares very favorably with the RBS results of 1.10e14 Sb/cm². These numbers correspond to an average of 1.6 antimony atoms per 40 nm thick silicon column at the peak of the profile.

EELS supports the ADF measured profiles. The antimony M_{45} edge at 528 eV energy loss was used to confirm that the ADF results are measuring the antimony profile. The M_{45} edge has a delayed maximum and appears only as a change in slope. Without the HF dip, a significant oxygen peak is seen at 532 eV with a sharp onset. Figure 2 shows EELS results from two points of a line scan across the delta layer. One point shown is from the inside of the delta layer and

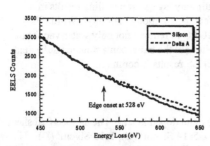

Figure 2. EELS spectrum of the antimony M_{45} edge from just off and centered on a delta doped layer. When surface oxides are present, a sharp onset oxygen edge can be seen at 532 eV.

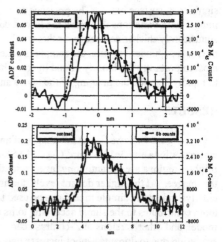

Figure 3. Two EELS and ADF linescans across delta layers. A: 1e14 Sb/cm² and B 5e14 Sb/cm². Both methods show the same profile.

another point off the layer. A detector background has been subtracted and the silicon only point has been scaled to overlay the pre-edge region of the delta layer spectrums. This silicon only signal is subtracted from the delta layer spectrums and the result integrated over a 150 eV range from the edge. This integrated signal for two line scans (1e14 and 5e14 Sb/cm² specimens) is shown in figure 3 with the corresponding ADF line scans. In both cases, the EELS and ADF signals give the same profile.

While many of the terms required to convert the EELS data to antimony numbers are not well known or routinely measured, estimates of their value can be made. For antimony the cross-section is on the order of 10^{-21} cm² if summing over a 150 eV window from the edge onset[6]. The approximate microscope numbers for the Cornell STEM are: electron flux ~ 10^{24} cm⁻²s⁻¹, efficiency ~ 1, and typical collection times for the PEELS is 10 s. This gives on the order of 10^4 counts for a single antimony atom and supports the ADF results of near single dopant atom sensitivity.

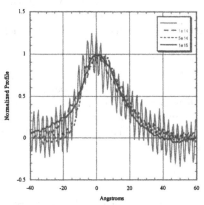

Figure 4. Delta layer profiles for all four antimony sheet concentrations. Peaks normalized to overlap the profiles. No significant difference in profile widths are seen over this wide range of doping.

These delta layer profile results have revealed a dramatic difference in electrical activity between 2-D and 3-D dopant distributions at the same effective volume concentrations. Comparison of 2-D and 3-D dopant concentrations requires accurate knowledge of the 2-D layer profile. Figure 4 show the ADF derived profiles, averaged along the layer, for the four images in figure 1. They have been scaled vertically to overlay and show that even though the sheet concentrations vary by two orders of magnitude, the profiles widths are essentially the same. Using the measured profiles to convert 2-D deposition and activation data to 3-D equivalents, the two data sets can be overlaid as in figure 5. While the 3-D distributions under go a deactivation process as the dopant level is increased the 2-D distributions do not. The various lines in figure 5 correspond to terms of a deactivation model involving the number of second nearest neighbor dopant atoms[2].

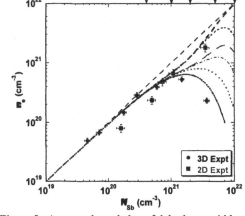

Figure 5. Accurate knowledge of delta layer widths allows conversion of 2-D dopant and carrier densities to their 3-D equivalent and comparison to bulk results. This reveals that 2-D distributions avoid the deactivation processes that occur in bulk doping.

CONCLUSIONS

Whereas the results shown here illustrate the ability to identify high Z atoms in lower atomic number substrates at near single atom sensitivity, the general application of these techniques will depend upon specific substrates, dopants, and preparation techniques. Silicon is perhaps an ideal substrate because of its low atomic number, crystalline perfection, wide availability, and ease of TEM specimen preparation.

ACKNOWLEDGMENTS

Support was received from Lucent Technologies and Air Force grant # F49620-95-1-0427. The Cornell STEM was acquired through the NSF (grant # DMR-8314255) and is operated by the Cornell Center for Materials Research (NSF grant # DMR-9632275). Support and helpful discussions with Earl Kirkland and Mick Thomas are gratefully acknowledged.

REFERENCES

1. H.-J. Gossmann and E.F. Schubert, Critical Reviews in Solid State and Materials Sciences, **18**, 1-67 (1993).

2. P.H. Citrin, D.A. Muller, H.J. Gossmann, R. Vanfleet, and P.A. Northrup, Phys. Rev. Lett. **83**, 3234 (1999).

3. E. J. Kirkland, M. Thomas, A High Efficiency Annular Dark Detector for a VG HB-501 STEM, Ultramicroscopy, **62** (1996) 79-88.

4. S.J. Klepeis, J.P. Benedict, and R.M. Anderson in *Specimen Preparation for Transmission Electron Microscopy*, Edited by J.C. Bravman, R.Anderson, and M.L. McDonald (Mater. Res. Soc. Proc. **115**, Pittsburgh, PA 1988), p. 179.

5. R.R. Vanfleet, M. Robertson, M. McKay, and J. Silcox, Prospects for Single Atom Sensitivity Measurements of Dopant Levels in Silicon, in "Characterization and Metrology for ULSI Technology: 1998 International Conference" edited by D.G. Seiler, A.C. Diebold, W.M. Bullis, T.J. Shaffner, R. McDonald, and E.J. Walters, (AIP press, 1998).

6. Calculated with data in: Electron Energy-Loss Spectroscopy in the Electron Microscope, R.F. Egerton, Plenum Press, New York 1996.

IN-SITU ANNEALING TRANSMISSION ELECTRON MICROSCOPY STUDY OF Pd/Ge/Pd/GaAs INTERFACIAL REACTIONS

F. RADULESCU, J.M. McCARTHY and E. A. STACH [*]
Department of Materials Science and Engineering, Oregon Graduate Institute of Science and Technology, P.O. Box 91000, Portland, Oregon, 97291

[*] National Center for Electron Microscopy, Lawrence Berkeley National Laboratory, Berkeley, California, 94720

ABSTRACT

In-situ TEM annealing experiments on the Pd (20 nm) / a-Ge (150 nm) / Pd (50 nm) / GaAs ohmic contact system have permitted real time determination of the evolution of contact microstructure. As-deposited cross-sectional samples of equal thickness were prepared using a focused ion beam (FIB) method and then subjected to *in-situ* annealing at temperatures between 130-400 °C. Excluding Pd-GaAs interactions, four sequential solid state reactions were observed during annealing of the Pd:Ge thin films. First, interdiffusion of the Pd and Ge layers occurred, followed by formation of the hexagonal Pd_2Ge phase. This hexagonal phase then transformed into orthorhombic PdGe, followed by solid state epitaxial growth of Ge at the contact / GaAs interface. The kinetics of the solid state reactions, which occur during ohmic contact formation, were determined by measuring the grain growth rates associated with each phase from the videotape observations. These data agreed with a previous study that measured the activation energies through a differential scanning calorimetry (DSC) method. We established that the Ge transport to the GaAs interface was dependent upon the grain size of the PdGe phase. The nucleation and growth of this phase was demonstrated to have a significant effect on the solid phase epitaxial growth of Ge on GaAs. These findings allowed us to engineer an improved two step annealing procedure that would control the shape and size of the PdGe grains. Based on these results, we have established the suitability of combining FIB sample preparation with *in-situ* cross-sectional transmission electron microscopy (TEM) annealing for studying thin film solid-state reactions.

INTRODUCTION

In the last decade, the Pd-Ge contact system emerged as the most promising ohmic contact replacement of the Au-Ni-Ge alloy, still used by most of today's GaAs technology. Its better electrical properties stem from the fact that the contact formation is based on a series of solid state reactions and no melting of the metal thin films and GaAs substrate take place during annealing. Previous research studies by Marshall et al. [1] demonstrated that thermally stable contacts with low resistivities could be achieved by low temperature annealing of Pd and Ge thin films. The behavior of the Pd-GaAs interface, when subjected to annealing temperatures lower than 400 °C, was studied in great detail by Sands et al. [2] Ottaviani et al. [3] studied the Pd-Ge thin film system and demonstrated that, at low temperatures, Pd_2Ge forms first, followed by transformation to PdGe at higher annealing temperatures. The microstructure of the entire contact system, Ge/Pd/GaAs, was investigated by Marshall et al. [4], who showed the predominant reaction takes place between Pd and Ge to first form Pd_2Ge and then PdGe. In a previous study [5], we determined the transition temperatures and the activation energies associated with these solid state transformations by combining DSC with cross-sectional TEM.

Mat. Res. Soc. Symp. Proc. Vol. 589 © 2001 Materials Research Society

The purpose of this work is to characterize the kinetics of the solid state reactions as they evolve during the annealing process of the Pd/Ge thin films on GaAs. By using *in-situ* annealing TEM, it is possible to obtain a continuous and complete record of the phase and microstructure evolution.

EXPERIMENT

In this study, Pd (20 nm)/Ge (150 nm)/Pd (50 nm) layers were deposited by alternate electron beam evaporation of Pd and Ge onto <001> GaAs. Electron transparent TEM cross-sections were produced by mechanical grinding and focus ion beam (FIB) milling. This TEM sample preparation technique was described in greater detail by Bassile et al. [6] The advantage of using this sample preparation technique for *in-situ* TEM experiments is two fold. First, it provides a large, uniform thickness cross-sectional electron transparent area which renders a more extended view of the thin films stack. Secondly, it answers a concern regarding the accuracy of all *in-situ* TEM experiments in general, the effect of the thin foil surface on the reaction sequence. The thickness of the transparent foil can be controlled, allowing one to determine the surface effects on the observed experiments by analyzing the same reaction on foils with different thicknesses.

A set of ten TEM samples was prepared from the same as-deposited wafer. These samples were loaded into a Gatan double tilt heating holder and then annealed *in-situ* in a JEOL 200CX TEM equipped with a Gatan intensified video camera. Isothermal as well as constant heating rate annealing between 130 °C and 400 °C was performed on this set of samples. The recorded temperature correction varied from –9 °C to –24 °C and was established by direct comparison with a DSC scan of equal heating rate.

The kinetics of the solid state reactions that occur during ohmic contact formation were determined by measuring the grain growth rates associated with each phase from the videotape observations. Images captured from the videotape provided a suitable way of performing these measurements.

RESULTS AND DISCUSSION

Figure 1 represents a sequence of video-captured images of an ohmic contact thin film structure subjected to 100 °C/min. heating ramp rate followed by a 10 minute holding time at 318 °C. This procedure resembles rapid thermal annealing (RTA) which is compatible with typical integrated circuit processing requirements.

The first phase transformation that takes place during annealing is shown in Figure 1a). We observe the nucleation and growth of hexagonal Pd_2Ge until the whole initial Pd layer is consumed. Both Pd layers reacted with the a-Ge and the newly formed germanide phase was identified by TEM microdiffraction.

The orthorhombic PdGe phase nucleates at the Ge-Pd_2Ge interface, as demonstrated by Figure 1b) and, afterwards, grows by consuming a-Ge and Pd_2Ge according to:

$$Pd_2Ge + Ge = 2PdGe \tag{1}$$

This reaction requires that Ge diffuses through the PdGe layer and reacts with Pd_2Ge. At 307 °C, a small amount of Pd_2Ge is still present at the interface with GaAs, see Figure 1c). As we increase the temperature, the Pd_2Ge layer is completely consumed and the PdGe surface makes direct contact with the GaAs substrate. In the next stage, the excess a-Ge diffuses through the

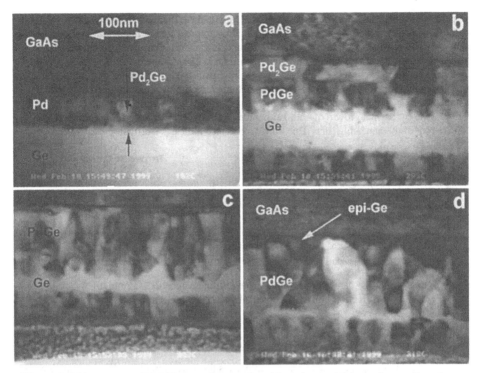

Figure 1. Video-captured TEM image of a sample subjected to rapid thermal annealing at
a) 182 °C, b) 293 °C, c) 307 °C, d) 318 °C for 10 min.

PdGe grain boundaries and grows on top of the GaAs substrate by a solid phase epitaxy
mechanism. This is clearly illustrated in Figure 1d).

The evidence presented so far, in regard to the microstructure evolution of the Pd-Ge
ohmic contact on GaAs, demonstrates the first phase change is the Pd_2Ge formation. This
hexagonal phase then transforms into orthorhombic PdGe followed by solid phase epitaxial
(SPE) growth of Ge at the GaAs interface. In the final stages of ohmic contact formation there
are two competing reactions that determine the extent of Ge epitaxial growth on GaAs - PdGe
formation and Ge crystallization. They both require excess a-Ge and, as shown by the recorded
videotapes, the Pd_2Ge has to be consumed entirely before Ge SPE growth on GaAs occurs.

These findings agreed with the results presented in an earlier study [5] that determined
the phase formation by combining TEM with the DSC method. They also indicate that, in the
overall thin film stack system, the dominant solid state reactions take place between Pd and Ge.

In order to analyze the kinetic parameters that describe the formation of Pd_2Ge and PdGe,
an Arrhenius temperature dependence was assumed. For this type of analysis, the accuracy of the
results relies on precise temperature measurements. Since the thermocouple configuration of the
typical TEM heating holder does not allow recording the actual temperature at the specimen
surface, one needs to calibrate the system. This is a common problem associated with any TEM
in-situ heating experiment and a unique fit-for-all temperature calibration procedure has not yet
been developed. In this study, the temperature was calibrated by comparing the DSC analysis
results with the real-time videotape recordings. The same heating rate was used in both

experiments to measure the temperature associated with three distinctive events in the microstructure evolution of the contact. In the DSC scan presented in Figure 2a) the start and the end of Pd_2Ge formation are measured at 178 °C and 238 °C, respectively. In addition to these two data points, the end of PdGe formation, measured at 280 °C, could also be used as a good indicator because it is a well-defined stage in the microstructure evolution of the contact. Assuming the same linear tendency, the temperature correction was extrapolated to lower temperatures, as demonstrated in Figure 2b).

The isothermal growth of both germanide phases was noted to be proportional to the square root of the annealing time. This time dependence indicates a diffusion-controlled

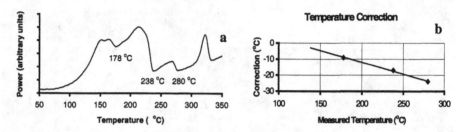

Figure 2. a) DSC scan of Pd/Ge/Pd multilayer film heated at 12 °C/min. b) Temperature correction plot as established by the DSC measurements.

mechanism as described by:

$$X^2 = 4 \cdot D_0 \cdot e^{-E_a/kT} \cdot t \qquad (2)$$

where X is the thickness, t is the annealing time, E_a is the activation energy, D_0 is a temperature independent coefficient, T is the temperature and k is the Boltzmann constant.

The kinetics of the solid state reactions were determined by measuring the isothermal grain growth rates associated with each phase from the videotape observations. By plotting $\ln[X(dX/dt)]$ vs. $1/kT$, one could obtain D_0 and E_a by measuring the intercept to the y axis and the slope, respectively. Figure 3 shows the Arrhenius plots associated with the evolution of both germanide phases. For both germanide phases, all the kinetic parameters were calculated from the Arrhenius plots and it was found that:

$X^2 = 2.94 \cdot 10^{15} \cdot e^{-0.98eV/kT} \cdot t$ describes the Pd_2Ge growth and

$X^2 = 6.89 \cdot 10^{17} \cdot e^{-1.28eV/kT} \cdot t$ describes the PdGe growth, where

X is the thickness in angstroms, T is the temperature in K and t is the time measured in minutes.

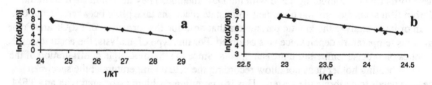

Figure 3. Arrhenius plots determined by measuring the isothermal grain growth of a) Pd_2Ge and b) PdGe

Based on the analysis presented above, the activation energy of Pd_2Ge formation was found to be 0.98 eV. In our previous study [5], a 1.12eV activation energy was calculated by using the DSC Kissinger plot method. Scott et al. [7] studied the kinetics of the Pd/c-Ge interactions and obtained a 1.08 eV activation energy for the growth of the Pd_2Ge hexagonal phase. Even though their analysis started with an initial single crystal Ge phase, as opposed to the amorphous Ge used in this study, the results show very close agreement. Ottaviani et al. [3] calculated an activation energy of 1.5 eV for the same thin film system by using backscattering spectroscopy and X-ray diffraction. Their study included a-Ge and c-Ge and the same activation energy was observed for both PdGe and Pd_2Ge. Our *in-situ* TEM findings for PdGe formation is 1.29 eV, which is significantly less than their value. The DSC analysis for PdGe [5], revealed a 1.33 eV activation energy for this phase growth, which agrees well with the *in-situ* TEM results.

In addition to the kinetic analyses of germanide growth, these *in-situ* TEM experiments allowed us to observe other features of the ohmic contact microstructure evolution. In the final stages of the ohmic contact formation, the Ge diffuses through the PdGe grain boundaries and grows epitaxially on top of the GaAs substrate. Although the exact role played by this Ge layer in the charge transport mechanism is not yet clearly established, there is some consensus that Ge transport to the GaAs surface provides an ohmic contact with low resistivity [1,8]. Since the Ge transport is achieved by grain boundary diffusion through the PdGe layer, it implies that the PdGe grain size would affect this transport mechanism. One way of controlling the shape and size of the PdGe grains is by modifying the annealing procedure of the ohmic contact. The videotape recordings demonstrated that the PdGe nucleates at the a-Ge:Pd_2Ge interface and then grows, primarily in a direction perpendicular to the interface, until the whole Pd_2Ge phase is consumed. In the first stage of the PdGe formation, nucleation takes place at the interface. The second stage is represented by a columnar grain growth and finally, in the third stage, PdGe coarsening combined with the formation of an equaxial grain structure. Clearly, the most favorable configuration for the Ge transport is given by the columnar grain structure because of the high grain boundary density. One way to achieve this microstructure is by a two step annealing process and Figure 4 clearly illustrates just that. The microstructure in Figure 4a) was obtained by a fast ramp rate followed by holding at 318 °C for 5 minutes, a typical annealing method employed in the integrated circuit fabrication line. In Figure 4b) the columnar PdGe grain microstructure resulted from a two step annealing procedure that consisted of fast ramping to 210 °C and holding for 5 minutes at this tempearture, followed by a second holding period at 320 °C for 5 minutes. The low temperature isothermal provided a high density of PdGe nuclei at

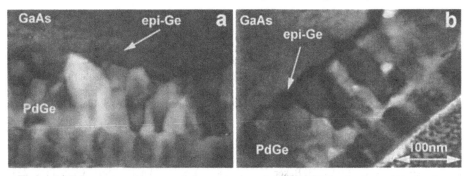

Figure 4. Video-captured images of a) sample annealed at 318 °C for 5 min. and b) sample annealed at 210 °C followed by fast ramp rate and hold at 320 °C for 5 min.

the a-Ge:Pd$_2$Ge interface without 2D coarsening while the second rapid ramp rate step contributed to the columnar growth of these nuclei. A rapid ramp rate anneal through the PdGe nucleation period renders a lower nuclei density, which in turn favors the subsequential lateral grain growth. This translates into a lower density of PdGe grain boundaries and a non-uniform Ge transport to the GaAs surface.

CONCLUSIONS

In this study, the solid state reactions that occur during the Pd-Ge ohmic contact formation on GaAs were studied by *in-situ* TEM analysis. These results agreed with the findings of earlier work on the phase formation sequence, and the activation energy values were in very close agreement with those determined by DSC. Furthermore, these data agreed with other studies presented in the literature that measured the activation energies through other indirect methods such as backscattering and x-ray techniques.

A new temperature calibration method for the *in-situ* TEM measurements was also presented in this study. The DSC information was correlated with distinct events that were easily observed in the *in-situ* TEM experiments. Because the temperature could be determined very accurately in the DSC cell, this technique provides a relatively precise method of calibration that could easily be adapted to study phase transformations in other systems.

As a result of these findings we demonstrated the possibility of controlling the Ge SPE growth and the morphology of the interface by employing a two step annealing procedure.

The use of FIB sample preparation method proved to be essential for obtaining electron transparent foils with appropriate thicknesses that minimized the effects of surface diffusion-driven reactions. Also, in comparison with more traditional techniques, this method provided a larger, more uniform cross-sectional area.

ACKNOWLEDGEMENTS

The authors would like to thank Dr. Andy MacInnes, Rick Morton and David Pye of TriQuint Semiconductor for their support. This work was funded by TriQuint Semiconductor and the State of Oregon. The work at NCEM was supported by the Director, Office of Science, Office of Basic Energy Sciences, Division of Materials Sciences of the U.S. Department of Energy under Contract No. DE-AC03-76SF00090.

REFERENCES

1. E. D. Marshall, B. Zhang, L. C. Wang, P. F. Jiao, W. X. Chen, T. Sawada, S. S. Lau, K. L. Kavanagh and T. F. Kuech, J. Appl. Phys. **62**, 942 (1987)
2. T. Sands, V. G. Keramidas, A. J. Yu, K-M. Yu, R. Gronsky and J. Washburn, J. Mater. Res. **2**, 262 (1987)
3. G. Ottaviani, C. Canali, G. Ferrari, G. Majni, M. Prudenziati and S. S. Lau, Thin Solid Films **47**, 187 (1977)
4. E.D. Marshall, S.S. Lau, C.J. Palmstrom, T. Sands, C.L. Schwartz, S.A. Schwarz, J.P. Harbison, and L.T. Florez, Mater. Res. Soc. Symp. Proc. **148**, 163 (1989)
5. F. Radulescu and J.M. McCarthy, J. Appl. Phys. **86**, 995 (1999)
6. D.P. Basile, R. Boylan, B. Baker, K. Hayes and D. Soza, Mat. Res. Soc. Symp. Proc. **254**, 23 (1992)
7. D. M. Scott, C. S. Pai and S. S. Lau, SPIE Conf. Proc. **463**, 40 (1984)
8. C.J. Palmstrom *et al.*, J. Appl. Phys. **67**, 334 (1990)

INCOHERENT HIGH-RESOLUTION Z-CONTRAST IMAGING OF SILICON AND GALLIUM ARSENIDE USING HAADF-STEM

Y. KOTAKA*, T. YAMAZAKI**, Y. KIKUCHI*, AND K. WATANABE***
*FUJITSU LABORATORIES LTD., Atsugi, Japan.
**Department of Physics, Science University of Tokyo, Tokyo, Japan.
***Tokyo Metropolitan College of Technology, Tokyo, Japan.

ABSTRACT

The high-angle annular dark-field (HAADF) technique in a dedicated scanning transmission electron microscope (STEM) provides strong compositional sensitivity dependent on atomic number (Z-contrast image). Furthermore, a high spatial resolution image is comparable to that of conventional coherent imaging (HRTEM). However, it is difficult to obtain a clear atomic structure HAADF image using a hybrid TEM/STEM. In this work, HAADF images were obtained with a JEOL JEM-2010F (with a thermal-Schottky field-emission) gun in probe-forming mode at 200 kV. We performed experiments using Si and GaAs in the [110] orientation. The electron-optical conditions were optimized. As a result, the dumbbell structure was observed in an image of [110] Si. Intensity profiles for GaAs along [001] showed differences for the two atomic sites. The experimental images were analyzed and compared with the calculated atomic positions and intensities obtained from Bethe's eigen-value method, which was modified to simulate HAADF-STEM based on Allen and Rossouw's method for convergent-beam electron diffraction (CBED). The experimental results showed a good agreement with the simulation results.

INTRODUCTION

High-angle annular dark-field scanning transmission electron microscopy (HAADF-STEM) is now an accepted high resolution imaging approach in materials research [1,2]. There are two distinctive characteristics of this method. One is the spatial resolution, which is comparable to high-resolution coherent imaging (HR-TEM) [1]. The other is the intensity of high-angle scattered electrons which increases in proportion to Z squared where Z is the atomic number. Therefore, this technique is called Z-contrast. With this technique, we obtain crystal structure and chemical composition information at the same time.

High resolution images of dumbbell structures in Si and GaAs have been observed with an HB603U, which is a dedicated STEM [2]. Therefore, the resolution reached 0.136 nm. The spatial resolution limit of a microscope for Z-contrast imaging depends on the objective lens spherical aberration coefficient (C_s), accelerating voltage and source brightness. For the HB603U, the C_s and accelerating voltage are 1 mm and 300 kV, respectively [2].

Recently, it was shown that the capability of a hybrid TEM/STEM approaches that of the dedicated STEM. The optical conditions, for example, convergence semi-angle, detector angle and collection angle, can be set up easily with the fluorescent screen in a hybrid TEM/STEM. However, it is not yet known well if we can change the lens conditions over a wide range [3].

The resolution of a JEM-2010F in the probe-forming mode was demonstrated to be 0.14 nm in HAADF-STEM images of Si single crystal when the C_s and accelerating voltage were 0.5 mm

185

and 200 kV, respectively [4]. The resolution reached 0.19 nm using a $SrTiO_3$ single crystal in a JEM-2010F with $C_s = 1.0$ mm [5]. For HR-TEM, the point resolution of coherent imaging is only 0.23 nm with $C_s = 1.0$ mm. Moreover, the distance between Si atoms dumbbell is elongated in the coherent phase contrast imaging. Therefore we could not obtain the exact structure image of Si and GaAs dumbbells [6][7].

It is the purpose of this study to confirm the Si dumbbell structure for high resolution HAADF-STEM imaging, and to identify atomic species of Ga and As in a GaAs single crystal through atomic-resolution Z-contrast imaging using a hybrid TEM/STEM.

EXPERIMENT

Experimental procedure

The specimens used in this experiment were a silicon single crystal for high-resolution imaging, and a gallium arsenide single crystal for atomic resolution imaging. The atomic numbers of Ga and As are 31 and 33, respectively. The specimens were thinned by using the cross-section Ar ion milling method. The [110] zone was observed in both experiments.

The HAADF-STEM images were obtained by a JEOL JEM-2010F with the $C_s = 1.02$ mm and thermal-Schottky field-emission gun in probe-forming mode at 200 kV. The incident electron beam conditions were determined by calculating the probe function [8]. The probe function indicates the effective electron beam spot size in HAADF-STEM imaging. The appropriate calculation parameters were determined by experimental measurement. The probe function is optimum when the convergence semi-angle is 11 mrad and the defocus value (Δf) is −51 nm, corresponding to the Scherzer focus. A schematic diagram of HAADF-STEM imaging is shown in figure 1. The detector position from inner (β_1) to outer (β_2) angles was determined by calculation of the elastic and inelastic scattering angles. Scattering factors by Weckenmeir and Kohl were used to calculate the real crystal potential and thermal diffuse scattering [9].

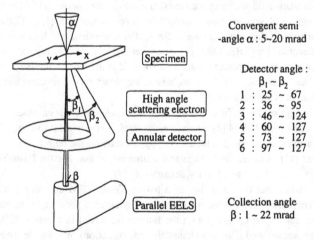

Fig. 1   Schematic diagram of HAADF-STEM imaging in this study.

The detector angle was set to prevent elastically scattered electrons reaching the annular detector. Detector positions of 36-95 and 73-127 mrad were used for Si, and 46-124 and 73-127 mrad for GaAs. The specimen thickness was measured by electron-energy-loss spectroscopy (EELS), taking into consideration the mean free path [10]. Image processing was made by fast Fourier transform filtering.

Calculation

In order to confirm the exact structure and atomic column height intensity, the experimental images were analyzed and compared with the atomic position and intensities calculated from Bethe's eigen-value method in which a wave function is written by three-dimensional Bloch waves. Image intensity was calculated using equation (1). Equation (2) gives a wave function of Bloch state, which is integrated into the probe. The lens aberration function is given by equation (3) [8].

$$I^{HA}(R_0) = \sum_{i-atom} \sigma_i \int_0^t \left[\left| \sum_j A^j (R_i - R_0, z) \right|^2 \right] dz \tag{1}$$

$$A^j (R_i - R_0, z) = \int_{probe} dK_\parallel \phi^j (R, z, K_\parallel) \cdot \exp[-iK_\parallel \cdot (R_i - R_0) + iW(K_\parallel)] \tag{2}$$

$$W(K_\parallel) = \frac{\Delta f \cdot |K_\parallel|^2}{2} + \frac{C_s \cdot K_\parallel^4}{4\chi^3} \tag{3}$$

The calculation was modified to simulate HAADF-STEM based on Allen and Rossouw's method for CBED [11]. Weickenmeier and Kohl's absorptive scattering factors for high angle scattering [9] were used in this calculation.

RESULTS

High-resolution imaging

Figure 2 shows a high resolution HAADF-STEM image of Si [110]. Clear Si dumbbell structure corresponding to (004) spacing 0.136 nm is exhibited. However, it is impossible to image such kind of Si dumbbell structure by HR-TEM with a C_s of 1.02 mm.

Figure 3 illustrates line profiles from the experimental image and also the simulated results along the [001] direction, depending on specimen thickness and detector positions. The dumbbell structure was observed with the detector angle setting at the position of 36-96 mrad except for the specimen with a thickness of 40 nm. When we set the detector position at 73-127 mrad, the dumbbell

Fig. 2   Z-contrast image along Si [110]

structure was only observed at a thickness of 70 nm. The intensity of simulated line profiles increased with increasing the specimen thickness at both detector positions. The structure image was dependent on the detector position.

Figure 4 compares experimental and simulated Si [110] images and line profiles of atomic column intensity along the [001] and [$\bar{1}$10]. The specimen thickness and detector position used in the experiments and calculation were 70 nm and 73-127 mrad, respectively. The atomic column positions of the experimental results agree with the simulated results. Thus we obtained the projected exact structural image of Si [110]. However, the FWHM of the peaks in the experimental image were broader than in the simulation, because the spot size of the electron probe actually used in the experiments was bigger than the effective beam spot size calculated by the probe function.

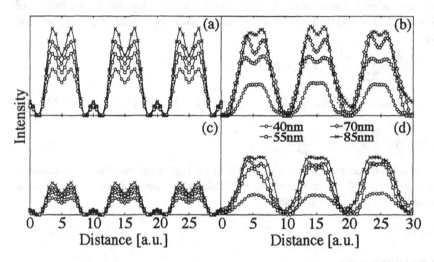

Fig. 3 Line profiles of atomic column intensity along Si [001] : (a) simulation, detector position is from 36 to 96 mrad ; (b) experiment, detector position is from 36 to 96 mrad ; (c) simulation, detector position is from 73 to 127 mrad ; (d) experiment, detector position is from 73 to 127 mrad.

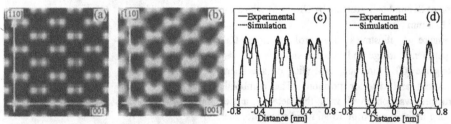

Fig. 4 Comparison of the experimental and simulated results : (a) simulated image along Si [110] ; (b) experimental image ; (c) line profile along [001] ; (d) line profile along [$\bar{1}$10].

Atomic-resolution imaging

Figure 5 shows line profiles of the experimental and simulated results along GaAs [001] depending on specimen thickness and detector position. The dumbbell structure was observed for both detector positions and specimen thicknesses of 55 and 70 nm. The simulated intensity increases with increasing specimen thickness. However, the experimental results did not depend on specimen thickness.

Figure 6 shows experimental and simulated images of GaAs [110] and line profiles showing their difference between gallium and arsenic. By comparing the line profiles of experimental and simulation, it is evident that the atomic positions of Ga and As in the experimental image fit to the simulation results. Furthermore, the atomic column intensity of experimental and calculated results are also in good agreement. Therefore, Ga and As in GaAs single crystal are distinguished in this experiment.

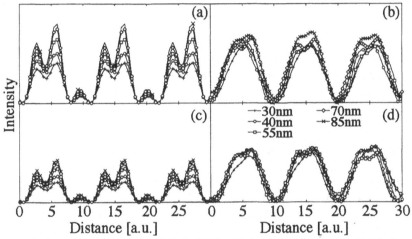

Fig. 5 Line profiles of intensity of GaAs [001] : (a) simulation, detector position is from 46 to 124 mrad ; (b) experimental, detector position is from 46 to 124 mrad ; (c) simulation, detector position is from 73 to 127 mrad ; (d) experimental, detector position is from 73 to 127 mrad.

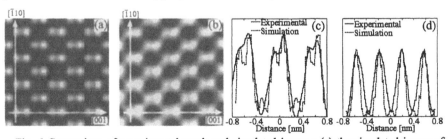

Fig. 6 Comparison of experimental result and simulated image : (a) the simulated image of GaAs [110] ; (b) the experimental result ; (c) the line profile along [001] ; (d) the line profile along [$\bar{1}$10].

CONCLUSIONS

We optimized the experimental conditions for HAADF-STEM imaging in a JEM-2010F, i.e. convergence semi-angle, detector position and specimen thickness. The dumbbell structure of Si was observed in high-resolution HAADF images. Furthermore, gallium and arsenic were distinguished in Z-contrast images. The experimental and simulated results showed good agreement in the effective thickness. Thus it is possible to identify the projected atomic species in HAADF-STEM imaging using a hybrid TEM/STEM.

ACKNOWLEDGMENTS

The authors would like to thank Dr. M. Kawasaki and Mr. K. Ibe of JEOL Japan for their advice on conducting this work, and T. Miyajima of FUJITSU LABORATORIES LTD. for his valuable discussions on experimental issues.

REFERENCES

1. P.D. Nellist and S.J. Pennycook, Phys. Rev. Lett. **81**, 4156 (1998).
2. P.D. Nellist and S.J. Pennycook, Ultramicroscopy **78**, 111-124 (1999).
3. E.M. James, N.D. Browning, A.W. Nicholls, M. Kawasaki, Y. Xin and S. Stemmer, J. Electr. Microsc. **47**, 561-574 (1998).
4. E.M. James and N.D. Browning, Ultramicroscopy **78**, 125-139 (1999).
5. M. Kawasaki, T. Yamazaki, K. Watanabe and M. Shiojiri, to be pubrished.
6. Y. Kikuchi, Philo. Mag. B**57**, 547-556 (1988).
7. K. Watanabe, Y. Kikuchi and H. Yamaguchi, phys. Stat. Sol. a**98**, 409-416 (1986).
8. S. J. Pennycook and P. D. Nellist, Ultramicroscopy **37**, 14-38 (1991).
9. Weckenmeir and Kohl, Acta Cryst. A**47**, 590-597 (1991).
10. R. F. Egerton, *Electron Energy Loss Spectroscopy in the Electron Microscope*, Plenum New York, 302-307 (1981).
11. L. J. Allen and C. J. Rossouw, Phys. Rev. B**39**, 8313-8321 (1989).

THE ATOMIC STRUCTURE OF MOSAÏC GRAIN BOUNDARY DISLOCATIONS IN GaN EPITAXIAL LAYERS

V. POTIN, G. NOUET*, P. RUTERANA* and R.C. POND**
Groupe d'Etudes des Semiconducteurs, Université de Montpellier II, Place Eugene Bataillon, 34095 Montpellier cedex 5, France.
*Laboratoire d'Etudes et de Recherches sur les Materiaux, UPRESA 6004 CNRS, Institut des Sciences de la Matière et du Rayonnement, 6 boulevard Maréchal Juin, 14050 Caen Cedex, France. ruterana@lermat8.ismra.fr
**Department of Materials Science and Engineering, the University of Liverpool, Liverpool L69 3 GH, England.

ABSTRACT

The studied GaN layers are made of mosaïc grains rotated around the c-axis by angles in the range 0-25°. Using high-resolution electron microscopy, anisotropic elasticity calculations and image simulation, we have analyzed the atomic structure of the edge threading dislocations. Here, we present an analysis of the $\Sigma = 7$ boundary using circuit mapping in order to define the Burgers vectors of the primary and secondary dislocations. The atomic structure of the primary ones was found to exhibit 5/7 and 8 atom cycles.

INTRODUCTION

Due to their large direct bandgap, which covers most of the visible spectrum, and due to their related promising application potential, the III-V nitride semiconductors have been extensively studied, leading to the fabrication of high brightness light emitting diodes and laser diodes [1,2]. In contrast with other semiconductors like gallium arsenide, for which the worst material that can be used has 10^4 defects cm^{-2}, these devices are made from layers which may contain a high density of defects (10^8-10^{10} cm^{-2})[3,4].

Among them, we have analyzed the atomic structure of the threading dislocations which form low and high-angle grain boundaries. They are due to the mosaic growth mode; they form at the coalescence of adjacent grains misoriented by rotation mainly about the c-axis in the range 0-25°. The interfacial structures can be described in terms of the coincidence site lattice (CSL). The CSL model is deduced by considering two identical interpenetrating lattices. If one lattice is rotated through a common lattice point, for certain angles, sites come into coincidence, forming the so-called coincidence lattice. Σ corresponds to the ratio of the volume of the CSL unit cell and the primitive lattice.

EXPERIMENTAL DETAILS

The GaN layers were grown on the (0001) sapphire surface by NH_3 gas source molecular beam epitaxy (MBE). The GaN layers were deposited at 800°C on top of a low-temperature (550°C) GaN buffer layer of 40 nm thickness. The plan view samples were prepared by mechanical grinding and ion milling. HREM experiments were carried out along the [0001] zone axis on a Topcon 002B microscope operating at 200 kV.

CRYSTALLOGRAPHY

In the topological theory of line defects in interfaces developed by Pond[5], it was shown that it is possible to determine all the admissible defects between crystallographically equivalent surfaces. In this formalism, one crystal is designated as white (λ) and the other as black (μ). The configuration formed by the interpenetrating lattices and crystals are called dichromatic pattern and complex, respectively[6]. The Burgers vectors of some admissible interfacial dislocations are given by:

$$b_{ij} = t\,(\lambda)_j - P\,t\,(\mu)_i$$

where $t\,(\lambda)_j$ and $t\,(\mu)_i$ represent the j-th and i-th translation vectors in the λ and μ crystals, respectively and P is a matrix which re-expresses $t\,(\mu)_i$ in the λ coordinate frame. The cores of such defects are associated with steps whose heights are given by $h\,(\lambda) = n_\lambda \cdot t\,(\lambda)$ and $h\,(\mu) = n_\mu \cdot t\,(\mu)$ where n are unit vectors normal to the interface, orientated towards the λ crystal.

These interfacial defects can be characterized in HREM images by circuit mapping as proposed by Pond[7]. A closed circuit is constructed on the micrograph with segments $c\,(\lambda)$ and $c\,(\mu)$ in the λ and μ crystals, respectively. When the circuit is mapped into a reference space, any closure failure is equal to the total defect content and, using the RH/FS convention, the defect is given by $c\,(\lambda, \mu)^{-1}$. If the circuit is mapped into a single crystal, the primary dislocations are exhibited, whereas in the dichromatic pattern, secondary dislocations are determined. In the simplest case, we obtain

$$b_{ij} = -c\,(\lambda, \mu) = -c\,(\lambda)_j - P\,c\,(\mu)_i.$$

RESULTS

The GaN layers are made of mosaïc grains rotated around the c-axis by angles in the range of zero to 25°. In the best layers, the misorientation is less than 1° and the size of the grains can reach several hundreds of nanometers[8]. These low-angle grain boundaries are made of threading dislocations which can be considered as isolated. Their atomic structure has been analyzed and two configurations have been reported[9]. The first one is made of one cycle with 8 atoms and the second one of two neighboring cycles of 5 and 7 atoms (fig. 1).

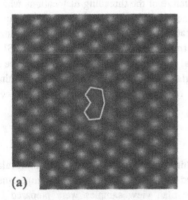 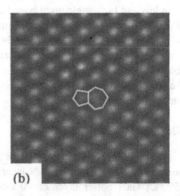

(a)　　　　　　　　　　　　　　　　　(b)

Fig. 1 : Atomic configurations for the 1/3 <11$\overline{2}$0> edge threading dislocations observed along [0001]: a) 8 and b) 5/7 atom cycles.

Besides these low-angle grain boundaries, high-angle grain boundaries are present. Adjacent grains are rotated around [0001] and the interfacial structures can be described in terms of coincidence site lattices (CSL): $\Sigma = 7$ ($\theta = 21.79°$), $\Sigma = 19$ ($\theta = 13.17°$), $\Sigma = 31$ ($\theta = 17.90°$)[10,11]. Here, we present an analysis of the $\Sigma = 7$ boundary using circuit mapping in order to define the Burgers vector of the dislocations. The diffraction pattern corresponding to the studied grain boundary indicates a misorientation equal to $\theta = 22° \pm 0.2°$ and very close to that of $\Sigma = 7$ in the CSL notation. The grain boundary is symmetric and we note that one step is present (fig. 2a). The white spots correspond to the tunnels at a defocus value close to -24 nm; the estimated thickness is 6 nm.

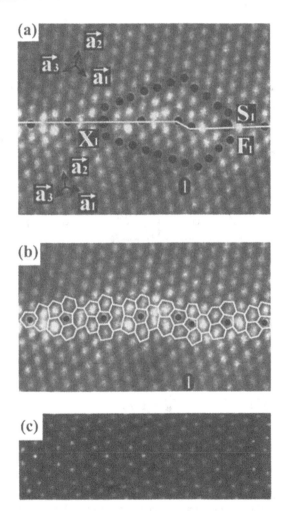

Fig.2: Symmetric $\Sigma = 7$ boundary where a step is present. a) the periods are defined by the black dots along the interface ; b) reconstruction of the same grain boundary made of 5/7 and 8 atom cycles ; c) simulated image (htickness: 6 nm, defocus: -24 nm).

A circuit corresponding to one period gives a closure failure of $\bar{a}_2 = 1/3 \, [\bar{1} \, 2 \, \bar{1} \, 0]$ when reported in the λ crystal. This circuit corresponds to $c \, (\lambda) = -2 \, \bar{a}_1 + \bar{a}_3 - \bar{a}_2 = 1/3 \, [\bar{4} \, \bar{1} \, 5 \, 0]_\lambda$ and $c \, (\mu) = - \, \bar{a}_2 + \bar{a}_1 - 2 \, \bar{a}_3 = 1/3 \, [5 \, \bar{1} \, \bar{4} \, 0]_\mu$. The normal to the interface was determined by making the cross product of $c \, (\lambda)$ by [0001] and [0001] by $c \, (\mu)$, leading to $n_\lambda = [\bar{2} \, 3 \, \bar{1} \, 0]_\lambda$ and $n_\mu = [\bar{1} \, 3 \, \bar{2} \, 0]_\mu$. Therefore, the interface plane is along $(\bar{2} \, 3 \, \bar{1} \, 0)_\lambda \, / \, (\bar{1} \, 3 \, \bar{2} \, 0)_\mu$: it corresponds to a side of the CSL unit cell of $\Sigma = 7$ (fig. 3a).

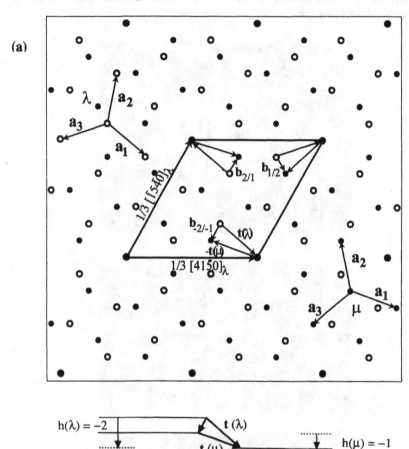

Fig. 3: a) Dichromatic pattern corresponding to $\Sigma = 7$ in the CSL notation; the unit cell of the CSL has been drawn. The smallest translation vectors $t(\lambda)$ and $t(\mu)$ and the corresponding Burgers vector $b_{-2/-1}$ are drawn. b) Schematic illustration of the step in fig.2.

If a circuit is mapped into the dichromatic pattern rather than into the lattice of a single crystal, secondary dislocations can be determined. For the circuit $S_1X_1F_1$, with $c\ (\lambda) = S_1X_1 = 1/3\ [\overline{1}4\ \overline{2}\ 16\ 0]_\lambda$ and $c\ (\mu) = X_1F_1 = 1/3\ [17\ \overline{4}\ \overline{13}\ 0]_\mu$, we calculate $b = -c\ (\lambda, \mu) = -c\ (\lambda) - P_7\ c\ (\mu) = 1/21\ [1\ \overline{5}\ 4\ 0]_\lambda$. P_7 is the 4×4 matrix which reexpresses $c\ (\mu)$ in the λ coordinate frame:

$$P_7 = \frac{1}{21}\begin{pmatrix} 20 & \overline{4} & 5 & 0 \\ 5 & 20 & \overline{4} & 0 \\ \overline{4} & 5 & 20 & 0 \\ 0 & 0 & 0 & 21 \end{pmatrix}$$

This Burgers vector corresponds to an interfacial edge dislocation with the smallest magnitude in the dichromatic pattern and its angle with the interface plane is 60°. The step heights are $h(\lambda) = -n_\lambda \cdot c\ (\lambda)\ /\ d(\overline{2}\ 3\ \overline{1}\ 0)_\lambda = -2$ and $h(\mu) = n_\mu \cdot c\ (\mu)\ /\ d(\overline{1}\ 3\ \overline{2}\ 0)_\mu = -1$ leading to a $b_{-2/\ -1}$ interface dislocation. This step relates equivalent surfaces in the λ and μ crystals by translations $t(\lambda)$ and $t(\mu)$ whose difference is equal to the Burgers vector of the secondary dislocation. In order to characterize completely the defect, it is necessary to determine the values of these translations. In theory, a multiplicity of configurations is possible because translation vectors of the CSL unit cell can be added to $t(\lambda)$ and $t(\mu)$ without modifying the Burgers vector. However, the resulting step configuration is different and only a comparison with the experimental image would allow determining the exact configuration of the defect. For Burgers vector $b_{-2/\ -1} = 1/21\ [1\ \overline{5}40]_\lambda$, the associated smallest translations vectors are $t\ (\lambda) = \bar{a}_1 = 1/3\ [2\overline{1}\ \overline{1}\ 0]_\lambda$ and $t\ (\mu) = \bar{a}_1 = 1/3\ [2\overline{1}\ \overline{1}\ 0]_\mu$, leading to the same step heights (fig. 3a). Thus, these translation vectors are compatible in theory. A diagram reproducing the configuration of the step with these values of $t(\lambda)$ and $t(\mu)$ has been compared to the experimental image (fig. 3b and 2a). They are consistent: the interface is deviated downwards and the two terraces are connected by a facet equal to $\bar{a}_1$.

In order to reconstruct the core of the primary dislocations, we have used the configurations generated for the analysis of the isolated threading dislocations (fig. 1). A careful examination of the boundary shows that different contrasts are exhibited: 5/7 atom cycles are present as well as 8 atom cycles. The whole boundary can be regarded as being composed of these two configurations, as shown in fig. 2b. A good agreement is observed between the experimental and the simulated images (fig. 2b and c).

SUMMARY

A $\Sigma = 7$ GaN grain boundary has been analyzed using circuit mapping. In this boundary, we have identified the periodic structure of the $\Sigma = 7$ symmetric interface ; it is based on only one 1/3 <11$\overline{2}$0> dislocation core. One secondary dislocation has been completely analyzed and the corresponding step has been reconstructed in agreement with the experimental image. Two configurations made of 5/7 and 8 atom cycles have been observed, just as for isolated threading dislocations.

REFERENCES

1. S. Nakamura, M. Senoh, S. Nagahama, N. Iwasa, T. Yamada, T. Matsushita, M. Kiyoku and Y. Sugimoto, Jpn. J. Appl. Phys. **35**, L74 (1996).
2. S. Nakamura, M. Senoh, S. Nagahama, N. Iwasa, T. Yamada, T. Matsushita, M. Kiyoku, Y. Sugimoto, T. Kozaki, H. Umemoto M. Sano and K. Chocho, Appl. Phys. Lett. **6**, 832 (1998).
3. A.H. Herzog, D.L. Keune and M.G. Craford, J. Appl. Phys. **43**, 60 (1972).
4. V. Potin, P. Vermaut, P. Ruterana and G. Nouet, J. Electron. Mater. **4**, 266 (1998).
5. R.C. Pond, "Dislocations in solids", Vol. **8**, Ed F.R.N. Nabarro, North Holland Publ. Co., Amsterdam/ New York, p. 1 (1989).
6. R.C. Pond and D.S. Vlachavas, Proc. Roy. Soc. London A **386**, 95 (1983).
7. R.C. Pond, Interface Science **2**, 1 (1995).
8. P. Ruterana , V. Potin and G. Nouet, Mat. Res. Soc. Symp. Proc. Vol. **482**, 459 (1998).
9. V. Potin, P. Ruterana and G. Nouet, Mat. Sci. Eng. B **50**, 29 (1997).
10. V. Potin, P. Ruterana, G. Nouet and R.C. Pond, Phys. Stat. Sol. B **216**, (1999).
11. V. Potin, P. Ruterana, G. Nouet, R.C. Pond and H. Morkoç, Phys. Rev. **B**, in press

INTERFACE STRUCTURE AND Zn DIFFUSION IN THE Cd Te/ZnTe/Si SYSTEM GROWN BY MBE

S.-C. Y. Tsen*, David J. Smith*, P. A. Crozier*, S. Rujirawat**, G. Brill**, S. Sivananthan**
*Arizona State Univ, Center for Solid State Science, Tempe, AZ.
**Univ. of Illinois at Chicago, Microphysics Laboratory, Department of Physics, Chicago, IL.

ABSTRACT

Two CdTe/ZnTe/Si samples were grown on Si (211) and Si(111) substrates by molecular beam epitaxy for use as lattice-matched substrates for HgCdTe growth. An As precursor was used before growing ZnTe. In order to understand the interface structure, the Zn diffusion problem and any effect due to As passivation, high resolution bright-field images, dark-field images and energy-dispersive X-ray spectra were studied. Fourier-filtered images have revealed details of the atomic arrangements and the dislocation defects at the interface. Local lattice parameters such as (111) d-spacings at different distances from the interface were measured to determine the Zn concentration based on Vegard's law. The Zn concentration profiles were consistent with measurements from energy-dispersive X-ray spectroscopy. The As-passivated interface showed vacancy-type defects.

INTRODUCTION

Substrates of CdTe/Si have been developed for use as alternatives for growth of HgCdTe hybrid infrared focal-plane arrays. CdTe(111)B (Te termination) grown on Si(001) has the advantage of requiring the least Hg flux for the growth of HgCdTe. However, the formation of twinning defects in this growth direction is a serious problem. Recent research has focused on growing CdTe(211)B on Si(211) since it is less sensitive to the formation of microtwins.

It has been reported that by depositing a thin buffer layer, with certain annealing conditions, the defects can be confined to within 0.1 μm of the interface and the orientation of the substrate can be preserved[1]. In this study, two CdTe/ZnTe/Si samples grown on either Si(211) or Si(111) substrates were studied. The best CdTe(111)B grown on As-passivated Si(111) with etch-pit-density of 2×10^{5} cm^{-2} was found to be better than the best CdTe(111)B film grown directly on vicinal Si(001) substrates.

The use of ZnTe thin buffer layers and As passivation can play important roles in affecting the quality of the film[1, 2]. An interface model with the outermost layer of Si atoms substituted by As atoms has been proposed [2]. The diffusion of Zn was also reported [3]. However, the interface structure and the Zn diffusion profile have not been characterized completely. It is essential to further investigate on these topics in order to optimize the conditions for growing better quality films.

EXPERIMENTAL DETAILS

Film Growth

The CdTe/ZnTe/Si samples studied here were grown by molecular beam epitaxy (MBE) in a Riber OPUS 45 MBE system at the Microphysics Laboratory. Reflection-high-energy electron diffraction (RHEED) at 10 keV was used *in situ* to monitor the growth. Three-inch Si(111) substrates (or Si(211) in the case of sample B) were cleaned by a modified RCA process. The substrate was preheated to around 400°C in a separate preparation chamber prior to growth.

Oxide removal was achieved by heating to ~850°C while monitoring with RHEED until a clean unreconstructed surface appeared. High purity sources of solid CdTe, ZnTe, As, and Te were used for growth. After oxide removal, the wafer was then cooled down under As$_4$ flux (partial pressure ~5x10^{-7} torr) to ~450 °C, which resulted in a Si(111):As 1x1 surface. The growth sequences for two particular samples identified as A and B are summarized in the following :

Sample A: CdTe/ZnTe/Si (111)

A thin ZnTe buffer layer was deposited at ~220°C to a thickness of ~200Å. The ZnTe buffer was then annealed to ~310°C under Te$_2$ (partial pressure ~5x10^{-7} torr) and CdTe fluxes for 10 minutes and a weak surface reconstruction was visible, indicating the smoothness of the single-domain (111)B surface. The wafer was then exposed to Te$_2$ flux for 5 minutes, and the growth of CdTe was initiated at ~310°C using a stoichiometric flux. The growth rate was about 1.0 μm/h.

Sample B : CdZnTe/CdTe/ZnTe/Si(211)

A ZnTe buffer layer was deposited at ~220°C to a thickness of ~70Å. It was annealed to ~240°C, then 310 °C under Te$_2$ and ZnTe fluxes for 30 s and 90 s. The CdZnTe layer was deposited on a ~2μm CdTe buffer layer which was grown at 310 °C. Initially, the CdZnTe was graded from x=0 to x~4% over about 1 μm. Then, CdZnTe was deposited with x~4% to a thickness of ~5μm.

High-Resolution Electron Microscopy

Cross-sectional TEM specimens were prepared by the standard rod and tube method. The specimen preparation procedure involved mechanical polishing to a thickness of about 110~150 μm, followed by dimpling to about 20 to 50 μm, and argon ion-beam milling to perforation using a liquid-nitrogen-cooled sample stage at an angle 15° with a voltage of 5 kV. The samples were finished with 12° and 1.5~2 kV to remove the amorphous surface layer.

TEM observations were made with a JEM-4000EX microscope having an interpretable resolution limit of about 1.6Å at the operating voltage of 400 kV. Typical observation directions are those which are perpendicular to Si{110} planes. Images were recorded at close to the optimum objective lens defocus (- 480Å).

Both bright-field and dark-field images of the interface region were studied. The contrast of high-resolution bright-field images was enhanced by Fourier filtering to reveal the atomic arrangement at the interface. The analysis of (002) and (004) dark-field images was based on the fact that the structure factors for (002) or (004) reflections are given, respectively, by a difference or sum of the scattering factors of the elements [4].

Energy Dispersive X-ray Spectroscopy

High spatial resolution energy-dispersive x-ray spectroscopy (EDX) was performed with a VG HB501 scanning transmission electron microscope (STEM) operating at 100 kV with a windowless Link EDX detector. This instrument is equipped with a field-emission gun and is capable of forming small electron probes approximately 1 nm in size. The STEM is interfaced to an EMiSPEC digital image and spectral acquisition system. The EMiSPEC system was used to acquire a series of spectra from the Si/ZnTe/CdTe layers. Up to 18 spectra were recorded in 100Å steps using an acquisition time of 5 s per spectrum. This acquisition time was long enough to give spectra with adequate counting statistics and at the same time short enough to minimize specimen drift during the acquisition of the series.

Digital Image Processing

Continuous areas along the growth direction in one micrograph were digitized using a light table and CCD camera (with a pixel size about 0.275 Å). Several small areas (~17.6Å x 17.6Å) were selected on each digitized image and analyzed using Gatan's DigitalMicrograph program with a script to measure the local lattice parameters. The script essentially takes a Fourier transform of a selected region of the image, generates a digital diffractogram and determines the lattice spacing and relative angle between the lattice planes. A reference micrograph with d-spacing of CdTe (111) planes of 3.74Å was used to give an independent calibration for the lattice spacing determination.

EXPERIMENTAL RESULTS AND DISCUSSION

1. Zn Diffusion

The EDX results of the elemental profiles for Cd, Zn and Te across the Si/ZnTe/CdTe interfaces have indicated the interdiffusion of Cd and Zn [3]. If we assume X as the Zn concentration in the formula $Cd_{1-x} Zn_x Te$, and take Y as the local lattice constant, which can be calculated from d(111) spacings measured by using DigitalMicrograph program, then, from the equation $Y = -0.376X+6.477$, based on Vegard's law, we can obtain the Zn concentration for different distances from the interface. Figure 1 shows the resulting Zn concentration profiles.

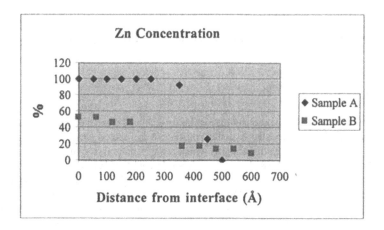

Fig. 1 Zn concentration profiles for sample A and B obtained from high-resolution images.

For sample A, the concentration of Zn is 100% from the interface region up to 250Å. This indicates that pure ZnTe exists around the interface, although the Zn concentration may vary between 96% to 100% from place to place horizontally. Annular dark-field images and bright-field images both gave the thickness of ZnTe to be around 250Å. Between 250Å and 500Å the Zn concentration drops continuously to zero. In sample B, around the interface region, the concentration of Zn is about 50%. At the thickness around 600Å , the concentration of Zn has dropped to ~10%. By comparing the Zn concentration profiles for samples A and B, it appears that the variation may be due to differences in ZnTe thickness and the growth directions, since the growth temperatures for ZnTe and CdTe are similar in both cases.

Figure 2 is a Zn Kα line intensity profile for sample A generated by scanning a nanometer sized probe over the interface and acquiring EDX spectra every 100Å. The scan was started before the interface and the approximate start of the interface is taken as the zero point. The profile shows a strong Zn signal rising up and peaking in the middle of the ZnTe layer. The detailed shape of the EDX profile is affected by local thickness variations and by spurious x-ray scattering from regions outside the primary irradiated area. However, despite these problems, Figure 2 suggests that Zn has diffused up to at least 400Å from the start of the interface and this profile is in reasonable agreement with the profile determined from Vegard's law (Figure 1).

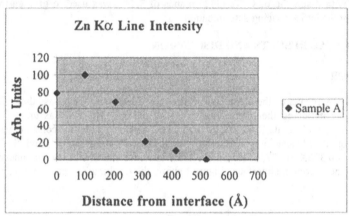

Fig. 2 Zn Kα line intensity profile for sample A obtained by EDX measurements.

2. Interface Structure

A. Dislocation

Figure 3a shows edge dislocations with regular intervals of 6 and 11 (111) d-spacings of Si in sample A. This result indicates that the interface was relaxed from a nonuniform strain condition. Figure 3b shows edge dislocations with a regular interval of 6 to 7 (111) d-spacings of Si in sample B. The spacing (~18.8Å) corresponds to a complete relaxation of the mismatched strain between the d-spacings of Si(111) (3.135Å) and ZnTe(111) (3.52Å) which is about 12.4%. However, the d(111) spacing measured near the interface for the ZnTe side was 3.63Å which corresponds to 51% Zn. This result indicated that the interdiffusion of Cd and Zn happened after relaxation of the strain at the interface.

B. As passivation

Based on data from energy dispersive X-ray spectroscopy, it is clear that As is present at the ZnTe/Si interface [3, 5]. We have studied both bright-field and dark-field images around the interface region for sample A.

Figure 4a shows one selected area of a 500KX image for sample A which has brighter contrast along the interface. The Fourier-filtered image shown in Figure 4b using 6 diffracted beams (no center beam) has been enlarged in three areas. In area 1 the atomic stacking across the interface is ABC type. If As is present in this region, it should fit into the zincblende structure. Area 2 and 3 both show vacancy -type defects which may relate to the use of As$_4$ flux.

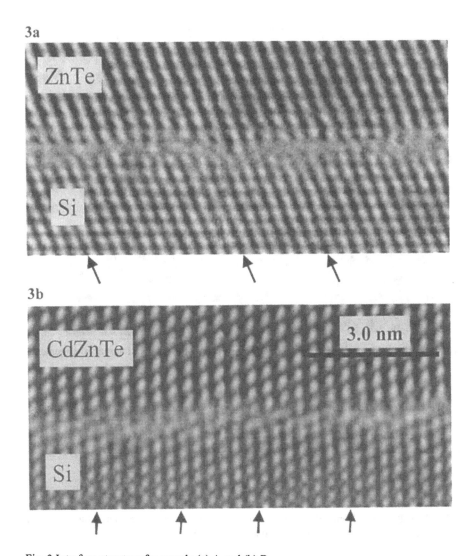

Fig. 3 Interface structure for sample (a) A and (b) B.

Although not shown here, the (002) and (00-2) dark-field images both showed dark contrast at the interface. Additionally, the (00-4) dark-field images also showed darker contrast. There are two possible explanations for this situation; one is vacancy defects, which were confirmed by the bright-field image with a clean crystalline interface; the other is a different crystal structure with smaller structure factors. Based on the d-spacing measurements in our study, the orthorhombic $SiAs_2$ and As oxides (such as As_2O_3) could be possible phases in some areas with visible interfacial layers (some areas showed a thicker disordered structure which may be due to the incomplete removal of Si oxides). Further investigations using other techniques are needed in order to explore the second possible explanation.

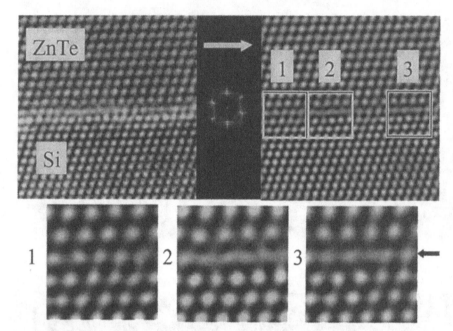

Fig. 4 Fourier-filtered image and three enlarged areas to show the details of atomic arrangement.

CONCLUSIONS

We have reported a new technique for studying the local Zn concentration profile in CdTe/ZnTe/Si samples. The result is supported by independent EDX measurements.

High-resolution bright-field and dark-field images have indicated that the As-passivated Si surface has introduced a vacancy-type defect to the crystalline interface.

ACKNOWLEDGMENTS

This work is supported by the Air Force Research Laboratory under Contract F29601-98-C-0053 and a subcontract from EPIR Ltd. Electron microscopy was carried out in the Center for High Resolution Electron Microscopy at Arizona State University.

REFERENCES

1. G. Brill, S. Rujirawat and S. Sivananthan, Private communication.
2. S. Rujirawat, Y. Xin, N. D. Browing, S. Sivananthan, David J. Smith, S.-C. Y. Tsen, Y. P. Chen and V. Nathan; Appl. Phys. Lett. **74** (1999) pp. 2346-2348 .
3. David J. Smith, P. A. Crozier, S.-C. Y. Tsen, S. Rujirawat, G. Brill, and S. Sivananthan; to be published.
4. L. H. Kuo, K. Kimura, T. Yasuda, S. Miwa, C. G. Jin, K. Tanaka and T. Yao; Appl. Phys. Lett. **68** (1996) 2413.
5. Y. Xin, S. Rujirawat, G. Brill, N. D. Browning, S. J. Pennycook, and P. Sporken; Microsc. Microanal. **5** (Suppl2: Proceeding), 1999 pp. 724-725.

RELATIONSHIP BETWEEN STRUCTURE AND LUMINISCENT PROPERTIES OF EPITAXIAL GROWN Y$_2$O$_3$:Eu THIN FILMS ON LaAlO$_3$ SUBSTRATES

H-J. GAO, G. DUSCHER, X.D. FAN, and S.J. PENNYCOOK, D. KUMAR*, K.G. CHO*, P.H. HOLLOWAY*, R.K. SINGH*
Solid State Division, Oak Ridge National Laboratory, Oak Ridge, TN 37831-6030
*Department of Materials Science and Engineering, University of Florida, Gainesville, FL32611-6400

ABSTRACT

Cathodoluminescence images of individual pores have been obtained at nanometer resolution in europium-activated yttrium oxide (Y$_2$O$_3$:Eu) (001) thin films, epitaxially grown on LaAlO$_3$ (001) substrates. Comparison with Z-contrast images, obtained simultaneously, directly show the "dead layer" to be about 5 nm thick. This "dead layer" is the origin of the reduced emission efficiency with increasing pore size. Pore sizes were varied by using different substrate temperatures and laser pulse repetition rates during film growth. These films are epitaxially aligned with the substrate, which is always Al terminated.

INTRODUCTION

Doped Y$_2$O$_3$ (YO) thin films are of major interest for electroluminescence device applications.[1-12] A detailed microscopic understanding of the effect of specific defects on luminescence and their correlation with growth conditions is highly desirable for maximizing luminescent efficiency. Two major sources of defects can be identified; first, threading dislocations nucleated at the film/substrate interface, and second, voids distributed throughout the active film thickness. In this paper, we therefore first review a recent atomic-resolution study of the YO/LaAlO$_3$ interface structure,[13] then demonstrate the relationship between pore structure and the luminescent properties. Using Z-contrast scanning transmission electron microscopy (STEM)[14], we directly observe the nanometer-scale cathodoluminescence (CL) of the film, and show that the reduction in luminescent efficiency is due to a "dead layer" around each pore caused by strong recombination of electron-hole pairs at the internal pore surfaces.

Eu-activated YO thin films were deposited by laser ablation[15-17] on (001) LaAlO$_3$ (LAO) substrates. For details see ref.18. Rocking curve measurements indicate a full width half maximum (FWHM) of 0.1°. Cross-sectional slices were obtained by cutting the LAO along the [100] or [010] directions (using pseudocubic indexing) and then gluing face to face in the usual way. Both plan-view and cross-section specimens were prepared for transmission electron microscopy (TEM) and/or STEM observations by mechanical grinding, polishing, and dimpling, followed by Ar-ion milling.

TEM bright field images and electron diffraction patterns were recorded in a Philips EM-400 electron microscope operated at 100kV. High-resolution Z-contrast imaging was conducted in a VG HB603U STEM at 300kV,[14,19] while the cathodoluminescent (CL) imaging was carried out in a VG HB501 STEM at 100 kV. The CL emission was collected by a lens system and detected by a photomultiplier, as shown in the schematic of Fig. 1.

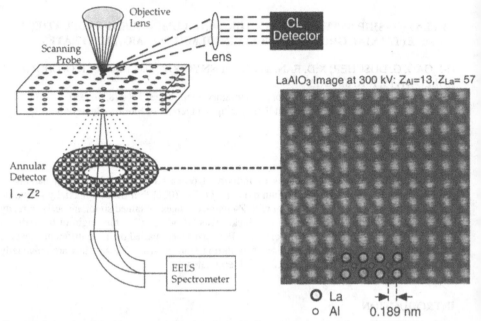

LaAlO₃ Image at 300 kV: $Z_{Al}=13$, $Z_{La}= 57$

O La
o Al   0.189 nm

Fig. 1 Schematic of the Z-contrast scanning transmission electron microscope (STEM) imaging together with the cathodoluminescent imaging system

FILM/SUBSTRATE INTERFACE STRUCTURE

It is well known that Y_2O_3 has a C-type rare-earth sesquioxide structure, closely related to the fluorite structure with a cell parameter a=1.060 nm and space group $T_h^7(Ia3)$.[20-23] In the fluorite lattice, each cation is surrounded by eight anions located at the corner of a cube. The C-type structure is derived by removing one-quarter of oxygen atoms and slightly rearranging the remaining ones.[24] For 75% of the cations the vacancies lie at the ends of a face diagonal, while for the other 25% they lie at the ends of a body diagonal. Therefore, each yttrium atom is surrounded by only six oxygen neighbors forming two different types of distorted octahedral structure in the unit cell, called S_6 and C_2.[25] Eight yttrium atoms have the S_6 symmetry and the other 24 atoms have the C_2 symmetry. From the crystallographic structure one can deduce that the distance of two neighboring Y atoms along the <100> direction of YO is 0.5302 nm, and along the <110> direction is 0.375 nm. LAO is a rhombohedral structure with lattice parameters $a = 0.378$ nm, $\theta \approx 90.5°$, very close to a cubic structure. The lattice mismatch with the <110> direction of the YO is therefore less than 0.8%, and so we would anticipate epitaxial growth of single crystalline YO thin films on the LAO (001) substrate to be feasible.

Figure 2 is a low magnification TEM micrograph and corresponding selected area diffraction (SAD) pattern of a plan view sample of the YO:Eu thin film. The diffraction pattern indicates an almost perfect single crystal film, but the image shows numerous small voids, suggesting an island growth mechanism with incomplete coalescence of the islands. An image of the cross section sample is presented in Figure 3 (a), showing the smooth surface, sharp interface and a uniform thickness of 300 nm maintained over the entire region. Figure 3 also shows SAD patterns of the film (b) and the LAO substrate (c), showing the orientation relationship to be [110]YO//[100]LAO and [-110]YO//[010]LAO. The columnar structure of the film is also apparent from the cross section image, with small rotations between neighboring grains giving the strong diffraction contrast. Each individual column however appears to be a good single crystal, which implies that the presence of the voids may avoid the need for a high density of dislocations between the grains to accommodate the rotations, and/or a high level of stress within the grains. The dominant direction of the voids is not crystallographic, suggesting that it is related to the deposition direction not being normal to the substrate.[26] The sample was not rotated during film deposition.

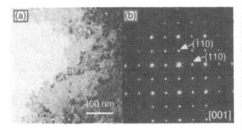

Fig. 2 Plan view TEM image (a) and corresponding electron diffraction pattern (b) of a laser-ablation-deposited YO:Eu thin film, showing the formation of a good single crystalline film containing numerous voids. The electron projection is along the [001] zone axis of the YO.

Fig. 3 Cross section TEM image (a) and corresponding SAD patterns of the as-grown YO:Eu film (b) and the LAO substrate (c) showing the orientation relationship to be [110]YO//[100]LAO and [-110]YO//[010]LAO.

In order to determine the detailed interfacial atomic structure, high-resolution Z-contrast STEM imaging of the samples was carried out. The Z-contrast image is a direct image with intensity highly localized about the atomic column positions and approximately proportional to the mean square atomic number (Z). Thus the La and Al columns in LAO, and the Y columns in YO, are directly distinguishable in a Z-contrast image taken along the [010] zone axis of the LAO substrate. Figure 4(a) is an atomic resolution Z-contrast STEM image of the film/substrate interface. The bright spots in the film are Y columns; in the substrate the brightest spots La columns, and the less-bright spots are Al columns. The O columns are not visible. Also shown in Fig. 4(b) is a higher magnification Z-contrast image that shows clearly the atomic structure of the interface. The substrate is seen to terminate with the Al plane, which matches directly onto the Y layer of the film as shown in the schematic.

Figure 5 is a schematic of two possible interface structures corresponding to the two possible terminations of the (001) substrate, either the $(AlO_2)^-$ or $(LaO)^+$ planes. Figures 5 (c) and (d) show these two planes, while (e) and (f) show the two (001) planes of YO, comprising pure Y and O. The full unit cell of the YO structure is four times the dimensions shown due to ordering of the O vacancies. For $(AlO_2)^-$ termination of the substrate, the four oxygen positions match almost exactly the oxygen positions in the YO. The Y atoms can sit over the center of the four O positions in Fig. 5(d), directly over the La site in the plane below, as seen in the image of Fig. 4(b). The interfacial Y is then coordinated by seven oxygen atoms instead of six, which may be compensated by some additional oxygen vacancies. In contrast, if the substrate is terminated by LaO, each Y sitting directly over one oxygen in the LaO plane, then each Y is substantially undercoordinated. This explains the observed termination and the fact that no single layer height steps were observed.

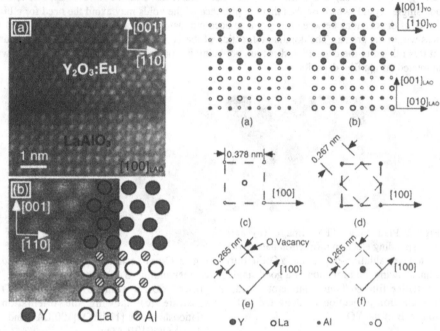

Fig. 4 (a) Z-contrast STEM dark field image showing the atomically abrupt interface, (b) higher magnification image showing clearly the Al terminated substrate, as shown in the schematic.

Fig. 5 Schematic interface structures for YO on the Al terminated surface (a), and on the La terminated surface of LAO (b). Atomic arrangements on the substrate (001) surface are shown for La termination (c), and Al termination (d). The atomic arrangements on the alternating oxygen and Y {004} planes in YO are shown in (e) and (f).

CORRELATION OF MICROSTRUCTURE TO LUMINESCENCE PROPERTIES

The variation of film microstructure with growth conditions is shown in Fig. 6. The size of the pores increases with lower substrate temperatures and higher pulse rates. The photoluminescent efficiency correlates directly with average pore size, as shown in Fig. 7, and we would expect the CL efficiency to be similar. The reason for this correlation is expected to be the existence of a "dead layer" near the specimen surface, resulting from non-radiative recombination of electron-hole pairs via surface states. Increasing pore size would create more internal surfaces per unit volume, resulting in lower overall efficiency.

Direct measurement of the "dead layer" at the surface of a pore has been achieved by comparing the Z-contrast image with the CL image obtained simultaneously. Fig. 8 (a) shows a Z-contrast image of a region of film containing some relatively widely spaced pores. The corresponding CL image (b) shows the variation in CL intensity. It is immediately clear that the edges of the holes in the CL image are much less sharp. Intensity profiles across the hole marked are shown in Fig. 8(c) and (d). The width of the pore in the Z-contrast image is 10 nm, whereas in the CL image it is more than doubled to 20 nm. Clearly if the grain size of a film is comparable to the extent of the dead layer, then the efficiency will be substantially reduced. This explains the order of magnitude reduction in efficiency for the sample shown in Fig. 6(a), in which the grain size is ~30 nm. It is clear that because of this "dead layer," porous structures can lead to greatly reduced emission efficiencies.

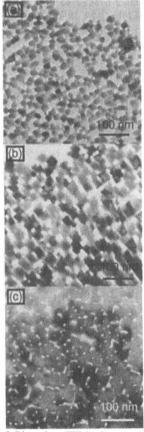

Fig. 6. Plan view TEM micro-graphs of Y_2O_3:Eu thin films grown at 735 °C (a, b) or 775 °C (c), with laser pulse frequency 10 Hz, (a,c) or 5 Hz (b).

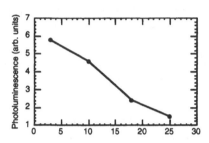

Fig. 7. Relationship between the average pore size and the photoluminescent efficiency.

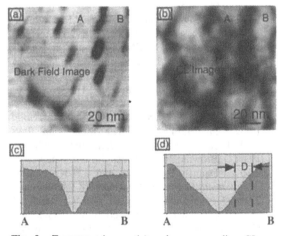

Fig. 8. Z-contrast image (a) and corresponding CL image (b) of the specimen shown in Fig. 6(c). Intensity profiles across the single pore from A to B are shown in (c) and (d) revealing the CL dead layer (D) to extend for about 5 nm.

ACKNOWLEDGEMENTS

The authors are grateful to A. Kadavanich for assistance with the CL detection system. The work at ORNL was sponsored by the Division of Materials Sciences, U.S. Department of Energy, under Contract No. DE-AC05-96OR22464 with Lockheed Martin Energy Research Corporation, and by appointment to the ORNL Postdoctoral Research Program administrated jointly by ORISE and ORNL. The work at University of Florida was supported by the Phosphor Technology Center of Excellence by DARPA Grant No. MDA972-93-1-0030.

REFERENCES

1. L. Manchanda and M. Gurvitch, IEEE Electronic Devices Lett. **9**, 180(1988).
2. T.S. Kalkur, Y.R. Kwor, and C.A. Paz de Araujo, Thin Solid Films **170**, 185(1989).
3. S.J. Duclos, C.D. Greskovich, and C.R. O'Clair, MRS Symp. Proc. **348**, 503(1994).
4. G.Blasse and B.C. Grabmaier, *Lum. Mater.* Springer, Berlin(1994).
5. S.C. Choi, M.H. Cho, S.W. Whangbo, C.N. Whang, S.B. Kang, S.I. Lee, and M.Y. Lee, Appl. Phys. Lett. **71**, 903(1997).
6. R.P. Rao, Solid State Communications **99**, 439(1996).
7. K.-I. Onisawa, M. Fuyama, K. Tamura, K. Taguchi, T. Nakayama, and Y.A. Ono, J. Appl. Phys. **68**, 719(1990).
8. A.F. Jankowski, L.R. Schrawyer, and J.P. Hayes, J. Vac. Sci. Technol. **A11**, 1548(1993).
9. W.M. Cranton, D.M. Spink, R. Stevens, and C.B. Thomas, Thin Solid Films **226**, 156(1993).
10. S.L. Jones, D. Kumar, R.K. Singh, and P.H. Holloway, Appl. Phys. Lett. **71**, 404(1997).
11. K.G. Cho, D. Kumar, D.J. Lee, S.L. Jones, P.H. Holloway, and R.K. Singh, Appl. Phys. Lett. **71**, 3335(1997).
12. K.G. Cho, D. Kumar, S.L. Jones, D.J. Lee, P.H. Holloway, and R.K. Singh, J. Electrochem. Soc. **145**, 3456(1998).
13. H-J. Gao, D. Kumar, K.G. Cho, P.H. Holloway, R.K. Singh, X.D. Fan, Y. Yan, and S.J. Pennycook, Appl. Phys. Lett. **75**, 2223(1999).
14. S. J. Pennycook, "STEM: Z-contrast", in *Handbook of Microscopy*, ed. By S. *Amelinckx, D. van Dyck, J. van Landuyt, and G. van Tendeloo*, VCH Publishers, Weinheim, Germany, pp. 595 (1997).
15. J. Fitz-Gerald, S.J. Pennycook, H. Gao, V. Krishnamoorthy, J. Marcinka, W. Glenn, R. Singh, Mat. Res. Soc. Proc. Spring 1998, **502**, (1998).
16. J. Fitz-Gerald, T. Trottier, R.K. Singh, P.H. Holloway, Appl. Phys. Lett. **72**, 1838(1998).
17. D. Kumar, J. Fitz-Gerald, R.K. Singh, Appl. Phys. Lett. **72**, 1451(1998).
18. H-J. Gao, et al., Appl Phys. Lett. (to be published).
19. N.D. Browning, M. F. Chisholm, S.J. Pennycook, Nature **366**, 143(1993).
20. J.L. Daams, P. Villars, and J.H.N. Vanvucht, *Atlas of Crystal Structure types for Intermetallic Phases*, P.6706. ASM International, Materials Park. OH(1994).
21. M.G. Paten and E.N. Maslen, Acta. Crystallogr. **19**, 307(1965).
22. B.H. O'Conner, T.M. Valentine, Acta Crystallogr. **B25**, 2140(1969).
23. M. Faucher, J. Pannetier, Acta Crystallogr. **B36**, 3209(1980).
24. W.v. Schaik and G. Blasse, Chem. Mater. **4**, 410(1992).
25. F. Jollet, C. Noguera, N. Thromat, M. Gautier, and J.P. Duraud, Phys. Rev. B **42**, 7587(1990).
26. E.S. Machlin, *Materials Science in Microelectronics*, Giro Press, Croton-on-Hudson (1995).

Partially Ordered and
Nanophase Materials

IMAGING HETEROPHASE MOLECULAR MATERIALS IN THE ENVIRONMENTAL SEM

B.L. THIEL, A.M. DONALD, D.J. STOKES, I.C. BACHE, N. STELMASHENKO
Cavendish Laboratory, University of Cambridge, Cambridge, UK

ABSTRACT

The Environmental Scanning Electron Microscope allows imaging of hydrated specimens, as well as fluid-containing or even fully fluid microstructures. This is possible so long as all of the phases can be thermodynamically stabilized. Experimental operating parameters include specimen temperature and chamber gas chemistry and pressure. However, it is also possible to alter the thermodynamic properties of phases by various chemical and physical means. These approaches have been used in ESEM studies on soft condensed matter such as emulsions, complex fluids and gels. By deliberately moving away from stabilizing conditions, it is also possible to perform dynamic experiments. A number of applications are illustrated.

INTRODUCTION

The Environmental Scanning Electron Microscope (ESEM) permits secondary electron imaging and spectroscopic analysis of liquid containing specimens at a spatial resolution and depth of field comparable to those obtained in a conventional SEM. Specimens do not require extensive preparation, fixing, dehydration, etc. nor do insulating specimens need to be given a conductive coating. Removing these restrictions greatly increases not only the range of specimens that can be examined with an SEM, but variety of *in situ* experiments that can be performed as well. These capabilities are chiefly due to two innovations:[1] First, a differential pumping system allows the specimen chamber to maintain a pressure of a few torr whilst the electron gun and optical column are held at high vacuum. This is achieved by effectively isolating portions of the electron column with pressure limiting apertures and applying increasingly efficient pumping methods for portions closer to the electron gun. A schematic of the basic vacuum system used in the FEI-Philips (previously ElectroScan) ESEM's is shown in Figure 1. For instruments operating with a tungsten thermionic source, either diffusion or turbomolecular pumps are sufficient. For LaB_6 or field emission sources, ion pumps are required to achieve the necessary vacuum.

The second critical innovation is the development of a secondary electron detector capable of operating with gas in the specimen chamber. Danilatos laid the foundations for the Environmental Secondary Detector in the early 1980's.[2] With this detector, an electrode is placed a few millimeters above the specimen, and in fact, forms the final pressure-limiting aperture. When a positive bias of a few hundred volts is applied, an electric field is created between the detector and specimen, as illustrated in Figure 2. Secondary electrons leaving the specimen are accelerated towards the anode by the electric field. Once their kinetic energy has exceeded the ionization threshold of the gas molecules, ionizing collisions occur. Each collision produces an additional electron, thus effectively amplifying the original signal. Depending on working distance, gas pressure and chemistry, and detector bias, gains of over 1000-fold are possible.[3,4] This cascade current is collected by the detector and is used to form an image. Positive ions are a beneficial by-product of the ionizing collisions. The detector field directs them back towards the specimen where they can neutralize specimen charging.

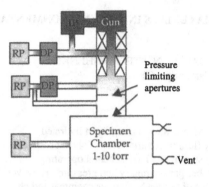

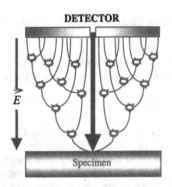

Figure 1. Differential pumping vacuum system in the ESEM. Ion (IP), diffusion (DP) and mechanical roughing (RP) pumps are used to create a pressure gradient in the column.

Figure 2. Gas cascade amplification of secondary electrons in the ESEM. A positive bias voltage on the detector creates an electric field with the specimen. Electrons are accelerated and have ionizing collisions in the gas.

The key to performing experiments on heterophase molecular materials is simultaneously stabilizing each of the phases present. Because these instruments typically operate with a few torr of water vapor in the specimen chamber, liquid water can be thermodynamically stabilized by cooling the specimen to a few degrees Celsius according to the phase diagram in Figure 3. While the most obvious application is to image fresh biological tissues, any fluid-containing or indeed fully fluid specimen may be examined, provided appropriate conditions are maintained. Furthermore, by deliberately moving into any of the single-phase fields in Figure 3, controlled hydration, dehydration or sublimation can be achieved. In general, the chamber gas can be considered to be at room temperature T_r, whereas the specimen temperature T_s is determined by the stage cooling. Under these conditions, Cameron & Donald show that the evaporation (or condensation) rate R (in molecules per unit area per second) for water is given as:

$$R = \left(\frac{kT_s}{2\pi m} \right)^{\frac{1}{2}} n_l \exp\left(\frac{-\varepsilon}{kT_s} \right) \left[1 - \left(\frac{T_r}{T_s} \right)^{\frac{1}{2}} \frac{P_w}{X(T_s)} \right] \qquad (1)$$

where k is Boltzman's constant, m is the mass of a water molecule, n_l is the number density of liquid water, ε is the heat of vaporization of a single water molecule, P_w is the partial pressure of water vapor in the specimen chamber, and $X(T_s)$ is saturated vapor pressure corresponding to temperature T_s. The evaporation rate as a function of chamber water vapor pressure is plotted is Figure 4 for a range of specimen temperatures. Equation 1 can be used to control solvent removal during dynamic experiments such as paint drying, film formation, freeze drying, etc. It should be remembered, though, that even when the evaporation rate is zero, the equilibrium is dynamic. Specifically, molecules are still striking and escaping from the specimen surface at a rate that increases with temperature.

It is not strictly necessary that a phase be thermodynamically stable in order to observe it in the ESEM. It is also acceptable if the rate of solvent gain/loss is negligible on the time-scale of an experiment. Many water-oil emulsions, for example, can be imaged readily because the evaporation rate of the oil is negligible. Thus, as long as the chamber conditions favor the stabilization of liquid water, the emulsion is effectively stable. Equation 1 gives insight into how the thermodynamics and kinetics of the experiment can be altered. The most influential factor in Equation 1 is the exponential dependence on the latent heat of vaporization. Any alteration to the system that increases this value will reduce the evaporation rate. Alternatively, the equilibrium vapor pressure of the liquid can be shifted. Addition of glycerol to water will reduce $X(T_s)$ and chaotropic salts can have an effect. Physical changes, such as the presence of a gel network or dissolved polymer will have an influence, as will geometric considerations such as capillary forces.

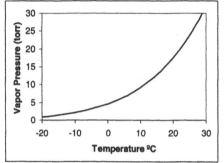

Figure 3. Saturated vapor pressure curve for water.

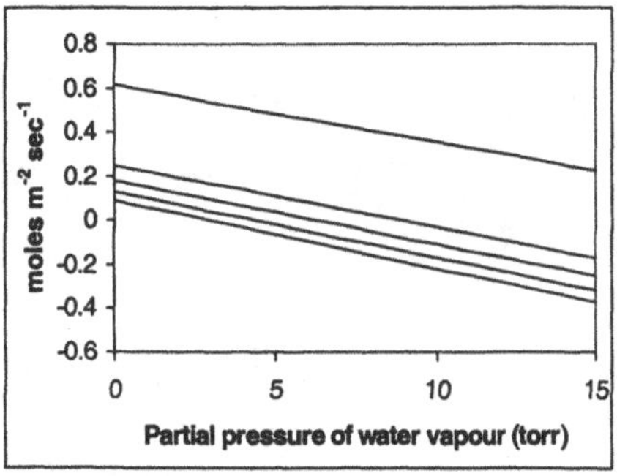

Figure 4. Evaporation rate for water at various partial pressures for several temperatures. Curves are for condensed phase temperatures of –5°C, 0°C, 5°C, 10°C and 25°C, with the lowest temperature being the bottom line.

EXPERIMENTAL

All experiments were performed on either Electroscan model E3 or 2010 ESEMs. The electron guns use LaB_6 and tungsten thermionic emitters respectively. Both instruments are equipped with thermoelectrically cooled specimen stages capable of operating over a range of –20°C to +40°C. Water vapor is used as the standard imaging gas, with a maximum pressure of 20 torr. Apart from the considerations of stabilizing specimens during observation, it is also necessary to avoid altering the specimen during the pumpdown from atmosphere. Accordingly, an optimized pumpdown sequence, as described by Cameron & Donald, was used for all specimens discussed here. This procedure cycles pumping and flooding (injecting water vapor into the chamber) steps in a manner than minimizes water gain or loss from the specimen.

RESULTS

Plant tissues are a very practical example of heterophase molecular materials. Fruit and vegetable parenchymal phloem (i.e., the 'fleshy' part) for instance, can mechanically be considered as 'closed-cell fluid-filled foams'. The mechanical properties depend on not only the geometric structure of the foam –the size and shape of cells- but on the filling fluid as well –the viscosity, permeability, etc. Food technologists continually struggle to relate complex textural properties of foods (e.g., mouth feel) to quantifiable mechanical properties such as elastic modulus. *In situ* dynamic mechanical testing of carrots in the ESEM was able to provide insight into the effects of cooking and aging on their mechanical behavior.[5,6] The series of images in Figure 5 shows the process of driving a scalpel blade through carrot tissue that has been boiled in water for 60 seconds. Boiling destroys the ability of the cell walls to retain fluid. When the blade is driven into the specimen, the entire cellular structure collapses, accordion-like, giving rise to a low modulus. Eventually, the few remaining cells that can still retain fluid are forced to bear the load, increasing the modulus as seen in stress-strain curve in Figure 5e. At the point of failure these cells burst, and the fluid is squeezed out (Fig. 5d) as if a sponge. This failure process is significantly different for carrots in the raw state, where high turgor pressure in the cells imparts a brittle response.

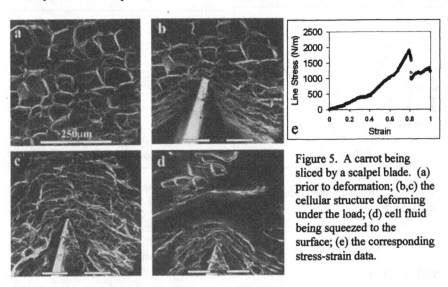

Figure 5. A carrot being sliced by a scalpel blade. (a) prior to deformation; (b,c) the cellular structure deforming under the load; (d) cell fluid being squeezed to the surface; (e) the corresponding stress-strain data.

Emulsions present some of the most striking examples of imaging complex fluid microstructures.[7] The micrograph shown in Figure 6 is a secondary electron image of a fully fluid vegetable oil and water emulsion. Oil is the darker, dispersed phase. Having demonstrated that it is possible to obtain such images of a multiphasic microstructure, the obvious question comes as to its interpretation. Specifically, what is the origin of contrast in the image? Results from confocal microscopy indicate that while there is some surface structure in these emulsions, it is not sufficient to account for the observed contrast. We have attributed the contrast to

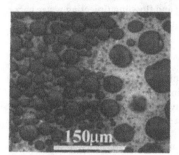

Figure 6. An ESEM micrograph of a protein stabilized vegetable oil-in-water emulsion. Oil is darker phase.

electronic structure differences between the molecules, and presented some general guidelines for predicting contrast between molecular liquids.[7] In this example, water is the bright phase.

Gel systems are an example of primarily liquid systems stabilized by a physical network of molecules.[8] The particular example shown in Figure 7 is of a dextran-gelatin gel, containing approximately 5% solids. This system is of commercial interest to the food industry as well as presenting significant theoretical challenges in polymer science. Depending on cooling conditions, a solution of this composition experiences a competition between spinodal decomposition and nucleation and growth transformations into dextran-rich and gelatin-rich regions. Furthermore, when the dextran-rich phase reaches a critical concentration, it undergoes a gelling transformation. This significantly alters the diffusion properties of the system, which in turn affects the kinetics first two transformation processes. Contrast in this system is believed to arise from the resonance bonds in the gelatin protein (dark phase) compared to the aliphatic bonds in the dextran (light phase).

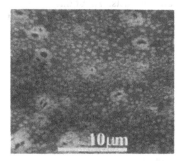

Figure 7. An ESEM micrograph of a phase segregated dextran-gelatin gel. The gel has approximately 9% solids content in water. The lighter regions are believed to be the dextran-rich phase. The length scales are consistent with those obtained by light scattering measurements.

Finally, the controlled drying of latex paint films is an example of an *in situ* dynamic hydration/dehydration experiment.[9] Figure 8 depicts a drying sequence of a latex suspension. Quality of the final film depends on the packing arrangement of the particles, the 'grain' size, crack formation, and the ability of the ordered regions to merge into a continuous film. These

Figure 8. Drying of a latex suspension. During the early stages of drying, water is held between the particles by capillary forces, giving the particles a bright halo effect. As drying proceeds, the particles form ordered domains.

factors are influenced by the relative humidity and the temperature (e.g., the solvent removal rate), making this an ideal system to be studies by ESEM. This system also illustrates the effect of capillary forces in retaining solvent in a microstructure well below the equilibrium vapor pressure. The first images in Figure 8 show water being retained at the interparticle necks. This water is gradually removed as the drying process continues.

CONCLUSIONS

The Environmental SEM is clearly more than just an imaging tool. Rather, it should be thought of as an experimental chamber in which the user has control over many thermodynamic variables. A wide variety of dynamic experiments can be performed on multiphasic materials. Control over the gas pressure and temperature provides a means for altering both the thermodynamics and kinetics of processes which can then be monitored at high spatial resolution and depth of field. However, these additional operating parameters also complicate experiments and allow much more scope for misinterpretation and introducing artifacts. All of this underscores the need for experiments to be well planned in order to obtain meaningful results. It is anticipated that these types of experiments will become more common in the future as additional, complimentary analytic techniques are incorporated into these microscopes.

REFERENCES

[1]G.D. Danilatos, "Foundations of Environmental Scanning Electron Microscopy," Adv. Elec. Electron Phys. **71**, 109-250 (1988).

[2]G.D. Danilatos, "Theory of the Gaseous Detector Device in the Environmental Scanning Electron Microscope," Adv. Elec. Electron Phys. **78**, 1 - 102 (1990).

[3]B.L. Thiel, I.C. Bache, A.L. Fletcher *et al.*, "An Improved Model for Gas Amplification in the Environmental SEM," J. Microscopy **187** (Pt. 3), 143-157 (1997).

[4]A.L. Fletcher, B.L. Thiel, and A.M. Donald, "Amplification Measurements of Potential Imaging Gases in Environmental SEM," J. Phys. D: Appl. Phys. **30**, 2249 - 2257 (1997).

[5]B.L. Thiel and A.M. Donald, "In-Situ Mechanical Testing of Fully Hydrated Carrots (Daucus carota) in the Environmental SEM," Annals of Botany **82**, 727-733 (1998).

[6]B.L. Thiel and A.M. Donald, "Microstructural Failure Mechanisms of Cooked and Aged Carrots," J. Text. Stud. submitted (2000).

[7]D.J. Stokes, B.L. Thiel, and A.M. Donald, "Direct Observation of Water-Oil Emulsion Systems in the Liquid State by Environmental SEM," Langmuir **14** (16), 4402-4408 (1998).

[8]B.L. Thiel and A.M. Donald, "The Study of Water in Heterogeneous Media Using Environmental Scanning Electron Microscopy," J. Molecular Liquids **80**, 207-230 (1999).

[9]N. Stelmashenko and A.M. Donald, "ESEM Study of Film Formation in Laticies Polymerised in the Presence of Starch," Microscopy & Microanalysis 4, supplement 2, 286-287 (1998).

A NEW APPROACH TOWARDS PROPERTY NANOMEASUREMENTS USING IN-SITU TEM

Z.L. WANG*, P. PONCHARAL**, W.A. DE HEER** and R.P. GAO*

* School of Materials Science and Engineering, ** School of Physics, Georgia Institute of Technology Atlanta GA 30332-0245.

ABSTRACT

Property characterization of nanomaterials is challenged by the small size of the structure because of the difficulties in manipulation. Here we demonstrate a novel approach that allows a direct measurement of the mechancial and electrical properties of individual nanotube-like structures by in-situ transmission electron microscopy (TEM). The technique is powerful in a way that it can directly correlate the atomic-scale microstructure of the carbon nanotube with its physical properties, providing an one-to-one correspondence in structure-property characterization. Applications of the technique will be demonstrated on mechanical properties, the electron field emission and the ballistic quantum conductance in individual nanotubes. A nanobalance technique is demonstrated that can be applied to measure the mass of a single tiny particle as light as 22 fg (1 f = 10^{-15}).

INTRODUCTION

Characterizing the properties of individual nanostructure is a challenge to many existing testing and measuring techniques because of the following constrains [1]. The size (diameter and length) is rather small, prohibiting the applications of the well-established testing techniques. The small size of the nanostructures makes their manipulation rather difficult, and specialized techniques are needed for picking up and installing individual nanostructure. Therefore, new methods and methodologies must be developed to quantify the properties of individual nanostructures. Among the various techniques, scanning probe microscopy (STM, AFM) has been a major tool in investigating the properties of individual nanostructures.

We have recently developed a novel approach which uses in-situ transmission electron microscopy (TEM) [2,3] as an effective tool for measuring the properties of individual carbon nanotubes. This is a new technique that not only can provide the properties of an individual nanotube but also can give the structure of the nanotube through electron imaging and diffraction, providing an ideal technique for understanding the property-structure relationship. The objective of this paper is to review our recent progress in applying in-situ TEM for characterizing the electrical, mechanical and field emission properties of carbon nanotubes, aiming at pointing out a new direction in nanomeasurements.

EXPERIMENTAL METHOD

TEM is a powerful tool for characterizing the atomic-scale structures of solid state materials. A powerful and unique approach could be developed if we could integrate the structural information of a nanostructure provided by TEM with the properties measured from the same nanostructure by in-situ TEM. Thus, an one-to-one correspondence can be achieved, providing a model system for comprehensively understanding nanomaterials. To carry out the property measurement of a nanotube, a specimen holder for an JEOL 100C TEM (100 kV) was built for applying a voltage across a nanotube and its counter electrode [4]. The nanotube to be used for property measurements is directly imaged under TEM (Figure 1), and electron diffraction patterns and images can be recorded from the nanotube. The information provided by TEM directly reveals both the surface and the intrinsic structure of the nanotube. This is a unique advantage over the SPM techniques. The static and dynamic properties of the nanotubes can be obtained by applying a controllable static and alternating electric field. The nanotubes were produced by an arc-discharge technique, and the as-prepared nanotubes were agglomerated into a fiber-like rod. The carbon nanotubes have diameters 5 - 50 nm and lengths of 1- 20 μm and most of them are nearly defect-free. The fiber was glued using silver past onto a gold wire, through which the electrical contact was made. The counter electrode can be a droplet of mercury or gallium for electric contact measurement or an Au/Pt ball for electron field emission characterization.

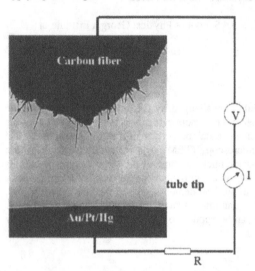

Fig. 1. TEM image showing carbon nanotubes at the end of the electrode and the other counter electrode. A constant or alternating voltage can be applied to the two electrodes to induce electrostatic deflection or mechanical resonance.

EXPERIMENTAL RESULTS

Bending modulus of a carbon nanotube

To measure the bending modulus of a carbon nanotube, an oscillating voltage is applied on the nanotube with ability to tune the frequency of the applied voltage. Resonance can be induced in carbon nanotubes by tuning the frequency (Figure 2). Resonance is nanotube selective because the natural vibration frequency depends on the tube outer diameter (D), inner diameter (D_1), the length (L), the density (ρ), and the bending modulus (E_b) of the nanotube [5]:

$$v_i = \frac{\beta_i^2}{8\pi} \frac{1}{L^2} \sqrt{\frac{(D^2 + D_1^2)E_\beta}{\rho}} \qquad (1)$$

where $\beta_1 = 1.875$ and $\beta_1 = 4.694$ for the first and the second harmonics. After a systematic studies of the multi-walled carbon nanotubes, the bending modulus of nanotubes was measured as a function of their diameters [2]. For nanotubes produced by arc-discharge, which are believed

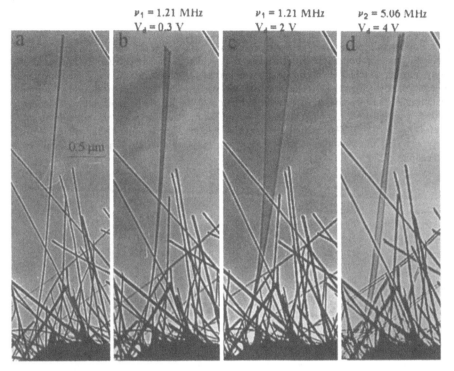

$\nu_1 = 1.21$ MHz $\nu_1 = 1.21$ MHz $\nu_2 = 5.06$ MHz
$V_4 = 0.3$ V $V_4 = 2$ V $V_4 = 4$ V

0.5 μm

Fig. 2. A selected carbon nanotube at (a) stationary, (b,c) the first harmonic resonance (ν_1 = 1.21 MHz) and (d) the second harmonic resonance (ν_2 = 5.06 MHz).

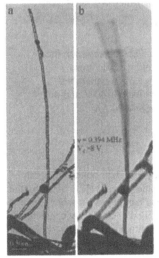

Fig. 3. Electro-resonance of a carbon nanotube produced by pyrolysis.

to be defect-free, the bending modulus is as high as 1.2 TPa (as strong as diamond) for nanotubes with diameters smaller than 8 nm, and it drops to as low as 0.2 TPa for those with diameters larger than 30 nm. A decrease in bending modulus as the increase of the tube diameter is attributed to the wrinkling effect of the wall of the nanotube during small bending. This effect almost vanishes when the diameters of the tubes are less than 12 nm.

Nanotubes produced by catalyst assisted pyrolysis contain a high density of point defects. Figure 3 shows the resonance of a carbon nanotube that exhibits bamboo-like structure. From the resonance frequency, the bending modulus is determined to be 0.03 TPa [6], which is almost 10 times smaller than that of a nanotube without defect, apparently demonstrating the effect of defects on mechanical properties.

Nanobalance of a single particle

In analogous to a pendulum, the mass of a particle attached at the end of the spring can be determined if the vibration frequency is measured, provided the spring constant is calibrated. If a very tiny mass is attached at the tip of the free end of the nanotube, the resonance frequency drops significantly (Figure 4). The mass of the particle can be thus derived by a simple calculation using an effective mass in the calculation of the inertia of momentum. This newly discovered "nanobalance" has been shown to be able to measure the mass of a particle as small as 22 ±6 fg (1f = 10^{-15}). *This is the most sensitive and smallest balance in the world.* We are currently applying this nanobalance to measure the mass of a single large biomolecule or a biomedical particle.

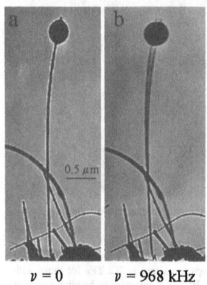

$v = 0$ $v = 968$ kHz

Fig. 4. A small particle attached at the end of a carbon nanotube at (a) stationary and (b) first harmonic resonance ($v = 0.968$ MHz). The effective mass of the particle is measured to be ~ 22 fg ($1 f = 10^{-15}$).

Fig. 5. In-situ TEM observation the electric field induced electron emission from carbon nanotubes. The applied voltage is 60 V and the emission current ~ 20 μA.

Electron field emission from individual carbon nanotubes

The unique structure of carbon nanotubes clearly indicates they are ideal objects that can be used for producing high field emission current density in flat panel display [7]. Most of the current measurements were made using a film of the aligned carbon nanotubes, in which there is a large variation in nanotube diameters and lengths, resulting in difficulty to clearly characterize the true switching-on field for electron field emission. Using the in-situ TEM setup we built, the electric field induced electron emission of a single carbon nanotube has been studied. Figure 5 shows an TEM image of the carbon nanotubes which are emitting electrons at an applied voltage of 60 V. The dark contrast near the tips of the nanotube is the field contributed by the charges on the tip of the nanotube and the emitting electrons. A detailed analysis of the field distribution near the tip of the carbon nanotube by electron holography is being carried out, which is expected to provide the threshold field for field emission and many other properties.

Electric transport properties

We have measured the electric property of a single multi-walled carbon nanotube using the set up of an atomic force microscope (AFM) [8]. A carbon fiber from the arc-discharge chamber was attached to the tip of the AFM, the carbon tube at the forefront of the fiber was in contact with a liquid mercury bath. The conductance was measured as a function of the depth the tube was inserted into the mercury. Surprisingly, the conductance shows quantized steps. The experiment had been repeated in TEM using the in-situ specimen holder. Figure 6 shows the contact of a carbon nanotube with the mercury electrode. The conductance of G_0 was observed for a single nanotube. It is interesting to note that the contact area between the nanotube and the mercury surface is curved. This is likely due to the difference in surface work function between nanotube and mercury, thus, electrostatic attraction could distort the mercury surface.

Fig. 6. Conductance of a carbon nanotube measured using the set up in TEM.

To directly find out if some nanotubes are conducting while some are not, we have observed the response of the nanotubes before and after applying a large voltage. Shown in Figure 7 is a case in which there are several nanotubes being in electrical contact at a small voltage/current (Figure 7a). Altering applying a larger voltage of about 10 V, all of the nanotubes but one were burnt out due to the heat generated, indicating that nanotube was non-conducting. This experiment showed the co-existence of conducting and non-conducting/semiconducting nanotubes, in agreement with theoretical prediction.

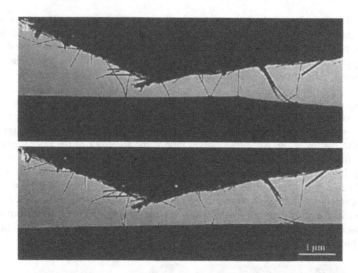

Figure 7. Electrical contact between carbon nanotubes (upper) and the Hg electrode (lower) (a) before and (b) after applying a 10 V potential, showing the existence of semiconducting nanotubes.

In summary, the approaches demonstrated here are a new direction in nanoscience of using in-situ TEM for measurements the electrical, mechanical and field emission properties. In this technique the properties measured from a single carbon nanotube can be directly correlated with the intrinsic atomic-scale microstructure of the nanotube, providing an one-to-one correspondence in property-structure relationship. This is an exciting field towards nanoscience and nanotechnology.

ACKNOWLEDGEMENT

Thanks for the financial support of NSF grants DMR-9971412 and DMR-9733160. Thanks to the Georgia Tech Electron Microscopy Center for providing the research facility.

REFERENCES

1. Z.L. Wang (ed.), *Characterization of Nanophase Materials*, Wiley-VCH, New York (1999).
2. P. Poncharal, Z.L. Wang, D. Ugarte, and W.A. de Heer, *Science* **283**, 1516 (1999).
3. Z.L. Wang, P. Poncharal, and W.A. de Heer, *J. Phys. Chem. Solids*, in press (2000).
4. Z.L. Wang, P. Poncharal, and W.A. de Heer, *Microscopy and Microanalysis*, in press (2000).
5. L. Meirovich, *Element of Vibration Analysis*, McGraw-Hill (New York) (1986).
6. R.P. Gao, Z.L. Wang, Z.G. Bai, P. Poncharal, and W.A. de Heer, *Phys.l Rev. B*, in press (2000).
7. W.A. de Heer, A. Chatelain, and D. Ugarte, *Science* **268**, 845 (1995).
8. S. Frank, P. Poncharal, Z.L. Wang, and W.A. de Heer, *Science* **280**, 1744 (1998).

ELECTRON MICROSCOPY OF SINGLE MOLECULES

D.E. LUZZI AND B.W. SMITH
Department of Materials Science and Engineering, University of Pennsylvania, Philadelphia, PA
19104-6272, luzzi@lrsm.upenn.edu, bwsmith@seas.upenn.edu

ABSTRACT

We report on the imaging of single C_{60} molecules adsorbed on the surface of carbon nanotubes. In the course of the work, the efficacy of carbon nanotubes as substrates for the high signal-to-noise observation of molecules is examined. The stability of nanotubes as a function of temperature, chemical environment and irradiation environment is explored. Carbon nanotubes are found to have a threshold energy for ballistic irradiation damage of approximately 85 keV. Finally, the use of nanotubes as *in-situ* reaction chambers and for other *in-situ* experimentation is demonstrated.

INTRODUCTION

Due to the small total scattering, the study of molecular structure has been a challenge. Techniques that rely on the study of crystals, for example x-ray diffraction, often require great effort and time in the production of specimens. Other techniques cannot provide the level of resolution necessary to see fine structure and/or require inferences from indirect evidence. The latest generation of transmission electron microscopes (TEMs) provide levels of brightness, energy stability, stage stability, detector sensitivity, lens aberration and electro-optic stability that should enable the imaging and structural analysis of single molecules.

One of the biggest challenges to overcome in the application of electron microscopy to the study of single molecules is the low signal from the molecule of interest, which is often swamped by a strong signal from the substrate. This is an especially egregious problem with organic molecules that often contain no elements that strongly scatter electrons. In this paper, we describe the use of carbon nanotubes as one-dimensional substrates for the support and study of single molecules. The imaging of the fullerene C_{60} is used as a proof of concept. The thermal, chemical and beam stability of the nanotubes is then addressed. Finally, *in-situ* experiments with nanotubes are demonstrated and some exciting possible applications of carbon nanotubes within the electron microscope are shown.

EXPERIMENTAL DETAILS

The starting material for this study was acid-purified carbon nanotubes produced by the pulsed laser vaporization (PLV) technique. The PLV material had been synthesized by the laser ablation of a graphitic target impregnated with 1.2 at% each Ni/Co catalyst. This raw nanotube "felt" was refluxed in HNO_3 for 48 hours, rinsed and neutralized, suspended in surfactant, and filtered to form a thin paper [1]. Such wet chemical etching is known to open the ends of multi-wall carbon nanotubes (MWNTs) [2] as well as attack the side walls of single-wall carbon nanotubes (SWNTs) [3]. This material is known to contain some C_{60} as an impurity in the synthesis process.

High temperature anneals in vacuum were used to produce well-ordered, clean nanotubes. These nanotubes were subjected to temperatures between 90 K and 1173 K *in-situ* under vacuums ranging from 20-40 μPa and *ex-situ* to 1473 K under similar vacuum levels. Nanotubes were also annealed *ex-situ* in Ar gas to 873 K. All *in-situ* experiments were carried out in a JEOL 2010F field-emission-gun transmission electron microscope (FEG-TEM). During in-situ anneals, temperature was monitored continuously via thermocouples. Only a few minutes were required to ramp between temperatures due to the small thermal mass of the heater.

A JEOL 4000EX high-resolution TEM (HRTEM) was used to investigate the response of the nanotubes to electron beams of energies 80, 100, 150, 200, 300 and 400 keV. Electron flux to the specimen was approximately $3.4(10)^{19}$ electrons cm^{-2} s^{-1}. The structure of both in-situ and ex-situ specimens was examined via TEM phase contrast imaging in either the FEG-TEM or the HRTEM. Magnification was determined using polyaromatic carbon shells present in the specimen, which originate from the decomposition of carbide crystals. It is known that the strong lattice fringes from these turbostratic shells have a well-defined spacing of 0.34 nm. Microscopy specimens were prepared from nanotube paper by tearing away a small sliver and fixing it inside an oyster TEM grid, thereby forgoing additional chemical or thermal processing. Imaging of individual molecules of C_{60} was performed in the JEOL 2010F at 100 keV.

RESULTS

As-received SWNTs are coated with surfactant as a result of the purification process. In the TEM, they appear as shown in Figure 1. A single SWNT is extended between two bundles of SWNTs oriented approximately perpendicularly at the top and bottom of the figure. A thin coating of disordered material is present on all surfaces.

In Figure 2, the same specimen as shown in Figure 3 is imaged after annealing in vacuum at 225°C for 63 hours. We have found that baking for at least 24 hours under these conditions removes the surfactant without modifying the nanotube material. In the figure, the nanotubes are seen to be free of most impurities. The walls of the nanotubes appear as dark parallel lines and are seen to have breaks and disclinations that are the result of acid attack (see arrow on the figure). SWNTs are stable in non-oxidizing and mildly oxidizing environments. However, oxidizing environments, as well as strong sonic energy, as produced in high intensity ultrasounds, are known to damage the walls of SWNTs and to cut SWNTs.

Figure 1. *A FEG-TEM image of a SWNT in as-purified PLV material showing the coating of surfactant.*

Vacuum anneals of damaged nanotube material has been found to repair most of the damage to the walls. The annealing induced recovery of defects begins to occur at temperatures above 400 °C. At higher temperatures, larger defects are seen to anneal out and the expected structure of SWNTs is recovered, that of a wrapped graphene sheet of hexagonally arranged carbon atoms, in the form of a cylindrically-wrapped chicken wire. An image of a SWNT that was annealed in vacuum at 1100°C is shown in Figure 3. Under phase imaging conditions, nanotubes can be considered as weak phase objects. Images of weak phase objects will be two-dimensional projections along the electron beam direction of the three dimensional specimen potential convoluted with the point transfer function of the electron microscope. With the resolution of the microscope significantly better than the finest scale detail in an image, the image can be considered to be a direct magnification of the carbon shells. The intra-shell structure of the wrapped graphene sheet is below the resolution limit of the microscope and appears as a uniform contrast level (gray). Since the maximum scattering potential of the

nanotube exists where the structure lies tangent to the electron beam, the images will appear as a pair of parallel lines.

At the level of column vacuum typical in the FEG-TEM, 20-40 μPa, the sublimation temperature of solid C_{60} is 325°C. Since SWNTs are stable to much higher temperatures, SWNTs will be surrounded by a gas of C_{60} molecules at temperatures above 325°C. An adsorbed C_{60} molecule on the surface of a SWNT has an increased coordination, will be more stable, and will therefore have a higher sublimation temperature. Thus, it can be expected that C_{60} molecules will transport from solid C_{60} to

Figure 2. *An FEG-TEM image of nanotubes in PLV material after annealing for 63 h at 225 °C. The damage due to the acid treatment is now visible. The arrow indicates a large defect in the sidewall of one SWNT.*

Figure 3. *HRTEM image of a SWNT after a 1 h vacuum anneal at 1100°C..*

the SWNT surface via the gas phase at these temperatures. In Figure 4, a FEG-TEM image of a SWNT sample held in-situ at a temperature of 350°C is shown. As can be seen, individual C_{60} molecules have adsorbed to the surface of the SWNTs. Some of these molecules are adsorbed to the side wall of a SWNT with respect to the electron beam and are therefore suspended over vacuum. This configuration is ideal for the study of single molecules as there is no substrate contribution to the image at the position of the molecule.

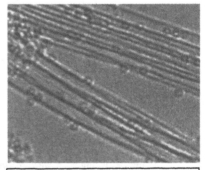

Figure 4. *An in-situ FEG-TEM image at 350 °C showing the adsorption of C_{60} molecules onto the surface of SWNTs.*

An important property of substrate materials for TEM experiments is stability under the electron beam. When considering the ballistic interaction of an electron beam with a nanotube, both the energy transferred to a carbon atom during a scattering event and the energy barrier that the atom must overcome to escape from the nanotube must be considered. To facilitate this discussion, the following geometry is defined: α is the angle between the normal to the nanotube surface and the direction of impulse of that atom and γ is the angle between the electron beam and the impulse direction. It can be shown that:

$$\Delta E = E_{transfer} - E_{escape} = \frac{2V(V + 2m_0c^2)}{m_Cc^2}\cos^2\gamma - f(\alpha) \qquad (1)$$

[4] where m_C is the mass of the carbon atom, V is the incident electron energy, and f(α) is a monotonically increasing function fit using a least squares method to established escape thresholds [5].

An ejection can occur for those (α, γ) having positive ΔE. In Figure 5, ΔE is plotted as a function of α and γ for a 100 keV electron beam. The maximized total angle from this curve predicts that only those atoms within ± 55° of the electron beam direction are susceptible to knock-on damage. The selective destruction of the top and bottom surfaces of a SWNT is thus predicted. The calculations further predict a threshold energy for the onset of irradiation damage of a nanotube of 86.3 keV. At energies above 140 keV, all carbon atoms can be displaced and complete destruction of the nanotube is predicted.

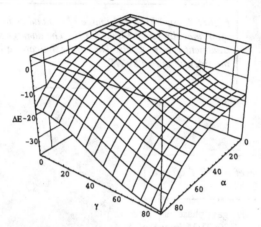

FIGURE 5. *ΔE versus (α, γ) for V = 100 keV. $\Delta E > 0$ only for small α, γ.*

The images in Figure 6 were recorded during irradiation of an isolated SWNT with 100 keV electrons. Figure 6a shows the unirradiated tube, whose walls are straight and parallel, suspended between two nanotube ropes. Figure 6b shows the tube after a moderate electron dose. The imaged walls are distorted, and their separation is no longer uniform. In this instance, the imaged walls of the tube retain strong contrast even after prolonged irradiation. This suggests that carbon atoms are not ejected from surfaces that are tangent with the electron beam, such that the tube is not damaged on its sides. The observed distortion is attributed to knock-on damage to the top and bottom of the tube, thereby destroying the cylindrical rigidity of the molecule. Consequently, the sides are less constrained and can adopt different conformations, with the sp^2 hybridized carbon atoms easily accommodating large out-of-plane bond angle distortions.

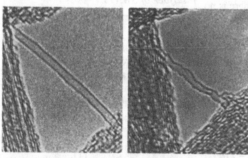

Figure 6. HRTEM images of an isolated SWNT (left) before irradiation, and (right) after irradiation with 100 keV electrons.

Different behaviors are observed if higher or lower energy electrons are used. Irradiation with 200 keV electrons produces an amorphous-like structure whose image has less contrast than that of the unirradiated tube. The imaged walls are seen to segment during irradiation, suggesting no anisotropic selectivity. Conversely, irradiation with 80 keV electrons produces essentially no change in the tube's image, even after high total electron dose. These experimental results are fully consistent with the theoretical model.

Knowledge of the behavior of SWNTs as a function of temperature and electron irradiation conditions is important in determining their use as substrates for experiments. It is also interesting to determine whether there could be interactions between the SWNTs and the molecules for study. In this case, the interaction of nanotubes with C_{60} was studied as a function of temperature. In Figure 7, an image of a bundle of SWNTs that have been filled with chains of C_{60} molecules is shown. The C_{60} molecules have a nominal spacing along the chain of 1 nm, which is the expected close-packed spacing of C_{60} molecules in solid C_{60}. However, there is some variability in the spacing that is not yet understood. Due to the unique microstructure of balls in a tube, these hybrid materials have been dubbed peapods[6]. This is the same sample as

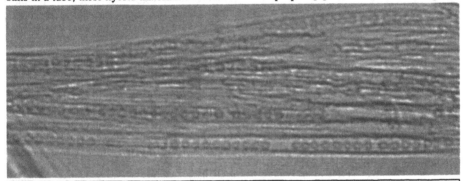

Figure 7. A FEG-TEM image of a bundle of SWNTs, many of which have been filled with chains of C_{60} molecules.

in Figures 1 and 2 after further annealing at 450 °C for two hours in vacuum. The walls of the SWNTs are partially healed. Many peapods are seen to be present indicating that the formation of peapods occurred during the annealing treatment following acid purification and the 225 °C cleaning anneal. These results indicate that C_{60} is entering the nanotubes during these relatively low temperature anneals. It is known from a number of analyses that acid purified PLV material contains residual crystallites of C_{60}. It is therefore likely that C_{60} enters the nanotube by transporting to open ends or side-wall defects by surface diffusion or in the gas phase.

Under a vapor phase transport mechanism, as the temperature is increased, the residence time of C_{60} on the surface of the SWNT, and thus the probability of entering the SWNT will decrease. As seen in the earlier results, at high temperatures, the defects in the SWNTs will anneal thereby blocking the entrance of C_{60} molecules into the tubes. Thus it would appear that there is a critical temperature window within which peapods can be efficiently formed. Some properties of these peapods have been studied by in-situ TEM. It is known that individual C_{60} molecules as well as short chains of molecules are mobile within a SWNT. Motion of these molecules can be driven by temperature or by inelastic interactions with the electron beam[7]. Under extended electron irradiation, C_{60} molecules will coalesce within the SWNTs forming short capsules[6].

This coalescence can also be induced by high-temperature anneals. In Figure 8, an image of a double tube is shown. Due to the special nature of these tubes, only being found as pairs with an inner 0.7 nm diameter nanotube encased in an outer 1.4 nm diameter nanotube, they have been named co-axial tubes (CAT)[7]. It is interesting to note that these features have been found in the 1100 °C annealed PLV material, but were never found in any specimen annealed at the lower temperatures of the experiments described above. In fact, peapods were found to be stable to at least 900°C under vacuum.

In order to determine the formation mechanism of the CATs, as-received PLV material was annealed in vacuum at 450 °C for two hours to form peapods and then immediately annealed at 1200 °C for 24 hours. In this specimen, only CATs and long capsules were found with no residual peapods. Thus it appears that CATs form through the coalescence of C_{60} molecules at high temperatures. These experiments show that nanotubes can act as nano-scale reaction chambers. In that role, they can provide two important functions. The constraining geometry of

Figure 8. A coaxial tube formed by the coalescence of C60 molecules at high temperature. The ends of adjacent inner tubes is seen at the left.

the nanotube can control the way in which molecules can interact. In the case of the C_{60} molecules, the surrounding SWNT insures that they collide at near-zero impact parameter. For high-aspect-ratio molecules, encapsulating SWNTs can increase the end-to-end reaction rate with respect to other reactions involving different geometrical orientations. The second function of the SWNT is to template the final reaction product. Upon the coalescence of two C_{60} molecules, a C_{120} molecule is formed in the shape of a hemispherically-capped cylinder. This is a metastable configuration of C_{120}; thus the SWNT determines the structure of the reaction product through steric constraint.

In conclusion, SWNTs are viable substrate materials for the investigation of single molecules by electron microscopy techniques. They provide the possibility of allowing the observation of molecules suspended over vacuum insuring no background noise from substrate scattering. The possibility of using SWNTs as *in-situ* reaction chambers for the observation of molecular reactions is also exciting. The authors gratefully acknowledge the support of the National Science Foundation through grant #DMR 98-02560 and the Office of Naval Research for supplying nanotube material through grant #N000149810893.

REFERENCES

1. Rinzler, A.G., *et al., Large Scale Purification of Single Wall Carbon Nanotubes: Process, Product and Characterization.* Applied Physics A, 1998. **67**: p. 29.
2. Tsang, S.C., *et al., A Simple Chemical Method of Opening and Filling Carbon Nanotbues.* Nature, 1994. **372**: p. 159.
3. Monthioux, M., B.W. Smith, and D.E. Luzzi, unpublished data, 1998.
4. Averback, R.S. and T.D.d.l. Rubia, *Solid State Physics,* . 1997, Academic Press: New York. p. 287.
5. Crespi, V.H. and et.al., Phys. Rev. B, 1996. **54**: p. 5927.
6. Smith, B.W., M. Monthioux, and D.E. Luzzi, *Encapsulated C60 in Carbon Nanotubes.* Nature, 1998. **396**: p. 323.
7. Smith, B.W., M. Monthioux, and D.E. Luzzi, *Carbon Nanotube Encapsulated Fullerenes: A Unique Class of Hybrid Materials.* Chem. Phys. Lett., 1999(in press).

NANOCRYSTAL THICKNESS INFORMATION FROM Z-STEM: 3-D IMAGING IN ONE SHOT

A.V. KADAVANICH†*, T. KIPPENY*, M. ERWIN*, S. J. ROSENTHAL*, S. J. PENNYCOOK†
†Oak Ridge National Laboratory, Solid State Division, Oak Ridge, TN 37831,
*Vanderbilt University, Department of Chemistry, Nashville, TN 37235

ABSTRACT

We have applied Atomic Number Contrast Scanning Transmission Electron Microscopy (Z-Contrast STEM) towards the study of colloidal CdSe semiconductor nanocrystals embedded in MEH-PPV polymer films.

For typical nanocrystal thicknesses, the image intensity is a monotonic function of thickness. Hence an atomic column-resolved image provides information both on the lateral shape of the nanocrystal, as well as the relative thickness of the individual columns.

We show that the Z-Contrast image of a single CdSe nanocrystal is consistent with the predicted 3-D model derived from considering HRTEM images of several nanocrystals in different orientations. We further discuss the possibility of measuring absolute thicknesses of atomic columns if the crystal structure is known.

INTRODUCTION

Nanocrystals are an interesting basic research problem and offer many potential applications in optics, electronics and catalysis. The interest is largely due to size-dependent electronic and thermodynamic properties arising from quantum effects or the large surface-to-volume ratio. For instance a 20Å diameter CdS nanocrystal has band gap that is 50% (0.8 eV) larger than the bulk crystal due to quantum confinement,[1] while the melting point is lowered by several hundred degrees Celsius due to the large surface area (~80% of the atoms are at the surface).[2]

A common problem in the study of nanocrystals is polydispersity. With a few exceptions [3,4] real nanocrystal samples are characterized by a distribution of sizes and shapes. While modern synthetic techniques have resulted in samples with less than 5% standard deviation on the average size,[5] this still implies that bulk characterization techniques provide only an average picture of nanocrystal specimens. Consequently techniques that can characterize individual nanocrystals are of great importance. A variety of electrical and optical measurements [6-11] have yielded insights into the physical properties of single nanocrystals. Theoretical interpretation of such results requires that the nanocrystal structure be known accurately.

At present, transmission electron microscopy (TEM) is the only technique that can yield atomically resolved structural information. However, due to the difficulty of tilting a nanocrystal specimen through a large angle, such measurements generally only yield a 2-dimensional picture of the nanocrystal structure. 3-dimensional models are then built up from analyzing different nanocrystals observed in different orientations. [12,13] Phase contrast HRTEM in conjunction with image simulations can provide the requisite 3-dimensional information directly if enough data exists to construct a realistic model. However, a more direct technique would be preferred for cases where little a priori information exists.

Z-Contrast STEM is an incoherent imaging process which results in images that are intuitively interpretable in terms of specimen structure and composition. [14,15] In this paper we demonstrate that for typical nanocrystal samples Z-STEM can provide information on the relative thickness of different parts of a nanocrystal. We also discuss the conditions necessary to achieve absolute thickness determinations and present the results of our initial attempts at accomplishing this.

We focus on CdSe nanocrystals as a test specimen since they are relatively easy to image by Z-STEM and the 3-dimensional structure is reasonably well-established. Furthermore, the binary nature is advantageous for absolute thickness determination, as discussed below.

EXPERIMENT

Sample Preparation

CdSe nanocrystals were prepared by the method of Murray [1,16] as modified by Peng [5] for

229

size-focussing. The TOPO surface ligands were exchanged with pyridine by heating in anhydrous pyridine for several hours. The nanocrystals were subsequently precipitated with hexanes and dissolved in chloroform. Poly (2-methoxy,5-(2'-ethyl-hexyloxy)-p-phenylenevinylene) (MEH-PPV) was prepared by the method of Wudl [17] and dissolved in chloroform. CdSe samples were stored in a glovebox until use, MEH-PPV was stored under argon in brown glass vials. TEM samples were prepared by mixing the MEH-PPV and CdSe solutions and spin-coating onto single-crystal NaCl substrates ({100} surfaces). Typical parameters were 20µl of 2 mg•ml^{-1} MEH-PPV/0.05 mg•ml^{-1} CdSe solution, spun at 2000 rpm. The films were removed by dipping into a water surface, whereupon the film floats onto the surface as the NaCl dissolves away. The floating films were picked up with lacey carbon coated copper TEM grids (Ted Pella Co.). Film thicknesses were typically in the range from 150-200Å as judged from the optical absorption of identical films spun onto glass slides. Specimens for Z-contrast imaging were prepared in air, stored under argon, and loaded in air.

STEM

Z-Contrast imaging was performed in a VG HB603 STEM operating at 300kV with a nominal resolution of 1.3Å. Bayesian image reconstruction techniques were used to remove noise from the image and deconvolute the resolution function due to the electron beam profile. Most of the analysis is based on a Maximum Entropy (MaxEnt) [18-20] algorithm running on a dedicated PC with a custom coprocessor card. The details are described elsewhere. [14] We have also analyzed the data using the Pixon™ method, a more recent algorithm for image restoration. [21, 22] For presentation, image brightness/contrast was adjusted in NIH Image 1.61, but intensities were measured on the unscaled images.

THEORY

Z-Contrast STEM

For normal TEM electron wavelengths, the high-angle scattering is largely incoherent, even for crystalline specimens, [15, 23, 24] and becomes more perfectly incoherent as the angle increases. In the limit of perpendicular scattering, coherence is destroyed completely, such that the only restriction on obtaining perfectly incoherent imaging is the ability to detect the scattered radiation. In practical terms, depending on the detector angle, the coherent contribution can vary up to 50% of the total detected intensity.

For a single atom, pure incoherent scattering is described by the Rutherford scattering formula, which predicts that the beam intensity scattered into a particular angle is proportional to the square of the atomic number (Z^2). This gives Z-STEM its elemental specificity.

It has been shown that for thin crystalline specimens in a channeling orientation, the scattering detected at high angles is predominantly from 1s Bloch states of the crystal. These states are highly localized with virtually no transverse coherence. [15, 23, 24]

Coherence does exist parallel to the beam direction and would result in intensity oscillations as a function of specimen thickness. However, vibrational motion of the atoms destroys the coherence, such that the effect is reduced to a thickness dependent modulation of the incoherent scattering intensity. The intensity of this thermally diffuse scattering is given by Equation 1, reproduced from Pennycook [25] and the thickness dependence is plotted for the Cd and Se columns of the wurtzite CdSe [100] zone axis orientation in Figure 1.

$$O_{TDS}(\mathbf{R},t) = \frac{Z^2 \varepsilon^{av^2}}{2\mu^{1s}\left(\xi^2\mu^{1s^2}+\pi^2\right)}\left[\pi^2\left(1-e^{-2\mu^{1s}t}\right) - \xi^2\mu^{1s^2}e^{-2\mu^{1s}t}\left(1-\cos\frac{2\pi t}{\xi}\right) - \pi\xi\mu^{1s}e^{-2\mu^{1s}t}\sin\frac{2\pi t}{\xi}\right]$$

Equation 1: O_{TDS} = Incoherent object function due to the 1s Bloch state as a function of $\mathbf{R}$ = position of an atomic column perpendicular to the beam direction and t = column thickness. Z = atomic number, ε^{av} = average excitation of the 1s Bloch state, μ^{1s} = absorption coefficient for 1s state, ξ = extinction distance.

For CdSe nanocrystals the relevant thickness range is the region below 100Å. As can be seen, the object functions for both Cd and Se are increasing monotonically in this region. Hence

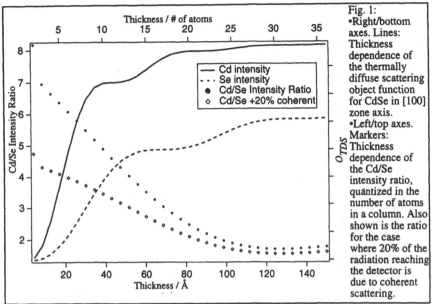

Fig. 1:
•Right/bottom axes. Lines: Thickness dependence of the thermally diffuse scattering object function for CdSe in [100] zone axis.
•Left/top axes. Markers: Thickness dependence of the Cd/Se intensity ratio, quantized in the number of atoms in a column. Also shown is the ratio for the case where 20% of the radiation reaching the detector is due to coherent scattering.

the relative intensities of adjacent columns in a Z-STEM image immediately reveal the relative thicknesses of these column.

The functions are plotted continuously, but for a crystal they are of course discrete, such that the intensities are quantized. More importantly, the ratios of Cd and Se intensities for columns of equal thickness—as one would expect to find for the dumbbell columns in [100] zone axis,—comprise a discrete, monotonic function of the thickness, and hence can be used to determine absolute thicknesses.

RESULTS/DISCUSSION

Z-STEM

The Z-STEM image of a nanocrystal near [100] zone axis orientation is shown in Figure 2. Panel (a) shows the raw image and panel (b) is the MaxEnt reconstruction of the object function with the point spread function of the microscope removed. Panel (c) shows a magnified view of the area in (b) indicated by the square. The dumbbell pairs of Cd and Se columns spaced 1.5Å apart are just resolved. The different intensities indicate that the Cd comprises the top right column of the dumbbell pair. Based on the contrast in the atomic columns the <001> lattice vector direction is assigned as up in the image.

Not all columns are as well resolved within this image. This could indicate strain in the lattice, but it could also be due to image noise. The maximum signal-to-noise ration (SNR) in the raw image is approximately 2. While the atomic centers are well separated (1.5Å) the 1s Bloch states have a diameter on the order of 0.8Å so some overlap between the Se and Cd states may be expected. The low SNR then makes it very difficult to accurately resolve the dumbbells. However, all the resolvable dumbbells support the assignment of the <001> direction

It is assumed that the Cd and Se columns are of equal thickness within each dumbbell column. One may reasonably expect them to differ by one atom at the entrance and exit surfaces each, depending on the exact nature of the surface termination. For the nanocrystal shown the thickness should be on the order of 15-18 atoms so the maximum error would be 13% which cannot account for the contrast difference observed. Hence the assignment of the elemental identities is clear-cut.

If we integrate the total intensity of each dumbbell pair we obtain a spatially resolved

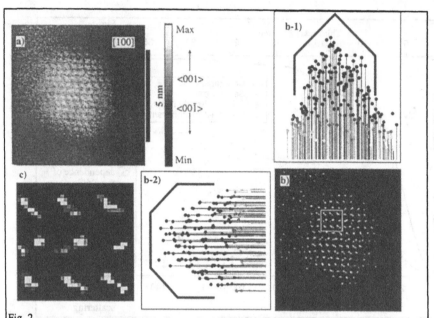

Fig. 2
a) Z-STEM image of a CdSe nanocrystal embedded in MEH-PPV, [100] zone axis orientation. The image has been aligned with the <001> lattice vector pointing up.
b) MaxEnt reconstruction of the image.
c) Closeup of the region in b) demarcated by the square, showing the different intensities in the Cd and Se columns comprising the dumbbells.
b-1) and b-2) Projections of the integrated dumbbell intensities in b) along the two axes of the image. The thick lines indicate the expected thickness profiles based on the HRTEM data of Shiang et. al.

thickness map, since the dumbbells are compositionally invariant. Projecting the thickness map along the two axes in the image we obtain the thickness envelopes depicted in panels (b-1) and (b-2) of Fig. 2 The thickness envelope corresponds to the expected shape based on previous HRTEM studies on such nanocrystals as indicated in the figure.[12] However, in this case the three-dimensional information is obtained in concert with the 2-dimensional projection, directly from the image.

We do not observe a discrete quantization of the image intensities as one would expect from Equation 1. This is likely due to the low SNR in the image which makes an accurate reconstruction difficult. Also, as discussed above, if the composition of the dumbbells deviates from 1:1 due to different terminations at the surfaces, the dumbbell intensities will not properly reflect the thickness.

We have calculated intensity ratios for the 14 resolvable dumbbell columns. The range of results varies from 4.1 to 1.0. Based on the intensity ratios in Figure 1 this corresponds to a thickness range of 50Å to hundreds of Å. While the low SNR results in a fairly large uncertainty in the ratios, estimated at approximately 0.5-1, this is clearly not reasonable for a nanocrystal whose major diameter in the image plane is 60Å. This discrepancy probably arises from residual coherent scattering reaching the detector. This can be seen from the second set of intensity ratios in Figure 1 which depicts the situation if 20% of the average intensity reaching the detector is due to coherent scattering. This reduces the lower thickness limit to 20Å, which is a reasonable value for this size of nanocrystal. The contribution of coherent scattering can be suppressed by increasing the HAADF collection angle at the cost of reduced SNR. At present, the overall low SNR makes this inadvisable. However, the addition of a spherical aberration corrector to the microscope should provide for adequate SNR, even at large scattering angles. The corrector is

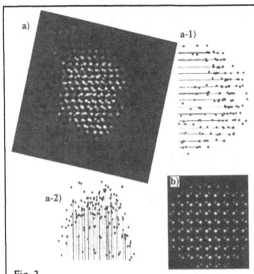

Fig. 3
a) Pixon reconstruction of the nanocrystal shown in Fig. 2
a-1), a-2) Projected intensities of dumbbell columns from the reconstruction in a)
b) Plot of the 1s Bloch states for the nanocrystal model of Shiang et. al. in [100] orientation.

scheduled to go online by 2002. In the meantime, improvements in specimen stability may allow us to integrate for longer times and thus keep the SNR within usable limits at higher angles.

The low SNR also makes it difficult to accurately identify the positions of surface atom columns. This is compounded by the inability of the MaxEnt to accurately fit the flat background signal from the polymer, giving rise to spurious column-like features in the region surrounding the particle. The Pixon™ method, a more recent image reconstruction algorithm [21] avoids this overfitting resulting in the smoother image shown in Figure 3. In addition to the smoother background, the image more accurately reflects the spatial extent of the Bloch states giving rise to the image. The relative thickness information obtained from this method matches that from the MaxEnt reconstruction of Figure 2 but the intensity ratios are generally lower. While this method seems promising, it is the first time the Pixon™ method has been applied to Z-STEM data so further testing will be necessary to ensure that the results can stand up to scrutiny.

CONCLUSIONS

Z-Contrast STEM is capable of resolving the lattice polarity in CdSe nanocrystals and can also provide thickness information directly from the image. By comparing intensities from compositionally similar columns, a map of relative thickness across a nanocrystal is obtained that matches the predicted shape to within the error of the measurement.

Absolute thickness can in principle be obtained from the Z-STEM images. However, this requires data of higher SNR than currently available.

FUTURE WORK

The Analysis of the Z-STEM data is crucially reliant on obtaining good SNR. In the short term this requires optimizing the specimen preparation to reduce drift and beam damage, allowing longer signal averaging. Presently, beam damage is the limiting factor as the MEH-PPV seems to degrade under the electron beam and cause contamination buildup. We are currently testing an alternate polymer system for use as a matrix.

In the long term the installation of a spherical aberration corrector will result in a beam profile with vastly improved imaging characteristics. The resolution will be improved to 0.5Å and the SNR is expected to improve by a factor of 7. The corrector will also reduce tailing in the beam profile, which makes it feasible to attempt column-resolved EELS measurements. The corrector is currently scheduled to be operational in early 2002 and should radically improve the capabilities of Z-STEM microanalysis.

ACKNOWLEDGEMENTS

The research presented here was funded by the Department of Energy, Basic Energy Sciences, Materials Sciences Division.

We wish to thank R. Puetter for performing the pixon reconstruction shown. We also wish to thank A. Yahil and Pixon LLC for a research use licence to the commercial Pixon™ code.

AVK gratefully acknowledges the assistance of P.D. Nellist, B.E. Rafferty and M. F. Chisholm in the operation of the STEM.

REFERENCES

1. Murray, C.B., Norris, D.J. & Bawendi, M.G. *Journal Of the American Chemical Society* **115**, 8706-8715 (1993).

2. Goldstein, A.N., Echer, C.M. & Alivisatos, A.P. *Science* **256**, 1425-7 (1992).

3. Vossmeyer, T., *et al. Science* **267**, 1476-9 (1995).

4. Herron, N., Calabrese, J.C., Farneth, W.E. & Wang, Y. *Science* **259**, 1426-8 (1993).

5. Peng, X.G., Wickham, J. & Alivisatos, A.P. *Journal Of the American Chemical Society* **120**, 5343-5344 (1998).

6. Blanton, S.A., Hines, M.A. & Guyot-Sionnest, P. *Applied Physics Letters* **69**, 3905-7 (1996).

7. Efros, A.L. & Rosen, M. *Physical Review Letters* **78**, 1110-13 (1997).

8. Empedocles, S. & Bawendi, M. *Accounts of Chemical Research* , 389-396 (1999).

9. Klein, D.L., McEuen, P.L., Bowen Katari, J.E., Roth, R. & Alivisatos, A.P. *Applied Physics Letters* **68**, 2574-6 (1996).

10. Nirmal, M., *et al. Nature* **383**, 802-4 (1996).

11. Tittel, J., *et al. Journal of Physical Chemistry B* **101**, 3013-3016 (1997).

12. Shiang, J.J., Kadavanich, A.V., Grubbs, R.K. & Alivisatos, A.P. *Journal of Physical Chemistry* **99**, 17417-17422 (1995).

13. Kadavanich, A.V., *et al.* in *Advances in Microcrystalline and Nanocrystalline Semiconductors - 1996* (eds. Collins, R.W.) 353-8.(Mater. Res. Soc, Boston, MA, USA, 1996).

14. Pennycook, S.J., Jesson, D.E., McGibbon, A.J. & Nellist, P.D. *Journal of Electron Microscopy* **45**, 36-43 (1996).

15. Nellist, P.D. & Pennycook, S.J. *Journal of Microscopy-Oxford* , 159-170 (1998).

16. Bowen Katari, J.E., Colvin, V.L. & Alivisatos, A.P. *Journal of Physical Chemistry* **98**, 4109-17 (1994).

17. Wudl, F. & Srdanov, G. in *United States Patent* (United States of America, 1993).

18. Burch, S.F., Gull, S.F. & Skilling, J. *Computer Vision, Graphics, and Image Processing* **23**, 113-28 (1983).

19. Skilling, J. & Bryan, R.K. *Monthly Notices of the Royal Astronomical Society* **211**, 111-24 (1984).

20. Skilling, J. & Sibisi, S. in *Invited and Contributed Papers from the Conference* (ed. Johnson, M.W.) 1-21 (IOP, Chilton, UK, 1990).

21. Pina, R.K. & Puetter, R.C. *Publications of the Astronomical Society of the Pacific* **105**, 630-637 (1993).

22. Puetter, R.C. *International Journal of Imaging Systems and Technology* **6**, 314-331 (1995).

23. Pennycook, S.J. & Jesson, D.E. *Physical Review Letters* **64**, 938 (1990).

24. Jesson, D.E. & Pennycook, S.J. *Proceedings of the Royal Society of London, Series A (Mathematical and Physical Sciences)* **441**, 261-81 (1993).

25. Pennycook, S.J. & Nellis, P.D. in *Impact of Electron Scanning Probe Microscopy on Materials Research* (eds. Rickerby, D., Valdrè, G. & Valdrè, U.) (Kluwer Academic Publishers, The Netherlands, 1999).

EPITAXY AND ATOMIC STRUCTURE DETERMINATION OF Au/TiO$_2$ INTERFACES BY COMBINED EBSD AND HRTEM

F. COSANDEY° and P. STADELMANN*

° Department of Ceramic and Materials Engineering, Rutgers University, Piscataway,
NJ 08854-8065, cosandey@scils.rutgers.edu
* Centre Interdepartemental de Microscopie Electronique, Ecole Polytechnique Federale
de Lausanne, CH-1015 Lausanne, Switzerland

ABSTRACT

We have studied the effects of deposition conditions on the epitaxial orientation of Au on TiO$_2$(110) and on the atomic structure of Au/TiO$_2$ interfaces by combined EBSD and HRTEM. Two experimental conditions were explored consisting of deposition of a 12 nm Au film at 300K followed by annealing at 770K and direct deposition of a 12 nm Au film at 770K. Deposition at 300K followed by annealing at 770K give rise to a (111)$_{Au}$//(110)$_{TiO2}$ epitaxial orientation relationship, while direct deposition at 700K temperature give rise to an epitaxial orientation relationship given by (112)$_{Au}$//(110)$_{TiO2}$. For both orientations, two epitaxial variants are observed which are twin related. The (112)$_{Au}$//(110)$_{TiO2}$ orientation has been found to minimize the interfacial lattice misfit while maximizing the number of Au-Ti bonds across the interface.

INTRODUCTION

Metals on oxide systems are important in numerous applications ranging from catalysts to gas sensors. The Au/ TiO$_2$ system is of particular interest because of its high activity for low temperature oxidation of CO and its good sensitivity as CO gas sensor [1, 2]. These beneficial effects are caused in part by a unique yet unknown synergistic effect between Au islands and the TiO$_2$ substrate. Therefore, recent studies have been conducted on model TiO$_2$(110) surfaces to understand the interaction mechanisms, growth mode and surface chemical reactions [3, 4]. In this study, we are reporting results on the effect of substrate temperature and preparation method on epitaxial orientation relationship of Au islands formed on TiO$_2$(110). Preliminary results on the atomic structure of the Au/TiO$_2$ interfaces are also presented.

EXPERIMENTS

The Au films were grown on TiO$_2$(110) substrates by vapor deposition in a UHV chamber. Details on TiO$_2$ surface preparation and deposition method have been presented elsewhere [3]. Two deposition conditions were studied. The first one consists of deposition of a 12nm thick Au film at 300K, followed by annealing at 775K for 1 hour (Sample A). The second one consists of direct deposition of a 12nm film on a TiO$_2$ substrate maintained at 775K (Sample B). The samples were then transferred in air to the LEO 982 Field Emission Scanning Electron microscope (FESEM) for imaging and diffraction. Electron Backscatter Diffraction (EBSD) patterns were taken from individual Au particles using the Opal system (Oxford Instruments). With this system, a spatial resolution of the order of 80 nm has been achieved for Au [5]. High resolution transmission electron microscopy (HRTEM) was done using a Philips CM-300 on samples viewed in cross-section.

Mat. Res. Soc. Symp. Proc. Vol. 589 © 2001 Materials Research Society

RESULTS AND DISCUSSION

Orientation of Au films deposited at 300K and annealed at 770K (Sample A)

A typical HRSEM image of a 12nm Au film deposited a 300K and annealed at 775K is shown in Fig.1 revealing a discontinuous Au film consisting of discrete Au particles with an average size of 250 nm. In this tilted SEM view, it can be seen that the islands are highly faceted with a high contact angle. In a previous study, a contact angle of 123° has been derived from quantitative analyses of particle size and density [3]. In order to determine the epitaxial orientation relationship of the Au islands with respect to the $TiO_2(110)$ surface, the orientation of individual islands has been determined by electron backscatter diffraction (EBSD). A typical EBSD patterns taken with incident electron beam energy of 20 keV is shown in Fig.2. In this EBSD patterns, the (111) surface normal of Au island is also represented. The direction in the islands parallel to the substrate normal is shown in Fig.3 in a stereographic triangle. It can be seen that all the islands have their (111) plane parallel to the $TiO_2(110)$ surface of the substrate. The {111} pole figure representing all four {111} planes from each island is shown in Fig.4. In this figure it can be seen that the Au particles are oriented according to two epitaxial domains rotated by 180° along a common [111] normal direction. These two orientations relationship are given by: $(111)_{Au}//(110)_{TiO_2}$ with $[1\bar{1}0]_{Au}//[001]_{TiO_2}$ (Orientation A-I) and $[\bar{1}10]_{Au}//[001]_{TiO_2}$ (Orientation A-II). In addition, it can be noted that although all the particles have their normal parallel to [111] to within 3° there is a large scatter in the azimutal in-plane orientation θ of 20°.

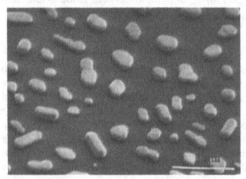

Fig.1 FESEM image of a 12nm Au film deposited at 300K and annealed at 770K

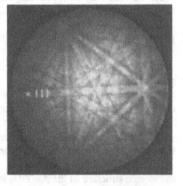

Fig.2 EBSD of an individual Au island with a (111) orientation

Orientation of Au films deposited directly at 770K (Sample B)

A typical HRSEM image of a 12 nm Au film deposited directly at 700K on the $TiO_2(110)$ surface is shown in Fig.5. As with the annealed sample A, the film consists of discrete Au islands but with a higher coverage of 62% as compared to 16%. In addition, the island size distribution is bimodal with average particle sizes of 67 nm and 11 nm for the two distributions. The small islands are the result of secondary nucleation in the area between the larger islands [3]. The particles are highly faceted and contain numerous planar defects aligned along the $[001]_{TiO_2}$ direction with re-entrant faceted edges typical of twinning planes. A high-resolution cross-section TEM image of a single Au island is shown in Fig.6 where numerous {111} twin planes

N

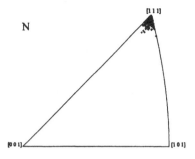

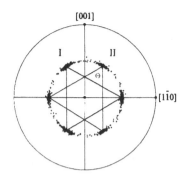

Fig.3 Stereographic representation of
Au cluster surface normal directions

Fig.4 {111} pole figure of the orientation
of Au clusters on TiO$_2$(110) surface
formed by annealing at 770K (Sample A)

are clearly visible. In addition, the top surface is bounded almost exclusively by {111} plane facets giving rise to a rough surface with numerous re-entrant edges visible also in the HRSEM image of Fig.5. The direction in the islands parallel to the substrate normal is shown in Fig.7 in a stereographic triangle. It can be seen that all the islands have their (112) plane parallel to the TiO$_2$(110) surface of the substrate. There is however, a distribution in this (112) orientation by a few degrees that can be observed in the HRTEM image of Fig.6. In this figure, the {111} twin interfaces are tilted away from the TiO$_2$ surface normal by 2°.

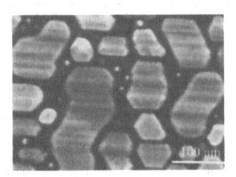

Fig.5 HRSEM image of a 12nm Au film
deposited at 770K
Fig.6 HRTEM image of a Au island revealing
the high density of (111) twin interfaces

The {111} pole figure taken from individual Au islands is shown in Fig.8 revealing a distribution in orientation of less than 2°. Also, the pole figure reveals two orientation variants that are rotated by 180° around the surface normal. These two orientation relationships are given by: $(112)_{Au}//(110)_{TiO2}$ with $[1\bar{1}0]_{Au}//[001]_{TiO2}$ (Orientation B-I) and $[\bar{1}10]_{Au}//[001]_{TiO2}$ (Orientation B-II). The numerous planar defects that are observed in Fig.5 and 6 are $(11\bar{1})$ twin planes separating epitaxial domains with these two orientations.

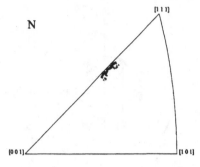

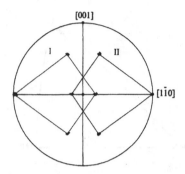

Fig.7 Stereographic representation of Au cluster surface normal directions

Fig.8 {111} pole figure of the orientation of Au clusters on $TiO_2(110)$ surface formed by deposition at 770K (Sample B)

The formation of epitaxial variants is a rather common observation in epitaxial growth studies and has been observed in numerous systems since the initial observations by Pashley and Stowell [6]. The requirements for the formation of epitaxial variants are that the symmetry of the 3D islands along the surface normal is different than the symmetry of the substrate and that the interfacial structures of each variant are identical [7]. For the $(111)_{Au}//(110)_{TiO2}$ epitaxial orientation (Orientation A), the Au islands have a 3-fold rotation axis along <111> while the $TiO_2(110)$ surface has a 2-fold rotation axis. Therefore, a rotation of 180° of the Au islands will lead to a different orientation while preserving the same atomic structure since the (111) plane has a 6-fold symmetry. The same argument holds for the $(112)_{Au}//(110)_{TiO2}$ epitaxial orientation (Orientation B) where the 3D Au islands have a 1-fold rotation axis along <112> while the (112) plane of Au has a 2-fold symmetry axis similar to the $TiO_2(110)$ substrate.

Geometrical models Au/TiO$_2$ interfaces and HRTEM images.

A schematic representation of the Au(111)//TiO$_2$(110) epitaxial orientation viewed along $[001]_{TiO2}$ direction is show in Fig.9. The lattice cells of Au and TiO$_2$ structures in contact with each other are also represented. The Au(111) cluster is oriented in a way that the <110> close packed direction in Au is parallel to $[001]_{TiO2}$. A characteristic feature of the TiO$_2$(110) surface is its termination which consists of rows of oxygen atoms called "bridging oxygen rows" aligned along the $[001]_{TiO2}$ direction. The Au(111) oriented islands lay on top of these bridging oxygen rows. Along the $[001]_{TiO2}$ direction, the misfit M(%) is 2.6%; which is expressed as $M(\%)=100\times2(P_M-P_O)/(P_M+P_O)$ where P_M and P_O are the lattice periodicity of the metal and oxide respectively. In the perpendicular $[1\bar{1}0]_{TiO2}$ direction, the misfit is 26.1%. This value is quite large but could be smaller if only individual <110>Au rows instead of the entire cell are considered. In this case, four <110>Au closed packed rows have a misfit of 14.3% with respect to bridging oxygen rows. A HRTEM image of the $(111)_{Au}//(110)_{TiO2}$ interface is shown in Fig.10. For the experimental conditions of thickness (3 nm) and defocus (57nm) the atomic columns in Au are white and the white contrast in the TiO$_2$ corresponds to the open channels formed by four oxygen column. At the present time, a quantitative description of atomic position across the interface is not known, but qualitatively, this experimental image is consistent with the model depicted in Fig.9. Further analysis and experiments are presently underway for quantitative analysis of atomic structure of the interface.

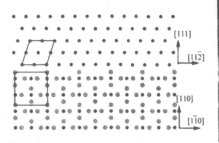

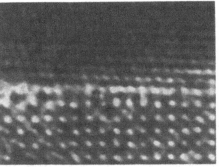

Fig.9 Schematic representation of Au(111)// Fig.10 HRTEM of Au(111)//TiO₂(110)
TiO₂(110) interface   interface

A schematic representation of the Au(112)//TiO₂(110) epitaxial orientation viewed along [001]$_{TiO2}$ direction is show in Fig.11. This orientations has the same [1$\bar{1}$0]$_{Au}$ direction parallel to [001]$_{TiO2}$ direction but the Au(111) plane is now rotated by 19.47°. Along the [001]$_{TiO2}$ direction, the misfit is the same as before at 2.6% but along [1$\bar{1}$0]$_{TiO2}$ it is now reduced to 8.4%. For the previous Au(111) orientation buckling of the close-packed (111) plane is expected, but upon tilting the rougher (112) plane inter-locks between the oxygen rows. In addition, Au is not expected to form strong bonds with oxygen, but this Au(112) plane configuration increases the number of Au-Ti bonds across the interface. A HRTEM image of the (112)$_{Au}$//(110)$_{TiO2}$ interface is shown in Fig.12 with similar experimental conditions as for the image depicted in Fig.10. This experimental image is consistent with the model described by Fig.11 but further analysis is presently underway for quantitative analysis. In addition, improved resolution to the Angstrom level will be required to resolve all the atoms, including oxygen, at the interface.

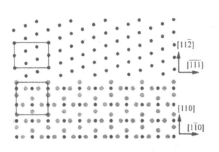

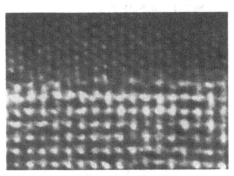

Fig.11 Schematic representation of Au(112)// Fig.12 HRTEM of Au(112)//TiO₂(110)
TiO₂(110) interface   interface

An interesting point to note here is that for the two Au(111) and Au(112) epitaxial orientations, the <110> close-packed direction in Au is parallel to the [001]$_{TiO2}$ direction. For this particular TiO₂(110) surface the rows of bridging oxygen along [001]$_{TiO2}$ direction structure might be controlling at least one specific direction for the orientation of the epitaxial layer. Indeed, an investigation on the initial growth of Ni by EXAFS [8] indicates that in the initial stages of growth, Ni forms two close packed rows of atoms in between the bridging oxygen rows thus forming the first step in the (111) close packed plane formation. In this regard, the common $[001]_{TiO2}$ and $<110>$Au directions play the role of a "door hinge" around which the (111) close-packed plane may rotate to produce the other Au(112) epitaxial orientation. The rotation angle between Au(111) and Au(112) orientations is 19.47°.

The Au(112) orientation and nearly perfect epitaxy formed at high temperature is surprising because of the combination of higher (112) surface energy, higher island coverage and large number of twin boundaries. This apparent higher energy configuration for (112) orientation as compared to (111) will be stable only if the interfacial energy between Au and TiO_2 for this orientation is low. Based simply on geometrical models, this low energy orientation could be explained from better lattice misfit between the Au overlayer, better inter-locking between the rougher Au(112) plane and TiO_2, and by the higher number of Au-Ti bonds across the interface.

SUMMARY-CONCLUSIONS

We have studied the effects of deposition conditions on the epitaxial orientation of Au islands on $TiO_2(110)$ surface by combined EBSD and HRTEM techniques. Two depositions conditions were used consisting of deposition at 300K followed by annealing at 775K (Sample A) and direct deposition at 775K (Sample B). For Sample A, the epitaxial orientation relationship is $(111)_{Au}//(110)_{TiO2}$ with $<110>_{Au}//[001]_{TiO2}$. For Sample B the orientation is $(112)_{Au}//(110)_{TiO2}$ with $<110>_{Au}//[001]_{TiO2}$. For both cases of epitaxial growth, two orientation variants are observed which are rotated by 180° around the $TiO_2(110)$ surface normal. In addition, extensive (111) twinning of the Au islands is observed which are the boundaries separating two orientation variants. The $(112)_{Au}//(110)_{TiO2}$ orientation has been found to reduce the lattice misfit while maximizing the number of Au-Ti bonds across the interface.

ACKNOWLEDGEMENTS

We would like to acknowledge numerous discussions with Ted Madey and Lei Zhang and thank them for the samples used in this study. Special thanks to Daniele Laub for her skill and patience in TEM sample preparation.

REFERENCES

1. M. Haruta, *et al.*, J. of Catalysis, **144**, 175 (1993)
2. T. Kobayashi, M. Haruta, S. Tsubota and H. Sano, Sens. and Actuators, **B1**, 222 (1990)
3. L. Zhang, F. Cosandey, R. Persaud and T.E. Madey, Surface Science, **439**, 73 (1999)
4. M. Valden, X. Lai and D.W. Goodman, Science, **281**, 1647 (1998)
5. F. Cosandey. Microscopy and Microanalysis, **3(2)**, 559 (1997)
6. D.W. Pashley and M.J. Stowell, Phil. Mag., **8**, 1605 (1963)
7. U. Dahmen, in *Boundaries and Interfaces in Materials: The David A. Smith Symposium*, R.C. Pond, *et al.*, Editors. TMS Publ., 225 (1998).
8. S. Bourgeois, *et al.*, Thin Solid Films, **304**, 267 (1997)

THEORETICAL EXPLANATION OF Pt TRIMERS OBSERVED BY Z-CONTRAST STEM

KARL SOHLBERG*, SOKRATES T. PANTELIDES**, and STEPHEN J. PENNYCOOK**
* Solid State Division, P.O. Box 2008, Oak Ridge National Laboratory, Oak Ridge, TN 37831-6031
** also, Department of Physics and Astronomy, Vanderbilt University, Nashville, TN 37235

ABSTRACT

First-principles quantum-mechanical calculations on γ-alumina have revealed a fascinating "reactive sponge" phenomenon. γ-alumina can store and release water, but in a unique, "reactive" way. This "reactive sponge process" facilitates the creation of aluminum and oxygen vacancies in the alumina surface. Earlier atomic-resolution Z-contrast STEM images of ultradispersed Pt atoms on a γ-alumina support showed the individual atoms to form dimers and trimers with preferred spacings and orientations that are apparently dictated by the underlying support[1]. In turn, the reactive sponge property of γ-alumina is the key to understanding the Pt clusters. Our calculations demonstrate that if three Pt atoms fill three vacancies created during the reactive sponge process, the resulting geometry precisely matches that of the Pt trimers observed in the Z-STEM images. Understanding the initial nucleation of small clusters on the complex gamma alumina surface is an essential first step in determining the origins of catalytic activity.

INTRODUCTION

γ-alumina is technologically one of the most important catalytic materials, serving both as a catalytic support[2] and as a catalyst[3, 4]. For example, catalytic reduction and oxidation of automotive pollutants such as nitric oxide (NO_x), carbon monoxide (CO), and hydrocarbons (HC) is accomplished with platinum (Pt) or rhodium (Rh) dispersed on a γ-alumina surface[5]. In refinery catalysis, the process is predominantly olefin cracking and it is carried out on catalysts formed by dispersal of transition-metal atoms on an insulating support surface[6], frequently a zeolite[7] or alumina[8].

It has long been recognized that electron microscopy is a powerful tool for investigating supported catalysts[9]. The ORNL STEM generates the smallest electron probe in the world (1.3 Å), and has revealed never-before-seen details about the distribution of the metal atoms and the structure of the metal clusters. It is capable of imaging individual catalyst atoms on the surface of γ-alumina and has revealed that Pt nucleates in the form of trimers and dimers[1].

Very recently, through quantum mechanical calculations, we have made some interesting discoveries about γ-alumina: It may exist over a range of hydrogen content, and this variable hydrogen content gives rise to a remarkable "reactive sponge" property[10]. In the reactive sponge process, water is absorbed and evolved much like in a kitchen sponge, but in a unique reactive way. Here we first review the reactive sponge process and then demonstrate how the filling of surface vacancies left behind by the reactive sponge process with Pt atoms leads to the formation of the Pt_3 clusters observed by Z-contrast STEM.

THEORY

The calculations were carried out with density functional theory[11] and employed the generalized gradient approximation (GGA) to the exchange-correlation energy[12], as described in the review by Payne $et\ at.$[13]. The electron-ion interactions were described with reciprocal-space pseudopotentials[14] in the Kleinman-Bylander form[15]. A plane wave basis was used to describe the electronic density with a cutoff energy of 1000 eV which was found to produce reliable results in earlier γ-alumina surface studies[16]. Integrations over the Brillouin zone employed a grid of k-points with a spacing of 0.1 Å^{-1} chosen according to the Monkhorst-Pack scheme[17].

The structural relaxation studies employed a unit cell consisting of a slab of γ-alumina 4 layers thick exposing the preferentially exposed[4] (110C) layer. The atoms in the bottom layer were frozen, as were the dimensions (a, b, c, α, β, γ) of the unit cell. The system was infinitely periodic with a vacuum spacing between slabs of 10Å. The starting structure from which the slab was generated was that of fully relaxed γ-alumina in its lowest energy form[10]. The starting Pt$_3$ was then generated by replacing two adjacent surface O atoms and a neighboring surface Al atom, (one in a nominal tetrahedral site) with Pt.

RESULTS

Despite its widespread use, two seemingly unrelated debates concerning the composition and structure of γ-alumina have long resisted resolution [4, 18, 19, 20]. First, it is widely held that γ-alumina has a defect spinel structure, with a fraction of the spinel cation sites vacant in order to maintain the balance of valence of a stoichiometric aluminum oxide. The distribution of the vacancies over the two cation sublattices has never been satisfactorily decided, however[4, 21, 22, 23, 24]. Second, numerous chemical analyses of γ-alumina have found non-negligible hydrogen content [25, 26, 27, 28, 29, 30, 31], but since none of these analysis techniques give structural information, the hydrogen was dismissed by many as "residual" from the incomplete dehydration of the boehmite precursor[32], or from incompletely removed surface adsorbed water.

We have carried out computational studies on γ-alumina and found that both controversies have a single resolution[10]. We have shown that that γ-alumina without bulk hydrogen[33] is merely the *terminus of a progression of hydrated forms*. Figure 1 depicts this progression. The starting point, boehmite[34] has the empirical formula $\overline{\text{H}}\text{AlO}_2$, which can be written $\text{Al}_2\text{O}_3 \cdot 1(\text{H}_2\text{O})$. Upon heating, boehmite loses water, progressing through various γ-alumina forms $\text{Al}_2\text{O}_3 \cdot n(\text{H}_2\text{O})$ until all of the water is driven off, leaving $\text{Al}_2\text{O}_3 \cdot 0(\text{H}_2\text{O})$ [35].

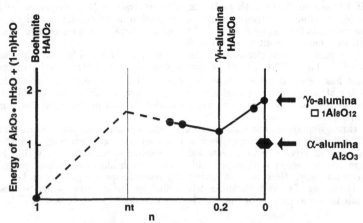

Figure 1. Energy diagram for the transition of boehmite to γ-alumina and, ultimately, corundum. All structures are fully relaxed.

The H atoms nominally occupy cation sites within the basic spinel structure, although they are displaced from the *ideal* cation sites, preferring to associate more closely with one of the neighboring oxygen atoms, thereby forming an OH group. There are two types of cation sites, tetrahedral and octahedral. We have computed vibrational frequencies for these OH groups and found excellent agreement with experimental results from IR spectroscopy[10]. Calculations of the material density versus hydrogen content and of the kinetics of hydrogen mobility are also in excellent agreement with experiments[10, 36].

It follows directly from the above analysis that γ-alumina has a very remarkable property which promises to unveil many of the mysteries associated with this material in catalytic processes[10]. From the observed water loss upon heating, and the observed variable hydrogen content, it follows that the material behaves as a sponge, in that it can store and subsequently release water, but in a unique, "reactive" way. When a water molecule arrives at the surface[37], it breaks up, H enters the bulk, and O stays at the surface. Al atoms countermigrate from the bulk to the surface where they recombine with the new O atoms and extend the crystal matrix. The ratios, determined by valence requirements, work out so that for every three H_2O molecules, six H move in, two Al move out, and the crystal extends by a stoichiometric Al_2O_3 unit. In the reverse process, H comes out from the bulk and combines with surface O to evolve as water while surface Al countermigrate into the bulk. The net result is an etching of the material as both Al and O atoms leave the surface. Of course the ratios work out the same way, so that the etching occurs by stoichiometric Al_2O_3 units. This unusual chemistry is a natural consequence of the fact that γ-alumina is not a single substance, but is a *sequence* of hydrogen-containing forms.

As noted above, the VG Microscopes HB603U STEM at ORNL produces an electron probe only 1.3 Å in diameter, sufficiently small and intense that individual heavy atoms scatter enough electrons to be detected with an annular detector placed around the transmitted beam. As the scattering cross section is proportional to atomic number (Z) squared, they are visible above the scattering of a light support. Figure 2a shows a close-up of the Z-contrast image of Pt atoms and clusters on γ-alumina[1]. Though noisy, single atoms are seen to form into dimers and trimers (example circled) with preferred spacings and orientations.

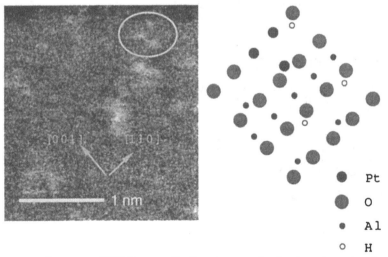

Figure 2. Z-contrast STEM image of a Pt trimer on the (110) surface of γ-alumina. Part b depicts how the reactive sponge mechanism produces the observed atomic configuration.

The reactive sponge property of γ-alumina now offers the key to understanding these clusters. The Z-STEM image (Fig. 2a) shows Pt_3 on a 110 surface of γ-alumina. We now know that a fraction of the O atoms in the 110C layer have an associated H atom in the HAl_5O_8 form of γ-alumina. In the reactive sponge process, for every three H that diffuse from the bulk, one Al counter-migrates from the surface into the bulk to maintain the balance of valence. When the three H reach the surface, they can combine with two surface O, and given that one of those surface O atoms already has an associated H, form $2H_2O$. When these two water molecules depart the surface, the surface is left with two O vacancies and one Al vacancy. If three Pt atoms fill these three vacancies as shown in Fig. 2b, a Pt trimer like that observed in the Z-STEM images is formed. Our calculations indicate that the resulting Pt_3 structures are not only stable, but the resulting optimized geometry is a near-perfect match to that of the Pt trimers observed in the Z-STEM images. The computed Pt-Pt distances are: 2.60, 2.61, and 2.89 Å. The DFT/pseudopotential calculations, however, are known to underestimate the lattice constant[10] in this system. The computed lattice constant for the cubic unit cell of γ-alumina is $a = 7.47$ Å. Experimentally, the lattice constant is not precisely established, but using the experimental density of γ-alumina as reported by Wefers and Misra[18] ($\rho = 3.2$ g/cc), we find $a = 8.18$ Å. Clearly the computed spacings should be scaled by a factor of about 1.087. The experimental Pt-Pt distances are: 2.88 ± 0.2, 3.21 ± 0.2, and 3.49 ± 0.2 Å [38]. The scaled computed distances: 2.83, 2.84, and 3.14 Å are in good agreement with the experimental values. Note also that while the Pt atoms in the starting structure for the calculations form an isosceles triangle, the optimized structure is distorted from this symmetry, just as in the Z-STEM images.

CONCLUSIONS

In summary, previous theoretical calculations have revealed an interesting reactive sponge process in γ-alumina which generated both O and Al vacancies in the surface. Our theoretical calculations find that if these vacancies are filled by Pt atoms such that 2 O + 1 Al are replaced by 3 Pt, the resulting Pt_3 structure formed on the γ-alumina surface has a structure matching that observed for Pt trimers on a γ-alumina surface by Z-contrast STEM.

ACKNOWLEDGEMENTS

This research was sponsored by the Laboratory Directed Research and Development Program (SEED) of Oak Ridge National Laboratory, managed by Lockheed Martin Energy Research Corp. for the U. S. Department of Energy under Contract No. DE-AC05-96OR22464, National Science Foundation grant DMR-9803768, by the William A. and Nancy F. McMinn Endowment at Vanderbilt University, and by financial support from Dupont through an ATE grant. Computations were partially supported by the National Center for Supercomputing Applications (NCSA) under grant CHE990015N to KS and utilized the SGI Origin2000 at NCSA, University of Illinois at Urbana-Champaign. KS was supported by an appointment to the ORNL Postdoctoral Research Program administered jointly by ORNL and Oak Ridge Associated Universities.

References

[1] P. D. Nellist and S. J. Pennycook, *Science*, **274**, 413 (1996).

[2] Satterfield, C. N., *Heterogeneous Catalysis in Practice*, §4.5 (McGraw Hill, New York, 1980).

[3] B. Shi and B. H. Davis, *J. Catal.*, **157**, 359 (1995).

[4] H. Knözinger and P. Ratnasamy, *Catal. Rev. - Sci. Eng.*, **17**, 31 (1978).

[5] B. C. Gates, *Chem. Rev.*, **95**, 511 (1995).

[6] M. Che and C. O. Bennett, *Adv. Catal.*, **36**, 55 (1989).

[7] R. Shah, M. C. Payne, M.-H. Lee, and J. D. Gale, *Science*, **271**, 1395 (1996).

[8] Z. Xu, F.-S. Xiao, S. K. Purnell, O. Alexeev, S. Kawi, S. E. Deutsch, and B. C. Gates, *Nature*, **372**, 346 (1994).

[9] S. J. Pennycook, A. Howie, M. D. Shannon, and R. Whyman, *J. Molec. Catal.* **20**, 345 (1983).

[10] K. Sohlberg, S. J. Pennycook, and S. T. Pantelides, *J. Am. Chem. Soc.*, **121**, 7493 (1999).

[11] W. Kohn and L. J. Sham, *Phys. Rev.*, **140A**, 1133 (1965).

[12] J. P. Perdew, *Phys. Rev. B*, **33**, 8822, (1986).

[13] M. C. Payne, M. P. Teter, D. C. Allan, T. A. Arias, and J. D. Joannopoulos, *Rev. Mod. Phys.*, **64**, 1045, (1992).

[14] Al pseudopotential: B. Winkler, V. Milman, B. Hennion, M. C. Payne, M. H. Lee, J. S. Lin, *Phys. Chem. Min.*, **22**, 461 (1995); H pseudopotential: CASTEP default reciprocal-space; O pseudopotential: B. Winkler, V. Milman, B. Hennion, M. C. Payne, M. H. Lee, J. S. Lin, *Phys. Chem. Min.*, **22**, 461 (1995); Pt pseudopotential: Troullier, N., and Martins J. L., *Phys. Rev. B*, **43**, 1993 (1991).

[15] L. Kleinman and D. M. Bylander, *Phys Rev. Lett.*, **48**, 1425 (1982).

[16] K. Sohlberg, S. J. Pennycook, and S. T. Pantelides, *J. Am. Chem. Soc.*, **121**, XXXX (1999).

[17] Monkhorst, H. J. and Pack, J. D., *Phys. Rev.*, **B 13**, 5188 (1976).

[18] Wefers, K. and Misra, C., *Oxides and Hydroxides of Aluminum* (Alcoa, 1987).

[19] Tsyganenko, A. A. and Mardilovich, P. P., *J. Chem. Soc., Faraday Trans.* **92**, 4843-4852 (1996).

[20] Henrich, V. E. and Cox, P. A., *The Surface Science of Metal Oxides* (Cambridge University Press, Cambridge 1994).

[21] A. J. Léonard, P. N. Semaille, and J. J. Fripiat, *Proc. Br. Ceram. Soc.*, **103**, 103 (1969).

[22] B. C. Lippens and J. J. Steggerda, in B. G. Linsen, (ed.) *Physical and Chemical Aspects of Adsorbents and Catalysts*, (Academic Press, London 1970).

[23] S-D. Mo, Y-N. Xu, and W-Y. Ching, *J. Am. Ceram. Soc.*, **80** 1193 (1997).

[24] M.-H. Lee, C-F. Cheng, V. Heine, and J. Klinowski, *Chem. Phys. Lett.*, **265**, 673 (1997).

[25] D. A. Dowden, *J. Chem. Soc.*, **1-2** 242 (1950).

[26] J. H. de Boer and G. M. M. Houben, *Proceedings of the International Symposium on the Reactivity of Solids*, **I**, 237 (1952).

[27] S. Soled, *J. Catalysis*, **81**, 252 (1983).

[28] V. A. Ushakov and E. M. Moroz, *React. Kinet. Catal. Lett.*, **24**, 113 (1984).

[29] R. M. Pearson, *J. Catal.*, **23**, 388 (1971).

[30] J. M. Saniger, *Mat. Lett.*, **22**, 109 (1995).

[31] A. A. Tsyganenko, K. S. Smirnov, A. M. Rzhevskij, and P. P Mardilovich. *Mat. Chem. and Phys.*, **26**, 35 (1990).

[32] S. J. Wilson, *J. Solid State Chem.*, **30**, 247 (1979).

[33] R-S. Zhou and R. L. Snyder, *Acta. Cryst.*, **B47**, 617 (1991).

[34] R. J. Hill, *Clays Clay Min.*, **29**, 435 (1981).

[35] C. S. John, N. C. M. Alma, and G. R. Hays, *Applied Catal.*, **6** 341 (1983).

[36] K. Sohlberg, S. J. Pennycook, and S. T. Pantelides, *Recent Research Developments in Physical Chemistry*, (submitted).

[37] For a detailed account of how H_2O breaks up on an alumina surface, see Ref.[39].

[38] P. D. Nellist, private communication.

[39] K. C. Hass, W. F. Schneider, A. Curioni, and W. Andreoni, *Science*, **282**, 265 (1998).

ION-IMPLANTED AMORPHOUS SILICON STUDIED BY VARIABLE COHERENCE TEM

J-Y. CHENG[1], J. M. GIBSON[2], P. M. VOYLES[3], M. M. J. TREACY[4] and D. C. JACOBSON[5]
[1]University of Illinois, Department of Materials Science and Engineering, Urbana, IL 61801, jcheng1@uiuc.edu
[2]Argone National Laboratories, Materials Science Division, 9700 S. Cass Ave, Argone, IL 60439
[3]University of Illinois, Department of Physics, Urbana, IL 61801
[4]NEC Research Institute, 4 Independence Way, Princeton, NJ 08540
[5]Lucent Bell Laboratories, 600 Mountain Ave, Murray Hill, NJ 07974

ABSTRACT

Amorphous silicon formed by ion-implantation of crystalline silicon is investigated with the use of VC-TEM (variable-coherence transmission electron microscopy). This technique is sensitive to medium-range-order structures. The results from high-energy Si implanted samples showed a striking similarity to sputtered amorphous silicon. We found that both ion-implanted and sputtered samples have paracrystalline structures, rather than the expected continuous random network (CRN). We also observed the structural relaxation of the ion-implanted amorphous silicon after ex-situ thermal annealing towards the random network. The more disordered structures are favored and a large heat of relaxation is released as the temperature increases. Finally, we show some preliminary results on the structural variation with the sample depth.

INTRODUCTION

Amorphous silicon is a well known system that has many device applications, such as flat panel displays and field effect transistors[1]. a-Si can be made by ion-implantation, reactive magnetron sputtering, vacuum evaporation, and by laser quenching methods. The ion-implantation method provides the cleanest interface between the amorphous layer and the substrate[2], but is more expensive and cannot make very thin films (<20nm). In the past, many studies of ion-implanted silicon focus on the dopant activation of silicon devices in integrated-circuit design. Most people believe the as-implanted amorphous structure is a continuous random network (CRN), due to the limitation of standard techniques, such as Raman spectroscopy and typical diffraction methods, to probe the short range ordering of the amorphous structure[3]. On the other hand, Treacy and Gibson have developed an alternative approach, called variable coherence transmission electron microscopy (VC-TEM)[4], that is able to distinguish between different amorphous structures in medium range order. Basically this technique is performed with hollow cone illumination, and statistically analyzes the dark field image. The quantitative results depend on the higher order atomic correlation, which reveals the medium range ordering (1-2nm) in amorphous materials. The evaporated a-Ge has been studied by VC-TEM[5], and the subtle change of amorphous structures in medium range order is illustrated by the variable coherence microscopy. Besides, the structural relaxation upon thermal annealing has been observed in the a-Ge by VC-TEM. The sputtered a-Si:H has also been studied by VC-TEM to understand light soaking effects on structures[6]. Nonetheless, ion implnatation is the cleanest method to study fundamental properties of a-Si.

EXPERIMENTS

A B-doped Si (100) wafer was self ion-implanted with doses and two ion energies: $5x10^{15}cm^{-2}$ at 700kV and $5x10^{15}cm^{-2}$ at 300kV at liquid-nitrogen temperature. This double implantation creates an amorphous layer about 1μm thick. The samples were dimpled, and chemically thinned by using HF/HNO$_3$ on the backside. In the pre-thinned samples, we first chemically etched the upside down to 700nm before backside thinning. The thickness of the thin areas around the central hole in the plan-view samples vary between 17nm to 35nm. This can be measured by the transmittance ratio of two bright field images: one is bight beam; the other has features, and corrected by the mean free path of 50nm[7]. The VC-TEM experiments are conducted in a Philips CM12 transmission electron microscope, which operates in the conical dark field mode at 120kV accelerating voltage. Images are recorded in the Gatan slow scan CCD camera, cooled down one hour before VC-TEM measurements. The spatial resolution of the images is about 1.1nm, associated with the objective aperture. The acquisition of dark-field images has been automated by the Digital Micrograph program to communicate with the instrument.

The "speckle" images, which reveal the fluctuation of scattered intensity from the amorphous sample, is Fourier transformed and filtered to give the relative variance

$$V(k) = \frac{\langle I(k)^2 \rangle}{\langle I(k) \rangle^2} - 1 = \int_{w_1}^{w_2} 2\pi w dw \left| \Delta \tilde{I}_i(k,w) \right|^2, \tag{1}$$

where k is the hollow cone scattering angle, nm^{-1}, $\Delta \tilde{I}_i$ is the Fourier transformation of ΔI, $\Delta I = I-\langle I \rangle$, which restored the incident intensity after dividing the recorded intensity by the modulation transfer function of CCD camera, as well as subtracting the shot noise of CCD camera[8]. ω_1 is the lower filtering frequency to remove the nonuniformity of illumination as well as the incoherent variation of thickness. ω_2 is the upper filtering frequency to remove the shot noise as well as to reduce false X-ray counts[9]. Under the kinematical assumption and the nonlinearity of dark field imaging[4], the relative variance is also expressed in term of atomic positions, j, l, m,n, with $r_{jl} = |r_j - r_l|$,

$$V(k) = N_0 \frac{\sum\limits_{j,l,m,n} A(r_{jn})A(r_{nl})A(r_{mn})F(r_{jl})F(r_{mn})}{\left| \sum\limits_{p,q} A(r_{pq})F(r_{pq}) \right|^2} - 1, \tag{2}$$

where N_0 is the average noise count, A(r) is the airy function of the objective aperture, and F(r) is the coherence function, which characterizes the partial coherence of hollow cone illumination. Experimentally, we can change the coherence function by tilting the hollow cone scattering angle to probe the structural fluctuation of dark field images in the sampling volume, 1/Q, Q is the angular size of objective aperture. In order to avoid beam damage effects and get reproducible results, the illumination scans over the cone surface on the well focused dark field regions with low to high and high to low k tilts. We acquired 33 images on each data series in k =2.5 to $7nm^{-1}$, and one more image at k = $14nm^{-1}$, for which the variance depends on incoherent variation with thickness only. Then we select 6-10 analyzed series of different thicknesses (20-30nm), offset the variance of each series at k =$14nm^{-1}$, and normalize the thickness of each series with 25nm to correct the error bar, assuming that V(k) is inversely proportional to thickness. More details on the theory of variable coherence microscopy are discussed by Treacy and Gibson[2].

Prior to the preparation of TEM plan-view samples, bulk amorphous silicon was annealed in an Omega furnace, with initial power output 100%, and initial ramping rate 8°C/min. When annealing approaches the set point temperature, 580°C and 600°C, the power output is set to 50% with the ramping rate 5°C/min, allowing the specimen to stay at 580°C for 2.5 mins, and at 600°C for 5 mins, then cooling down gradually to room temperature. The set point temperature 600°C with duration 5 mins was chosen to observe recrystallization of a-Si according to Washburn's regrowth experiments[10]. The recrystallized thickness can be measured from Rutherford BackScattering (RBS) spectrum. The RBS experiments are performed under Ar^+ ion bombardment at the beam energy of 2.0MeV, with channeling axis along the orientation of Si (100) substrate. The energy spectrum is calibrated by the standard sample (Ti,Au,Al). The thickness of a-Si is measured as 670nm by RBS; in other words, there is 330nm thick a-Si recrystallized. The RBS spectrum is shown in Fig. 1.

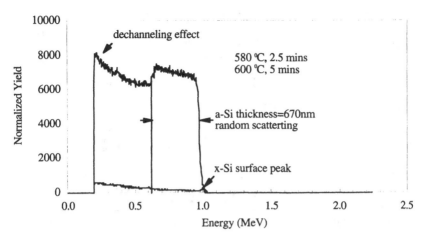

Fig.1. RBS spectrum for a-Si samples annealed at 580°C for 2.5 mins and 600°C for 5 mins.

RESULTS

We obtained several preliminary results for ion-implanted a-Si with variable coherence microscopy, as shown in Fig. 2. The positions of the two peaks in the variance curve are the same ring positions in the diffraction pattern of a-Si. The atomic correlation length can be extracted from the peak height[11]. In paracrystalline models[9], the variance was calculated for different paracrystalline structures of a-Si, and the simulated variance curve is similar to that of sputtered and as-implanted a-Si in Fig. 2. It follows that the ion-implanted a-Si and sputtered a-Si:H are paracrystalline structures, but the degree of paracrystallity is slightly different, though we examined the near surface area of the ion-implanted sample. We noticed the spatial resolution on sputtered a-Si:H data is 0.5nm (controlled by larger objective aperture), while the spatial resolution on as-implanted a-Si data is 2.2nm. Meanwhile, we have not corrected the variance offset on as-implanted data points. We need further experiments and more reproducible results to prove that the ion-implanted a-Si has smaller correlation length than sputtered a-Si:H.

We have succeeded in observing the structural relaxation in ion-implanted a-Si samples upon annealing. Gibson and Treacy have also observed the same relaxation phenomena on the evaporated a-Ge[5]. This is the direct evidence of relaxation to support the assumption of the paracrystalline model for the α-Si system. The paracrystal with high free energy, stored in terms of strain energy or heat, will transit towards the more disordered state, like a continuous random network, which is energetically favorable upon thermal annealing. The thermal relaxation behavior of α-Si has been confirmed by calorimetric experiments[12,13]. It is thought the unrelaxed and relaxed a-Si are made up of continuous random networks in short range order, and the degree of randomness, due to point defect annihilation, changed upon relaxation[3]. However, the failure of standard techniques to detect the subtle structural change, especially in medium range order, is obvious. As with X-ray diffraction experiments[3], the difference of the diffraction pattern between as-implanted and annealed a-Si is hardly noticed. The average distribution of the diffracted intensity is the Fourier transformation of the pair correlation function, which is only sensitive to the short range ordering. Intrinsically, diffraction techniques are not able to probe the atomic ordering in 2nm, which corresponds to the grain size of the paracrystalline structure, whereas variable coherence microscopy has demonstrated its capability at this point.

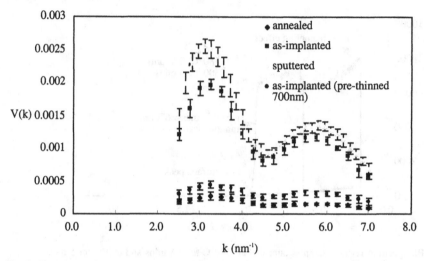

Fig. 2. Experimental variance curves of different amorphous materials: ion-implanted a-Si and sputtered a-Si:mid-H[6]. They both reveal the paracrystalline structures. Another variance curve from annealed ion-implanted a-Si shows the large structural relaxation. The heights of two pronounced peaks is diminished drastically. This result is consistent with the paracrystalline model[9]: a large heat of relaxation is released upon annealing so that atoms in the paracrystal rearrange to form the continuous random network. The last variance curve is the preliminary result of the pre-thinned (700nm) ion-implanted a-Si. We found the amorphous structures may vary from depth to depth (the large reduction of the peak height is observed in the curve) due to the formation of a-Si by double implantation.

The other interesting issue is that the degree of paracrystallity may vary in our ion-implanted sample. We thinned the amorphous layer down to 700nm, and examined the change on the variance curve of the deep amorphous region in the sample. We found the deep region has largely reduced peak heights, compared with the variance result from the near surface region. It follows that the deep region in ion-implanted samples is likely to be a continuous random network. The higher energy irradiation of the deep region than the near surface region upon double implantation may lead to the more disordered atomic configuration. We will examine our ion-implanted sample from depth to depth to understand the structural dependence on irradiation conditions, under which both ion dose and energy change with depth.

DISCUSSION

Treacy and Gibson derived the statistical fluctuation in variable coherence microscopy depends on three- and four- body correlation functions[9]. As yet it is not possible to convert the dark field imaging directly into pair-pair correlation functions because of the complicated mathematical manipulation needed to extract meaningful information from VC-TEM signals. However, the experimental variance curves can be compared with the calculated variance curves, in which the dark field images are acquired according to several simulated CRN and paracrystalline models from molecular dynamic methods. The variance curve of the annealed ion-implanted a-Si is in good agreement with the CRN-K1 model[9]. The diminished peak height signifies the large reduction of atomic ordering. The CRN is a lower energy state, and upon annealing, a large heat of relaxation is released from the ordered configuration of the paracrystal. During the energy drop, atoms have chances to rearrange themselves in the distorted region, such as grain boundaries, and rebond to each other to establish more stable, completely random configuration. The relaxation of the annealed a-Si is consistent with Donovan's calorimetric measurements[12], though they are not aware of structural transition upon thermal annealing effects. In Roorda's differential scanning calorimetric (DSC) results[13], the low temperature homogeneous heat of relaxation (integrated value ≈3.8kJ/mol) equals to one-third of heat of crystallization (integrated value ≈12kJ/mol). From the thermodynamic point of view, the Gibbs free energy (ΔG) decreases with increasing annealing temperature[14]. As annealing temperature approaches the crystallization temperature 600°C, like 580°C in our annealing experiments, ΔG drops drastically, i.e. an amount of heat greater than 5kJ/mol for a-Si is released at that temperature[3] so that we observed the thermal relaxation effect on the large reduction of the peak height, or on the large degree of structural relaxation. The structural relaxation of ion-implanted a-Si has not been characterized in in-situ experiments We believe that is because the accumulation of oxygen atoms in the near surface region of plan-view samples may hinder the amorphous/crystal interface moving, so that the maximum relaxation phenomena in thin areas of a-Si is hardly seen before homogeneous crystallization. In the near future, we will anneal thinned samples in the furnace at the same temperature, 580°C, and examine 20-30nm thick regions in hollow cone measurements to see if any difference on the variance curve is caused by oxygen effects.

Though the paracrystallity of ion-implanted a-Si is observed by variable coherence microscopy, it is unclear how amorphous silicon will form paracrystalline structures by ion implantation. One possible model to interpret the formation of ion-implanted a-Si is thermal spike effect[15]. By the collision cascade of ion irradiation, when the deposited energy density Θ exceeds a critical value, for which a phase change occurs (typically 1.0eV/atom in melting; 3.0-4.0eV/atom in evaporation), it is possible the thermal spike effect exists - some sort of collective "hot spot" occurs in a few picoseconds till the incident ion comes to rest. The energy loss mechanism is categorized in two types: atomic motion and electron excitation. The incident ion energy is equal to the nuclear collision loss energy plus the electronic excitation energy. Winterbon et. al.[16] developed an analytic procedure based on the transport equations to predict the mean volume of the collision cascade. In most heavy ion implantation cases, such as Si and Ge, most energy is deposited into atomic motion (if energy is equally shared among all atoms within the small cascade volume), and the high density cascade will correspond to a significant increase in thermal motion. It is noted that the quenching rate of the small cascade volume is of the order of 10nm in ≈10^{-12} sec, i.e. 10^3 to 10^{-4} m/sec, for which the amorphous regime can be formed because the solidification rate is faster than 15 m/sec. In other words, by large deposited energy density (≈1.0eV/atom) within a small cascade dimension (≈10nm), at quenching time ≈10^{-12} sec, the random liquid like (i.e. amorphous) distribution of atoms can be created in the spike. Though we have not succeeded to link the paracrystalline models with the thermal spike effect directly, analytical procedure of ion implantation will be made to understand thermal spike effects on the formation of paracrystalline structures as observed in ion-implanted a-Si.

CONCLUSIONS

We observed the structural relaxation of the ion-implanted a-Si by the variable coherence microscopy upon thermal annealing. Large heat of relaxation release is associated with the high free energy paracrystalline state transfer into an energetically stable continuous random network. We also found the structural variation with depth (≈700nm deep) in the ion-implanted sample. Further studies on the forming mechanism of a-Si by ion implantation would be continued with more VC-TEM experiments as well as irradiation experiments with systematic variations of ion dose, energy and species.

ACKNOWLEDGMENTS

We gratefully acknowledge the support of the Department of Energy under grant DEFG-96ER45439, the Center of Microanalysis of Materials under, and the National Science Foundation under grant DMR97-05440.

REFERENCES

1. G. L. Olson and J. A. Roth, Mat. Sci. Rep. **3**, 1 (1988).
2. E. F. Kennedy, L. Csepregi, J. W. Mayer and T. W. Sigmon, J. Appl. Phys. **48** (10), 4241 (1977).
3. S. Roorda, W.C. Sinke, J. M. Poate, D. C. Jacobson, S. Dierker, B. S. Dennis, D. J. Eaglesham, F. Spaepen and P. Fuoss, Phys. Rev. B **44** (8), 3702 (1991).
4. M. M. J. Treacy and J. M. Gibson, Acta Crystallogr. **A52** (2), 212 (1996).
5. J. M. Gibson and M. M. J. Treacy, Phys. Rev. Lett. **78** (6), 1074 (1997).
6. J. M. Gibson, M. M. J. Treacy, P. M. Voyles, H-C. Jin and J. R. Abelson, Appl. Phys. Lett. **73** (21), 3093 (1998).
7. J. M. Gibson, J-Y Cheng, P. Voyles, M. M. J. Treacy and D. C. Jacobson, Mat. Res. Soc. Symp. Proc. **504** (1999).
8. J. M. Zuo, Ultramicroscopy **66**, 21 (1996).
9. M. M. J. Treacy and J. M. Gibson, J. Non-Cryst. Solids **231**, 99 (1998).
10. R. Drosd and J. Washburn, J. Appl. Phys. **53** (1), 397 (1982).
11. J. M. Gibson and M. M. J. Treacy, Ultramicroscopy to be published.
12. E. P. Donovan, F. Spaepen, J. M. Poate and D. C. Jacobson, Appl. Phys. Lett. **55**, 1516 (1989).
13. S. Roorda, S. Doorn, W.C. Sinke, P.M. L. O. Scholte and E. van Loenen, Phys. Rev. Lett. **62** (16), 1880 (1989).
14. S. Roorda and W.C. Sinke, Mat. Res. Soc. Symp. Proc. **205**, 9 (1992).
15. J. S. Williams and J. M. Poate, *Ion Implantation and Beam Processing*, Academic Press, New York, 1984, p. 81-97.
16. K. B. Winterbon, P. Sigmund and J. B. Sanders, Kgl. Danske Vid. Selsk. Mat. Fys. Medd. **37** (14) (1970).

ANALYTICAL HIGH RESOLUTION TEM STUDY
OF Au/TiO$_2$ CATALYSTS

T. Akita, K. Tanaka, S. Tsubota, M. Haruta
Osaka National Research Institute, AIST, Midorigaoka1-8-31, Ikeda, Osaka, 563-8577, JAPAN,
akita@onri.go.jp

ABSTRACT

HRTEM(High-Resolution Transmission Electron Microscope), HAADF-STEM (High Angle Annular Dark Field Scanning Transmission Electron Microscope) and EELS(Electron Energy Loss Spectroscopy) techniques were applied for the characterization of Au/TiO$_2$ catalysts. HAADF-STEM provides precise size distributions for Au particles smaller than ~2nm in diameter. It was observed that many small particles under 2nm were supported on anatase TiO$_2$ having a large surface area. The HAADF-STEM method was examined as a way to measure the shape of Au particles. EELS measurements were also used to examine the interface between Au and TiO$_2$ support to study electronic structure effects.

INTRODUCTION

Gold is known to be chemically inert, however, it shows remarkably high activity as a catalyst when finely dispersed with a diameter smaller than 10nm on a metal oxide support[1]. Especially, interesting features have been observed in Au/TiO$_2$ systems. The selective oxidation of propylene to propylene oxide occurs only on anatase TiO$_2$ support, but not on rutile TiO$_2$. The reaction switches from epoxidation to hydrogenation depending on the amount of Au loading in the system of propylene in the presence of hydrogen and oxygen[2]. This phenomenon might depend on the size of the Au particles. Many details of the above characteristic behavior are not clearly understood.

We have studied the crystal structure and electronic structure of the Au/TiO$_2$ catalyst by HR-TEM, EELS and HAADF-STEM. Some basic features were found about Au/anatase TiO$_2$ and Au/rutile TiO$_2$. There was a preferred orientation between Au particles and anatase TiO$_2$, but not with rutile TiO$_2$[3]. This orientation relationship was consistent with the result predicted from geometrical calculation[4-6].

In order to study characteristic size effects of Au particles[7], precise size distribution measurements are indispensable. Therefore, the HAADF-STEM method was investigated to detect small particles under 2nm. HAADF-STEM is powerful for detecting small particles composed of heavy atoms, whereas it is difficult to measure size distributions of small particles by conventional TEM since phase contrast disturbs the detection of particles under 2nm[8].

The possibility of estimating the shapes of Au particles from the intensity of Z-contrast images

253

was examined. Because the activity of a gold catalyst is strongly dependent on the shape of the Au particles[9], it is very important to know the shape of particles. The intensity of a Z-contrast image reflects the number of atoms contained in a particle[10,11]. Therefore, even from the projected image, the shape may be estimated. A method to estimate particle shape statistically is proposed in this study.

EELS measurements in the STEM mode were also used to study the specific electronic structure of the TiO_2 support around Au particles.

EXPERIMENT

Au/TiO_2 (anatase and rutile) catalysts prepared by a deposition precipitation method[12]were used for this investigation. TiO_2 supports that have different surface area (Ishihara Sangyo Co. Ltd. TTO-55(rutile): 39 m²/g, ST-01(anatase): 279 m²/g, ST-21(anatase): 57.3 m²/g) were used. Au was loaded at 10wt%. Observation was performed with a JEM-3000F transmission electron microscope equipped with an annular dark field detector and Gatan Imaging Filter for EELS measurements. Bright field (BF) and HAADF-STEM images were digitally processed through a Gatan-Digiscan unit.

Intensity analysis of Z-contrast image

In a Z-contrast image, the integral image intensity will be proportional to the number of contained atoms. When (intensity)$^{1/3}$ is plotted versus the diameter of Au particles, each point will lie on straight line on condition that the particles have similar shapes. The slope should depend on the mean particle shape[10] as shown in fig.1.

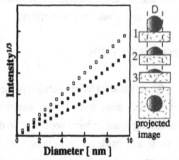

Figure.1 Calculated volume for three kinds of model particles. In Z-contrast image, volume directly corresponds to the Intensity.

For intensity measurements of Au particles, Z-contrast images of 512x512pixel were taken with 42.7sec. The scanned area was about 50nmx50nm. One pixel corresponds to 0.095nm. The inner collection angle of the annular detector was 65mrad. The intensity of Au particles in Z-contrast images was measured with Gatan DigitalMicrograph software. The intensity of individual particles was calculated following the literature[10].

RESULTS

Size distribution from Z-contrast images

Fig.2 shows BF and HAADF- STEM images of Au/anatase TiO_2 catalyst. In the HAADF image,

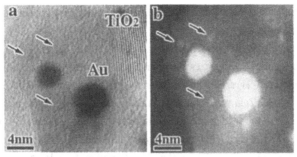

Figure 2 Bright Field STEM image and HAADF-STEM image of Au/TiO$_2$ catalyst. Small particles under 2nm were clearly seen in HAADF-STEM image.

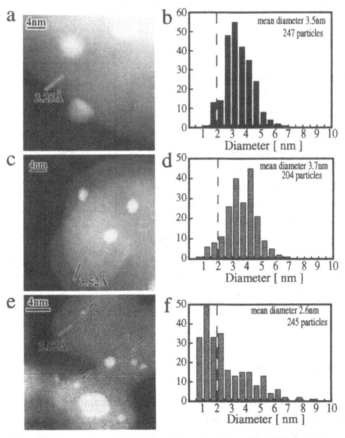

Figure 3 Z-contrast images and size distribution of Au/rutile TiO$_2$(TTO-55)(a,b),Au/anatase TiO$_2$(ST-21)(c,d) and Au/anatase TiO$_2$(ST-01)(e,f). Small particles under 2nm are seen on anatase TiO$_2$(ST-01).

small particles about 1nm are clearly seen, whereas in bright field they are not clear since contrast from the substrate disturbs the image.

Size distributions of Au/anatase TiO$_2$(ST-01 and ST-21) and Au/rutile TiO$_2$(TTO-55) were measured from the Z-contrast images, as shown in fig 3. The mean diameters of Au/rutile (TTO-55) and Au/anatase TiO$_2$(ST-21) are 3.5nm and 3.7nm, respectively. In the case of Au/anatse TiO$_2$(ST-01), small particles under 2nm were observed as shown in fig.3(e) and the mean diameter was 2.6nm. The percentage of the small particles under 2nm is 47% for Au/anatase TiO$_2$(ST-01), while only 6% for Au/rutile TiO$_2$. This might be caused by the difference in surface area (rutile TiO$_2$(TTO-55) 39.1m^2/g, anatase TiO$_2$ (ST-01) 279m^2/g) and number of surface defects which may prevent the aggregation of Au particles. The mean diameter shows almost no dependence on crystal structure; little difference was found in the size distributions shown in fig3b and 3d that have nearly same surface area (TTO-55: 39.1 m^2/g, ST-21: 57.3 m^2/g) but different crystal structures.

Intensity analysis of Z-contrast images

Preliminary experiments on intensity analysis of Z-contrast was done for Au on active carbon in order to confirm that the intensity of Z-contrast images depends on the volume of the Au particle. (Intensity)$^{1/3}$ was plotted versus diameter for individual Au particles. The points lie on straight lines as shown in fig.4. This means the Z-contrast intensity depends on the particle volume. The relation of (Intensity)$^{1/3}$ and particle diameter for Au on rutile TiO$_2$ is also presented in fig.4. A slight difference in slope between Au/rutile TiO$_2$ and Au/active carbon was seen. This result may indicate that the Au particles tend to have a more hemispherical shape on a TiO$_2$ support compared with Au particles on active carbon.

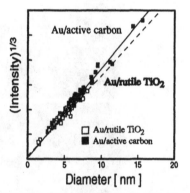

Figure 4 Intensity of Au particles in Z-contrast images in the case of Au on active carbon and Au on rutile TiO$_2$. The difference is seen in these slope.

But it may also reflect that particles tend to be more spherical when they are large since the size distribution of particles on active carbon is larger than for those on rutile TiO$_2$.

Line-Scan EELS measurement

EELS measurements made in STEM line-scan mode were examined for the Au/rutile TiO$_2$ catalyst in order to study the electronic structure of local areas such as the interface between Au and TiO$_2$ support. Fig.5 shows the HAADF-STEM image of rutile TiO$_2$ and corresponding EELS spectra. Ti-L and O-K edges are seen from the TiO$_2$ support at 458eV and 535eV. The positions of the line scans are shown in the images. Spectra from the interface between Au and TiO$_2$ and from

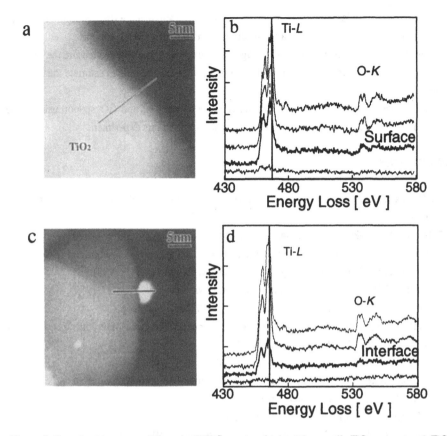

Figure 5 Z-contrast images and line scan EELS spectra obtained from rutile TiO_2 support and TiO_2 around Au particle. in both case, EELS spectra shows the feature of reduced titanium oxide. detectable specific feature caused by depositting Au was not seen in these spectra.

the surface of TiO_2 are different from the spectra for bulk TiO_2. Four peaks are observed at the Ti-L edge for bulk TiO_2, but only two are seen in surface and interface spectra. The features in surface and interface spectra are attributed to metallic titanium or reduced titanium oxide[13]. This might be caused by electron irradiation damage during observation since such areas are very thin and sensitive. Under this experimental condition, no specific feature was found at the interface between Au and TiO_2 support.

CONCLUSIONS

HAADF-STEM and EELS methods were applied for the characterization of Au/TiO_2 catalysts. Precise size distributions of Au particles under 2nm were measured and it was found that there were many small particles under 2nm on anatase TiO_2(ST-01) and that the size distribution depends on the surface area of support and not on the crystal structure of the TiO_2 support.

Intensity measurements of Z-contrast images were examined as a way to measure the shape of particles. Preliminary experiments showed that this method may be able to estimate the shape of particles.

EELS measurements were also done for the interface between Au and TiO_2 support using a line scan method, but detectable specific features were not found in this experiment.

REFERENCES

1. M. Haruta, S.Tsubota, T.Kobayashi, H.Kageyama, M.Genet and B.Delmon, J.Catal. **144,** 175(1993).
2. T.Hayashi, K.Tanaka and M.Haruta, J.Catal. **178,** 566(1998).
3. T.Akita, K.Tanaka, S.Tsubota and M.Haruta, J.Electro.Microsco.(1999)submitted
4. Y.Ikuhara and P.Pirouz, Microscopy Research and Technique **40,**317(1998)
5. S.Stemmer, P.Pirouz, Y.Ikuhara and R.F. Davis, Phys. Rev. Lett. **77,**1797(1996).
6. Y.Ikuhara and P.Pirouz, Material Science Forum, **207-209,**121(1996).
7. K.Tanaka, T.Hayashi and M.Haruta, Interface Science and Material Interconection Proc. Of JIMIS-8, The Japan Institute of Metals,p.547(1996)
8. M.M.J.Treacy and A.Howie, J.Catal. **63,** 265(1980).
9. S.Tsubota, D.A.H.Cunningham, Y.Bando and M.Haruta, Preparation of catalysts VI. edited by G.Poncelet et.al.(Elsevier, Amsterdam,1995)p.227.
10. M.M.J.Treacy and S.B.Rice, J. Microsco. **156,** 211(1989).
11. A.Singhal, J.C.Yang and J.M.Gibson, Ultramicroscopy **67,**191(1997).
12. S.Tsubota, M.Haruta, T.Kobayashi, A.Ueda and Y.Nakahara, Preparation of catalysts V. edited by G.Poncelt,et al. , (Elsevier, Amsterdam,1991)p.695.
13. R.D.Leapman, L.A.Grunes and P.L.Fejes, Phys.Rev.B **26,**614(1982).

HIGH-RESOLUTION SCANNING ELECTRON MICROSCOPY AND MICROANALYSIS OF SUPPORTED METAL CATALYSTS

JINGYUE LIU
Monsanto Corporate Research, Monsanto Company, 800 N. Lindbergh Blvd., St. Louis, MO 63167

ABSTRACT

The use of a high-brightness field emission gun and novel secondary electron detection systems makes it possible to acquire nanometer-resolution surface images of bulk materials, even at low electron beam voltages. The advantages of low-voltage SEM include enhanced surface sensitivity, reduced sample charging on non-conducting materials, and significantly reduced electron range and interaction volume. High-resolution images formed by collecting the backscattered electron signal can give information about the size and spatial distribution of metal nanoparticles in supported catalysts. Low-voltage XEDS can provide compositional information of bulk samples with enhanced surface sensitivity and significantly improved spatial resolution. High-resolution SEM techniques enhance our ability to detect and, subsequently, analyze the composition of nanoparticles in supported metal catalysts. Applications of high-resolution SEM imaging and microanalysis techniques to the study of industrial supported catalysts are discussed.

INTRODUCTION

The development of a commercially viable catalyst requires understanding the factors that enhance selectivity, increase activity, and improve lifetime. Most industrial catalysts are highly complex mixtures of different phases including active components, promoters, stabilizers, supports, etc. The understanding of the structure-performance relationship is crucial to successful development of industrial catalysts. The size and spatial distribution and the surface structure of metal or alloy particles are the most important parameters in determining the performance of supported metal catalysts. The morphology of the supports and the particle-support interactions, however, can profoundly affect the distribution and the nature of highly dispersed metal or alloy particles. A complete structural characterization of both the metal or alloy particles and the supports is essential for a full understanding of the catalyst's performance.

Traditionally, transmission electron microscopy or scanning transmission electron microscopy (TEM/STEM) has to be used to visualize highly dispersed metallic nanoparticles. A major limitation of TEM/STEM techniques is, however, the stringent requirement of samples that can be examined: useful information can be extracted from only very thin areas of a sample. It is very difficult, if not impossible, to use TEM/STEM techniques to extract information about the surface properties of bulk catalyst samples such as powders, cylinders, or beads that are most frequently used in industrial catalytic reactions.

With the use of a high-brightness field emission gun and novel electron detection systems in field emission scanning electron microscopes (FE-SEM), it is now possible to examine surface features of bulk samples on a nanometer scale [1]. Highly dispersed metal nanoparticles as well as detailed surface topography of catalyst supports can now be examined in FE-SEM instruments [2-3]. Secondary electron (SE) and backscattered electron (BE) signals can be used independently or in combination to give complementary information. The high stopping power of electrons at low energies significantly reduces the volume of electron-specimen interactions and the range of the electrons penetrating into the sample. Thus, high-resolution chemical microanalysis can be

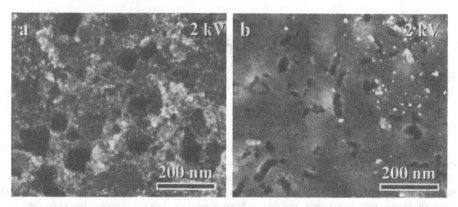

Figure 1. High-resolution and low-voltage SE images of a Pt/carbon catalyst revealing detailed surface topography and pore structure of the highly activated carbon powders.

realized at low beam voltages by collecting and analyzing the emitted characteristic X-ray photons. In this paper, we will demonstrate how these newly developed techniques can be used to provide high-spatial-resolution morphological, structural, and chemical information of supported metal catalysts.

HIGH-RESOLUTION SECONDARY ELECTRON IMAGING

Image resolutions of about 1 nm and 2.5 nm can be obtained at incident beam voltages of 15 kV and 1 kV, respectively, in the most recent generation FE-SEM instruments. At low voltages the electron-specimen interactions are restricted to the near surface regions of a specimen. Therefore, even the SE2 signal (secondary electrons generated by backscattered electrons) originates from a region near the sample surface and has a narrow spatial distribution. Compared to high-voltage SE images, low-voltage SE images are more surface-sensitive, especially at low magnifications.

Figure 1 shows high-resolution SE images of a Pt/carbon powder catalyst. Detailed surface morphology of the highly activated carbon powders is clearly revealed. The pore structure of the activated carbon is highly inhomogeneous and macro-pores, meso-pores, and micro-pores coexist. Quantitative information about the size, shape, and connectivity of the carbon pores can be extracted by detailed analyses of these high-resolution SE images.

Small metal particles located on top of, or very close to, the support surface can also be imaged with high contrast in high-resolution SE images, although the detailed contrast mechanism(s) is complicated and not well understood [4]. The visibility of metal nanoparticles in high-resolution SE images depends on many parameters including the size and location of the metal particles, the energy of the incident electrons, the size of the electron probe, and the surface cleanness of the metal nanoparticles. Figure 2a shows a high-resolution SE image of a Pt/carbon catalyst, clearly showing Pt nanoparticles as small as 2 nm or less in diameter with high contrast. The pore structure of the highly activated carbon support is also clearly revealed. The relative locations of the Pt nanoparticles with respect to the pores of the carbon support can also be extracted. Information about sintering of small metal particles and blocking of the pores of catalyst supports during catalytic reactions can be extracted from high-resolution SE images. This type of knowledge is crucial to understanding the deactivation processes of supported metal catalysts.

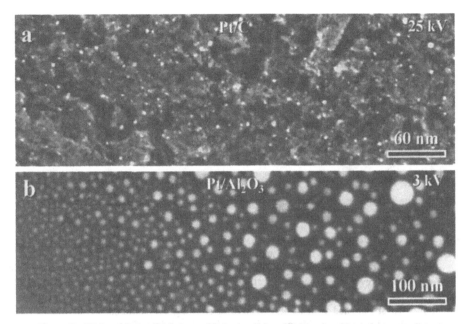

Figure 2. High-resolution SE images of Pt/carbon (a) and Pt/alumina (b) catalysts revealing
highly dispersed Pt nanoparticles with high contrast.

Because of electron charging effects, non-conducting materials are often coated with a
conductive layer prior to examination in conventional SEM instruments. One of the advantages
of low voltage FE-SEM is that there generally exists an electron energy at which the electron
beam induced charging is minimized or eliminated, even though the specimen is non-conducting.
This dynamic charge neutralization makes it possible to image metal particles dispersed onto non-
conducting catalyst supports. Figure 2b shows a low voltage (3 kV) SE image of a Pt/α-alumina
catalyst, clearly revealing the size and spatial distribution of Pt nanoparticles.

This type of information about size and spatial distribution of metal particles and support
morphology of bulk catalyst samples is extremely important for understanding and optimizing the
performance of supported metal catalysts. High-resolution and low-voltage SEM is the only
imaging technique now available for providing nanometer-resolution surface details of highly
heterogeneous powdered catalysts.

HIGH-RESOLUTION BACKSCATTERD ELECTRON IMAGING

The BE signal is one of the most useful signals generated inside a FE-SEM instrument.
Because of its high atomic-number contrast (or material contrast) high-resolution BE images can
be effectively used to identify metal or alloy particles highly dispersed in supported metal
catalysts. With high-energy electrons, the backscattering coefficient increases monotonically with
the atomic number of the sample. Below 5 keV, however, the backscattering coefficient no
longer increases monotonically with increasing atomic number. For example, the backscattering
coefficient of light-element materials increases with decreasing electron energy while that of
heavy-element materials decreases with decreasing electron energy. Therefore, low-voltage BE

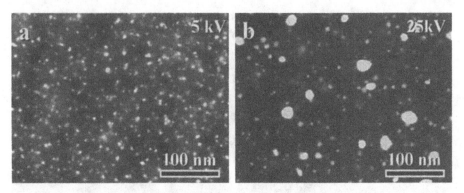

Figure 3. High-resolution backscattered electron images of Pt/carbon catalysts showing Pt nanoparticles with high atomic number contrast.

images need to be interpreted cautiously. At very low electron energies (< 1 keV), the BE signal becomes as surface sensitive as that of the SE signal.

Figure 3a shows a high-resolution BE image of a Pt/carbon catalyst obtained with 5 keV electrons, clearly revealing a uniform distribution of Pt nanoparticles with an image quality comparable to that of STEM annular dark field images. Figure 3b shows a BE image of a Pt/carbon catalyst obtained with 25 keV electrons, showing very different Pt dispersion. The contrast of Pt nanoparticles in BE images is complicated and depends on many parameters [4]. Generally, Pt particles located on the surface of the carbon support give higher image intensity than those located below the carbon surface. Because of the effect of beam broadening, Pt particles located deep inside the carbon are shown with a faint, diffuse contrast in figure 3b. With further improvement in the collection efficiency of BE detectors, high-resolution BE images will prove to be more and more useful for characterizing supported metal catalysts.

The combination of SE imaging with BE imaging can provide useful information on correlating the location of metal particles with respect to support topography. Figure 4 shows a set of images of the same area of a Pt/carbon catalyst obtained by collecting the SE and BE signals with 3 keV and 25 keV electrons. While the SE images clearly show the topography of the highly activated carbon support the BE images unambiguously reveal the size and spatial distribution of the Pt nanoparticles. Furthermore, by comparing the contrast variations in SE and BE images obtained with different electron energies we can extract information about the depth distribution of Pt nanoparticles. For example, most of the larger Pt particles shown in figure 4 are located on the external surface of the carbon support while some of the smaller Pt particles shown in figure 4d are definitely located deep inside the carbon support.

HIGH-RESOLUTION AND LOW-VOLTAGE X-RAY MICROANALYSIS

One of the most useful analytical techniques available in the FE-SEM is X-ray energy-dispersive spectroscopy (XEDS). The spatial resolution of XEDS depends on the range of the incident electrons and the nature of the electron-specimen interactions. In most solid materials, the electron range at high energies (>15 keV) is of the order of micrometers to tens of micrometers which makes high spatial resolution chemical analysis of bulk samples impossible. The tremendous decrease in the interaction volume at low electron energies, however, makes it possible to perform high-resolution chemical microanalysis of bulk samples [5].

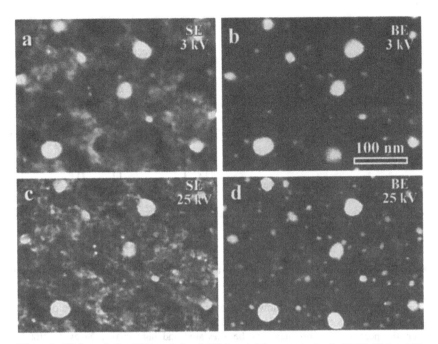

Figure 4. High-resolution SE (a and c) and BE (b and d) images obtained with 3 keV (a and b) and 25 keV (c and d) electrons, respectively.

Figure 5a shows a SE image of a Pt/α-alumina catalyst obtained under microanalysis conditions with 4 keV electrons. The Pt particles are revealed with extremely high contrast and some surface steps on the alumina crystal are also shown with observable contrast. Figure 5b shows XEDS spectra obtained from the alumina support (spectrum A) and from a Pt platelet (Spectrum B). There is no aluminum signal detected in spectrum B because the electron-specimen interactions are confined within the Pt particle. It is also interesting to note that the low-energy background signal (Bremsstrahlung X-rays) is significantly enhanced when the electron beam is positioned on the Pt particle. This has some practical implications in obtaining high-resolution elemental maps or X-ray spectrum-images. The enhancement of the carbon signal shown in spectrum B may be related to the preferential deposition of carbon on Pt particles.

Low-voltage microanalysis not only improves the spatial resolution but also significantly enhances the surface sensitivity of the XEDS technique. Monte Carlo simulations suggest that a layer of carbon as thin as 1 nm in thickness, deposited on alumina or Pt surfaces, can be detected in XEDS spectra obtained with electron energies < 2 keV. The high surface sensitivity of low-voltage XEDS is of great importance in studying catalyst deactivation processes. For example, in many hydrocarbon and related reactions, such as catalytic cracking, reforming, hydrodesulfurization, and various dehydrogenations, carbonaceous deposits (coke) are slowly formed on the surface of the catalyst. The deposition of coke causes a reduction in catalytic activity by blocking active sites. High spatial resolution surface analysis is required to understand the processes of coke formation. Low-voltage microanalysis has proved extremely useful in analyzing the deactivation processes of heterogeneous catalysts.

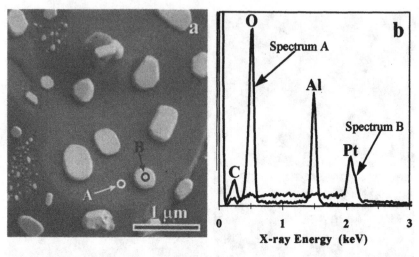

Figure 5. SE image of a Pt/alumina catalyst (a) and XEDS spectra (b) obtained from the alumina support and a Pt platelet.

One of the main limitations of low-voltage microanalysis is the peak overlap of low-energy X-ray photons. With the recent development of high energy-resolution X-ray spectrometers [6], however, the peak overlap problem may be alleviated and high quality X-ray spectra can be obtained for quantitative analyses of the sample composition.

SUMMARY

FE-SEM techniques are powerful high-resolution surface analytical tools for catalyst characterization. A major advantage of low-voltage FE-SEM is the ability to directly examine non-conducting catalyst supports without prior coating. The significant decrease in the interaction volume at low voltages makes it possible to perform high-resolution chemical microanalysis of bulk samples. Low-voltage SEM imaging and XEDS techniques enhance our ability to detect and analyze the chemical composition of nanoparticles in supported catalysts, surface coatings or deposits, small inclusions or precipitates, and a variety of biological samples with high surface sensitivity. The combination of high-resolution SE and BE imaging with low-voltage microanalysis will become more and more useful in characterizing supported metal catalysts and other inhomogeneous material systems.

REFERENCES

1. D. C. Joy and J. B. Pawley, Ultramicroscopy **47**, p. 80 (1992).
2. D. J. Smith, M. H. Yao, L. F. Allard, and A. K. Datye, Catalysis Letters **31**, p. 57 (1995).
3. J. Liu in *Electron Microscopy 1998, Vol. 2*, edited by H. A. Calderon Benavides and M. Jose Yacaman *(Proc. 14th ICEM*, Cancun, Mexico 1998) p. 399-400.
4. J. Liu, Microscopy and Microanalysis (2000), in press.
5. E. D. Boyes, Advanced Materials **10**, p. 1,277 (1998).
6. D. A. Wollman, K. D. Irwin, G. C. Hilton, L. L. Dulcie, D. E. Newbury, and J. M. Martinis, J. Microscopy **188**, p. 196 (1997).

ON-PARTICLE EDS ANALYSIS OF BIMETALLIC, CARBON-SUPPORTED CATALYSTS

Deborah L. Boxall,[a] Edward A. Kenik[b] and Charles M. Lukehart[a]
[a]Department of Chemistry, Vanderbilt University, Nashville, TN 37235
[b]Metals and Ceramics Division, Oak Ridge National Laboratory, Oak Ridge, TN 37831

ABSTRACT

Thermolysis of single-source molecular precursors has been used to prepare carbon-supported Pt_1Sn_1, Pt_1Ru_1, and Pt_3Mo_1 alloy nanoparticles. On-particle, high spatial resolution energy dispersive spectroscopy (HR-EDS) has been used to determine the extent of compositional variability between particles in the nanocomposite. The loss of volatile Ru and Mo oxides during the course of the HR-EDS analysis results in average compositions greatly enriched in Pt. Mathematical correction for loss of these volatile species results in average metal:metal ratios comparable to those measured by bulk elemental analysis.

INTRODUCTION

Optimization of the performance of direct methanol fuel cells (DMFCs) is an active area of research since DMFCs offer a means to generate electricity locally through the platinum-catalyzed oxidation of methanol, a readily available liquid fuel source. One of the limiting factors to achieving commercial feasibility of these devices is the poisoning of the active platinum sites on the anode by adsorbed CO.[1] However, it has been found by several groups that incorporation of a secondary metal such as Mo, Ru or Sn into the anode catalyst will ameliorate this poisoning effect.[2] The preparation of bimetallic catalysts with specific stoichiometries and surface compositions has therefore become an active area of research.

We have been exploring the use of heterometallic single-source molecular precursors as a means to prepare bimetallic nanocomposite materials useful in DMFCs.[3] We have determined that through judicious control of the thermal treatment conditions used to degrade the molecular precursor, it is possible to prepare bimetallic catalysts with the same overall metal stoichiometry as that initially present in the molecular precursor.

While the bulk stoichiometry of the metal catalyst can be determined via chemical analysis or powder x-ray diffraction, the compositional variability of individual particles remains ambiguous. High-resolution TEM coupled with EDS allows the elemental composition of individual metal particles in the composite material with diameters as small as 1 nm to be determined. We report the results of on-particle HR-EDS obtained from carbon-supported Pt_1Sn_1, Pt_1Ru_1 and Pt_3Mo_1 nanocomposites prepared using single-source molecular precursors. While the results from several different bimetallic catalysts indicate that there is some compositional size dependence, this dependence is consistent with loss of volatile oxide species from the surface of the metal particles. Application of a mathematical model to compensate for this presumed material loss results in calculated average metal:metal ratios within experimental error of those measured by bulk chemical analysis.

Mat. Res. Soc. Symp. Proc. Vol. 589 © 2001 Materials Research Society

EXPERIMENTAL

Nanocomposite preparation. A Pt_1Sn_1/Vulcan carbon nanocomposite of 26 wt% total metal (**1**) was prepared by sequentially depositing a total of 556 mg of *trans*-$(PEt_3)_2ClPtSnCl_3$ onto 251 mg of a high surface area (250 m^2/g), graphitic carbon powder (Cabot Corp. Vulcan Carbon, XC72R).[4] Following each deposition, the carbon was placed in an alumina boat in a Lindberg programmable furnace and heated to 250° C under air and then up to 650° C under getter gas (10%H_2:90% N_2) at a rate of 15°/min. Upon reaching 650° C, the atmosphere was switched to N_2 and held at 650° C for five minutes. Following the final deposition, the sample was annealed at 650° C under N_2 for an hour before cooling under flowing N_2. Elemental analysis of the final nanocomposite (wt%): C, 55.34; H, < 0.5; Pt, 18.53; Sn, 7.86; P, 0.48.

A Pt_1Ru_1/Vulcan carbon nanocomposite of 16 wt% total metal (**2**) was prepared by depositing $(\eta^2\text{-}C_2H_4)ClPt(\mu\text{-}Cl)_2Ru(\eta^3,\eta^3\text{-}C_{10}H_{16})Cl$ onto Vulcan carbon from an acetone solution[5] and then subjecting the treated carbon to a total of 100 s of microwave irradiation as previously described.[6] Elemental analysis of final nanocomposite (wt%): C, 78.63; H, < 0.5, Pt, 10.77; Ru, 5.19.

A 47 wt% total metal Pt_1Ru_1/graphitic nanofiber nanocomposite (**3**) was prepared by sequentially depositing 1.20 g of $(\eta^2\text{-}C_2H_4)ClPt(\mu\text{-}Cl)_2Ru(\eta^3,\eta^3\text{-}C_{10}H_{16})Cl$ onto 600 mg of graphitic nanofibers[7] from an acetone solution. Between each deposition, the treated carbon was heated to 350° C under getter gas at a rate of 15°/min and then cooled to ambient conditions under N_2. Following the final deposition, the carbon was heated to 350° C under air, then to 650° C under getter gas at a rate of 15°/min. The sample was annealed at 650 °C for 1 hour then cooled to room temperature under N_2. Elemental analysis of final nanocomposite (wt%): C, 47.45; Pt, 32.12; Ru, 14.85.

A Pt_3Mo_1/Vulcan carbon nanocomposite of 43 wt% total metal (**4**) was prepared by depositing 860 mg of $\{Pt_3Mo[(PPh_2)_2CH_2]_3(C_5H_5)(CO)\}[BPh_4]$ in two portions onto 250 mg of Vulcan carbon from a CH_2Cl_2 solution.[8] Between deposition steps, the vacuum dried treated carbon was heated in a Lindberg programmable tube furnace to 350° C under air. After the second deposition, the treated carbon was heated to 350° C under air, then to 650° C under getter gas at a rate of 15°/min. The sample was then annealed at a temperature of 650° C for one hour under N_2 before cooling to ambient conditions under N_2. The final stage of the heat treatment consisted of heating to 350° C under air at a rate of 30°/min and then cooling to room temperature. Elemental analysis (wt%): C, 31.13; H, 1.37; P, 6.91; Mo, 4.2; Pt, 25.1.

Due to the presence of oxygen-containing functionalities on the Vulcan carbon support[9] and the positive oxidation potentials of the alloying elements,[10] the balance of each elemental analysis was assumed to be oxygen.

General Methods. Samples for transmission electron microscopy imaging were prepared by placing a drop of a nanocomposite/CH_2Cl_2 suspension on a carbon-coated, copper grid. Single-particle HR-EDS was used to determine particle composition using a Philips CM200FEG 200kV TEM equipped with an Oxford light element EDS detector and an EMiSPEC Vision data acquisition system. The HR-EDS data were collected in scanning transmission electron microscopy (STEM) mode at a tilt angle of 35°, an acceleration voltage of 200 kV, and a collection time of 20 s with a 1.4 nm probe in the stopped-scan mode. Calibration of the integrated intensities from the $PtL_{\alpha1}$, $RuK_{\alpha1,2}$, $SnL_{\alpha1}$, or $MoK_{\alpha1,2}$ lines as appropriate was accomplished by collecting the emissions for 100 s from several 1 μm^2 areas of a sample of known composition (Pt_1Sn_1 and Pt_3Mo_1) or precursor-doped Vulcan carbon.[11] XRD scans were obtained using a Scintag X_1 θ-θ automated powder diffractometer with a Cu target, and a Peltier cooled solid-state detector. Bulk chemical analyses were performed by Galbraith Laboratories, Knoxville, TN.

RESULTS AND DISCUSSION

Powder XRD analysis of the four nanocomposites resulted in patterns consistent with the formation of the 1:1 niggliite Pt_1Sn_1 phase,[12] a Pt_1Ru_1 alloy[13] and a Pt_3Mo_1 alloy.[2a,14] The lattice constants calculated from the XRD peak positions for the four nanocomposites are shown in Table I.

Nanocomposite	M	XRD		EDS Pt_xM_1
		a_{Exp} (Å)	a_{Ref} (Å)	
1	Sn	4.119(6)	4.100	1.00
2	Ru	3.870(9)	3.862	1.64
3	Ru	3.852(4)	3.862	4.64
4	Mo	3.911(4)	3.908	4.11

Table I: XRD results and initial volume-weighted average particle compositions determined by on-particle EDS.

The volume-weighted average nanoparticle composition for the four nanocomposites obtained from EDS measurements are also shown in Table 1. Since it is reasonable to assume that the average particle composition should be comparable to the expected metal:metal ratios indicated by the powder XRD patterns, it is surprising to note that only nanocomposite **1** has an average composition consistent with the XRD evidence.

A comparison of the relationship between the integrated intensities for Pt and Sn, I_{Pt} and I_{Sn} respectively, in **1** and I_{Pt} and I_{Ru} for **2** is shown in Figure 1. As expected from the consistency between the XRD and the average EDS composition results, a plot of I_{Sn} vs. I_{Pt} reveals a linear relationship between the two elements. However, a similar plot, shown in Figure 1b for **2**, results in a parabolic relationship consistent with loss of volatile ruthenium oxides from the surface of the nanoparticles during the course of the EDS analysis.[6] Computationally adding back 0.5 monolayers of ruthenium to the outside of the particles results

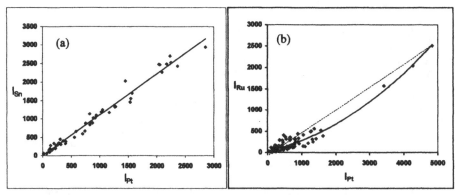

Figure 1: (a) Comparison of the integrated Sn and Pt intensity relationship for **1** and (b) the integrated intensity relationship for **2**.

in a volume-weighted average composition of $Pt_{1.08(3)}Ru_1$ as compared to the bulk elemental analysis composition of $Pt_{1.07}Ru_1$. The mean corrected Pt/Ru atomic ratio for **2** was 1.12 with a standard deviation of ±0.39 corresponding to a 1σ composition variation range of Pt_2Ru_3 to Pt_3Ru_2.

While the absence of any pure or greatly enriched Pt particles in **2** indicates that thermolysis of single-source molecular precursors affords stoichiometric control over the initial composition of the nanoparticles, the preservation of the stoichiometry imparted by the precursor depends upon judicious control of the thermal treatment used to decompose the precursor. The type of EDS results expected to be observed from phase separated nanoparticles is exemplified by nanocomposite **3**. An annular dark-field image of **3** is shown in Figure 2a. As can be seen, the nanocomposite contains particles of dramatically different sizes, ranging from *ca.* 3 nm for the smallest nanoparticles to many times that value for the larger particles. On-particle EDS measurements determined that the large particles were predominantly Ru with a maximum Pt content of about 20%. The integrated intensity plot for this nanocomposite, shown in Figure 2b, consists of two distinct regions with the large Ru-rich particles high on the I_{Ru} axis and the smaller Pt-Ru alloy nanoparticles lying along the abscissa. If the data points corresponding to Pt:Ru ratios less than 0.20 are discarded, a best fit line through the I_{Ru} vs. I_{Pt} plot has a roughly parabolic shape suggesting that there may have been some loss of Ru from the surface. After excluding the large particles, a computationally corrected, volume-weighted average nanoparticle composition of $Pt_{1.14(0.02)}Ru_1$ is obtained as compared to the bulk composition of $Pt_{1.12}Ru_1$. The close agreement between the bulk elemental analysis and the volume-weighted average composition for just the small particles indicates that while there is obviously some phase separation present, the amount of Ru used to form the large particles may not be significant.

The moderate scatter of the integrated Mo and Pt intensities collected from **4** indicates that there may have been a small amount of phase separation present in this sample (see Figure 3). However, even in the case of gross phase separation, mass conservation requires that the average nanoparticle composition should still be comparable to that obtained by chemical analysis. Instead, what is observed is a 40 atomic% Pt enrichment of the nanoparticles over the

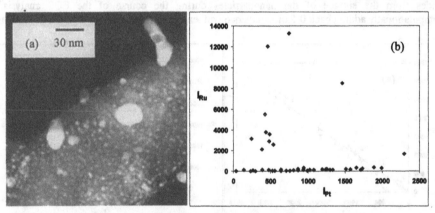

Figure 2: (a) Annular dark-field STEM image of nanocomposite **3** showing the large external Ru-rich particles and smaller Pt-Ru alloy particles. (b) Integrated Ru and Pt intensities for **3** collected during on-particle EDS.

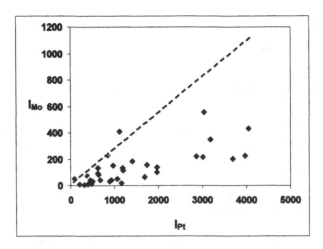

Figure 3: Integrated Mo *vs.* Pt intensity for **4**. Dashed line illustrates the linear relationship expected for a 3:1 Pt/Mo alloy in the absence of MoO_x volatilization.

bulk value. Considering the fact that Mo, like Ru, is also capable of forming volatile surface oxides,[15] it is likely that the discrepancy between the average EDS results and the bulk analysis is due to loss of volatile surface Mo containing species. Modifying the calculations to account for the different lattice dimensions of the Pt_3Mo_1 alloy and assuming an initial surface coverage of 25% Mo results in an average calculated composition of $Pt_{3.08(4)}Mo_1$ as compared to the bulk value of $Pt_{2.94}Mo_1$.

CONCLUSIONS

In the absence of volatile surface species, on-particle HR-EDS analysis provides an indication of <u>both</u> the stoichiometry of the metal nanoparticles and their compositional variability. If volatile surface species are present, the resulting average particle composition determined by EDS is enriched in the non-volatile element present. Mathematical corrections for loss of volatile surface species result in average metal:metal ratios comparable to those obtained by bulk chemical and XRD powder pattern analysis.

ACKNOWLEDGMENTS

Research support provided by the U.S. Army Research Office under grant numbers DAAH04-95-0146, DAAH04-96-1-0179, DAAH04-96-1-0302 and DAAG55-98-1-0362 is gratefully acknowledged. On-particle EDS and HR-TEM performed at the Oak Ridge National Laboratory Shared Research Equipment (SHaRE) User Facility was sponsored by the Division of Materials Sciences, U.S. Department of Energy, under contract DE-AC05-96OR22464 with Lockheed Martin Energy Research Corp. and through the SHaRE Program under contract DE-AC05-76OR00033 with Oak Ridge Associated Universities. We are indebted to Eve S. Steigerwalt and Krzysztof C. Kwiatkowski for preparing the PtRu/graphitic nanofiber and Pt_3Mo/Vulcan carbon nanocomposites, respectively.

REFERENCES

1. S. Wasmus and A. Küver, *J. Electroanal. Chem.* **461**, 14 (1999).

2. a) B.N. Grgur, N.M. Markovic, and P.N. Ross, *J. Electrochem. Soc.* **146**(5), 1613 (1999).
b) H.A. Gasteiger, N. Markovic, P.N. Ross, Jr., and E.J. Cairns, *J. Phys. Chem.* **98**, 617 (1994).
c) M.J. Gonzalez, C.T. Hable, and M.S. Wrighton, *J. Phys. Chem. B* **102**, 9881 (1998).

3. C.M. Lukehart, D.L. Boxall, J.D. Corn, M. Hariharasarma, W.D. King, K.C. Kwiatkowski, E.S. Steigerwalt, and E.A. Kenik, *ACS Fuel Chem. Div. Preprints* **44**(4), 982 (1999).

4. P.S. Pregosin, and S.N. Sze, *Helv. Chim. Acta* **61**(5), 1848 (1978).

5. Precursor prepared via a modification of that reported by K. Severin, K. Polborn, and W. Beck, *Inor. Chim. Acta* **240**, 339 (1995).

6. D.L. Boxall, E.A. Kenik, and C.M Lukehart, *Proceedings of the ASME: Advanced Energy Systems Division* **39**, 327 (1999).

7. N.M. Rodriguez, M.S. Kim, and R.T.K. Baker, *J. Catal.* **144**, 93 (1993).

8. K.C. Kwiatkowski and C.M. Lukehart, *Abstr. Pap. Am. Chem. Soc.* **218**(Pt. 1), 466-INOR (1999).

9. C. Prado-Burguete, A. Linares-Solano, F. Rodriguez-Reinoso, and C. Salinas-Martinez de Lecea, *J. Catal.* **115**, 98 (1989).

10. a) D.R. Lide, ed., *Handbook of Chemistry and Physics*, 71st ed. (CRC Press, Boston, 1990), pp. 8-18, 8-19. b) D.R. Rolison, P.L. Hagans, K.E. Swider, and J.W. Long, *Langmuir* **15**, 774 (1999).

11. D.B. Williams and C.B. Carter, *Transmission Electron Microscopy* (Plenum, New York, 1996), p. 605.

12. PDF card #25-614.

13. V. Radmilovic, H.A. Gasteiger, and P.N. Ross, Jr. *J. Catal.* **154**, 98 (1995).

14. W. Pearson, *A Handbook of Lattice Spacings and Structures of Metals and Alloys* (Pergamon Press, Oxford, 1958), p. 755.

15. M.S. Samant, A.S. Kerkar, S.R. Bharadwaj, S.R. Dharwadkar, *J. Alloys Compounds* **187**, 373 (1992).

Interfaces in Metals
and Ceramics

STRUCTURE REFINEMENT OF S-PHASE PRECIPITATES IN Al-Cu-Mg ALLOYS BY QUANTITATIVE HRTEM

R. KILAAS *, V. RADMILOVIC *,**, and U. DAHMEN *
*National Center for Electron Microscopy, Lawrence Berkeley Laboratory, University of California, Berkeley, CA, USA,
**University of Belgrade, Faculty of Technology and Metallurgy, Dept. of Physical Metallurgy, Belgrade, Yugoslavia

ABSTRACT

The crystal structure of the Al_2CuMg S-phase precipitate in an Al matrix has been determined by quantitative high resolution electron microscopy. This work combines techniques of image processing and quantitative comparison between experimental and simulated images with automatic refinement of imaging and structural parameters.

INTRODUCTION

Over the last few years much progress has been made in the field of quantitative high resolution transmission electron microscopy (HRTEM) based on pattern recognition techniques for image comparison and automatic structure determination/refinement [1-7]. These techniques are particularly important in the analysis of nanoscale structures, which may not be stable at larger size and are often inaccessible to X-ray analysis. The present work is an application of HRTEM to the study of Al_2CuMg intermetallic S-phase precipitates, and forms part of an ongoing study of the formation mechanisms, morphology and interface structure of S-phase precipitates in Al-Cu-Mg alloys [8].

Al-Cu-Mg based alloys are of significant interest for many structural applications, due to their low weight, mechanical strength and corrosion resistance. The mechanical properties of these alloys are based on a dispersion of S-phase precipitates which have been shown to alter the deformation mode [9]. S-phase precipitates have the composition Al_2CuMg and form as laths along $<100>_{Al}$, with $\{012\}_{Al}$ habit [10]. The following crystallographic orientation relationship is observed[11-13]:

$$[100]_S//[100]_{Al} \ (lath \ axis) \quad (001)_S//(021)_{Al} \ (habit \ plane) \quad [010]_S // [012]_{Al},$$

indicating the presence of 12 S-phase orientation variants in the Al-matrix. The crystal structure of this intermetallic phase has been studied using different diffraction techniques for more than five decades. Since the mechanical behavior of Al alloys strengthened by the S-phase cannot be fully understood without a clear understanding of its crystal structure, morphology and interface structure, quantitative characterization of these crystallographic and microstructural features is essential for future alloy development.

Several models have been proposed for the structure of the S-phase. The first model was given by Perlitz and Westgren (PW) [14] based on X-ray diffraction. The unit cell of the PW model is shown in Fig. 1. The structure is orthorhombic with unit cell dimensions $a = 0.4$ nm, $b = 0.923$ nm, and $c = 0.714$ nm, space group $Cmcm$, containing 16 atoms in the ratio Al:Cu:Mg = 2:1:1. Mondolfo [15] suggested a modified PW model with slightly different lattice parameters ($a = 0.4$ nm, $b = 0.925$ nm, and $c = 0.718$ nm). Cuisiat et al. [16] offered a model with a unit cell dimensions identical to Mondolfo's, but with space group $Im2m$ containing only 8 instead of 16 atoms. Yan et al. [17] proposed an orthorhombic structure with space group *Pmm2* (No. 25), lattice parameters $a = 0.4$ nm, $b = 0.461$ nm, $c = 0.718$ nm and four atoms per unit cell in the ratio Al:Cu:Mg = 2:1:1. However, X-ray diffraction [18] and electron diffraction [19] experiments support the models of PW and Mondolfo.

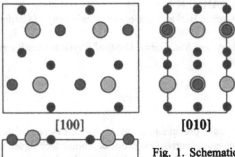

[100]   [010]

[001]

Fig. 1. Schematic drawing of the Perlitz and Westgren (PW) model of the S-Phase. The three principal orientations [100], [010] and [001] are shown; black – Al, light gray-Mg, dark gray-Cu.

EXPERIMENTAL PROCEDURE

Two alloys, Al-Cu-Mg and Al-Li-Cu-Mg (referred to as ACM and ALCM, respectively) with compositions given in Table I were solution treated at 550°C for 2 h, quenched into ice brine and aged for 72 h at 190°C (ACM) and for 16 and 100 hrs at 190°C (ALCM) to produce peak aged and over-aged conditions, respectively. The ALCM alloy was also deformed 3% prior to aging. The disks were thinned to electron transparency using a twin jet electropolisher with a solution of 30% nitric acid and 70% methanol below -25°C, at 15 V. High resolution electron microscopy (HREM) was performed using the Berkeley ARM operating at 800 kV and a Philips CM300 operating at 300 kV, while conventional diffraction contrast and microanalysis was done on various 200 kV instruments.

Table I - Alloy Compositions [wt. %]

Alloy	Cu	Mg	Li	Zr	Fe
ACM	2.01	1.06	-	0.14	0.08
ALCM	1.30	1.0	2.5	0.09	-

RESULTS AND DISCUSSION

Several HREM images of S-phase precipitates located near the edge of the foil, as is shown in Fig. 2, recorded along the $[100]_S$ and $[010]_S$ directions, were digitized from film and used for analysis. To reduce noise present in the as-recorded images (Fig. 3, column 1), these images were subjected to an automatic Wiener filter [20], unit cell averaging and crystallographic image processing [21] which imposes space group symmetries on the experimental image. These processed images (Fig. 3- columns 2 and 3) were then compared with image simulations from different models for a range of thickness and defocus conditions using the cross correlation coefficient (CCC) and chi-square (χ^2) as a measure of the goodness of fit. Based on the various models proposed for the S-phase, image simulations were carried out for a range of thickness (T), defocus (Δf) and convergence angle (α), and the theoretical images were compared quantitatively to the unit cell image obtained from the experimental image.

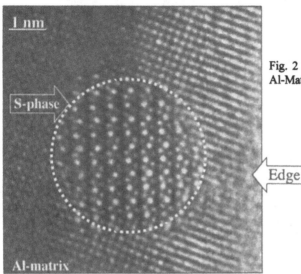

Fig. 2 - Al_2CuMg phase (S-Phase) in Al-Matrix at the edge of the foil.

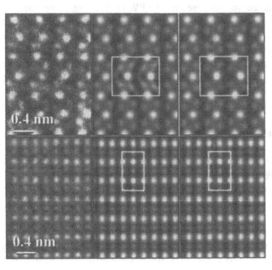

Fig. 3. Quantitative analysis of HREM images along $[100]_S$ (first row) and $[010]_S$ (second row). First column: as-recorded images; second and third columns: processed images with insets showing image simulations based on two different models, PW and RaVel, respectively. For the PW model, the best fit of CCC = 0.894 is obtained for T = 1.2 nm; Δf = -88 nm; α = 3.3 mrad. For the RaVel model the best fit of CCC = 0.98 is obtained for T = 7.2 nm; Δf = -77.3 nm; α = 1.5 mrad. Both models give the same fit of CCC = 0.99 for the image taken in $[010]_S$ zone axis, with T = 12 nm; Δf = -88.1 nm; α = 1.5 mrad.

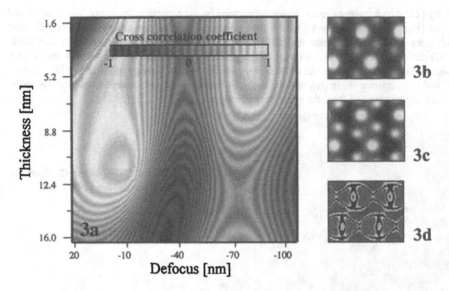

Fig. 4. Map of the cross correlation coefficient (3a) as a function of thickness and defocus for the PW model. Best agreement (CCC = 0.90) is found for a thickness of 11.2 nm and a defocus of −10 nm. Average experimental image (3b); corresponding "best" match simulated image (3c); difference image between experiment and simulation (3d).

Of the four proposed models, only the PW model gave results with reasonable similarity to the experimental image. However, simulated images based on the PW model (Fig. 3- second column), still exhibited some distinct differences with the observations. Fig. 4 shows the map of cross-correlation coefficients giving goodness-of-fit values for the investigated range of thickness and defocus for the PW model. Although the highest obtained value for CCC (0.90) indicates a relatively good fit between experiment and theory, both the value for χ^2 and the difference image given in Fig. 3 indicate definite discrepancies between experimental images and the PW model. In addition the indicated defocus is quite far from the defocus value of about −70 nm obtained from the experimental data. This led to a search for an alternative model for the structure of the S-phase.

A new model was introduced, referred to as the RaVel model, where the Cu and Mg exchange positions within the structure. In order to compare the fit between theory and experiment for the two models, the Cu/Mg sites were exchanged gradually by testing the structure as $PW_{1-x}Ravel_x$ where x = 0 gives the PW model (all Cu and Mg on the original sites) and x = 1 gives the RaVel model (all Cu and Mg atoms exchanged). Fig. 5 shows two measures of the goodness of fit, χ^2 and the cross correlation coefficient, as a function of exchange parameter x. As can be seen from a complete exchange of Cu and Mg gives the best fit.

The next step was to refine the foil thickness and the imaging parameters. This was done through an automatic refinement procedure employing simulated thermal annealing [20]. The parameters that were used in the refinement procedure were specimen thickness, objective lens defocus, and spread of defocus due to chromatic aberration and spatial incoherence of the electron beam (the convergence angle α).

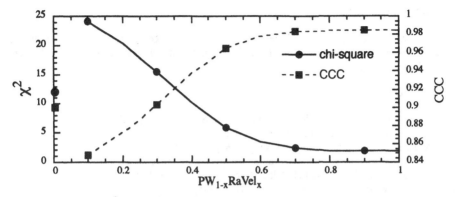

Fig. 5. Chi-square (χ^2) and Cross-Correlation Coefficient (CCC) plotted as a function of the degree x of exchange between Cu and Mg.

It should be noted that the value for defocus is in agreement with the value of -73 nm that was obtained by taking the Fourier transform of the amorphous region adjacent to the precipitate and estimating the defocus from the contrast transfer of the objective lens.

Once the imaging parameters were determined as accurately as possible, the fractional coordinates of the atoms in the unit cell were refined. Only three atoms are unique, with the rest being symmetry related. The z-coordinates of the Cu and Mg atoms are special and cannot be varied. Changing these would result in more atoms in the unit cell than physically possible. The initial positions for the RaVel model are taken from the coordinates given in the literature for the PW model [5]. These coordinates together with the refined ones are shown in Table II.

Table II - Initial P/W and refined RaVel atomic positions of Al_2CuMg.

Element	Initial-PW			Refined RaVel		
	x	y	z	x	y	z
Cu	0	0.072	0.25	0	0.765	0.25
Mg	0	0.778	0.25	0	0.074	0.25
Al	0	0.356	0.056	0	0.362	0.056

CONCLUSIONS

By careful analysis of experimental images taken of S-phase Al_2CuMg in aluminum and by comparing these images to image simulation using automatic techniques for refining structural parameters and imaging parameters, a new model was developed for the crystal structure of the S-phase. This model is based on the model given by Perlitz and Westgren, but differs in an exchange in the positions of the Mg and Cu atoms. The initial atomic coordinates have been further refined while preserving the symmetries of the space group.

ACKNOWLEDGMENT

This work is supported by the Director, Office of Energy Research, Office of Basic Energy Sciences, Materials Sciences Division of the U.S. Department of Energy under Contract No. DE-ACO3-76SFOOO98.

REFERENCES

1. H.W. Zandbergen, S.J. Andersen, and J. Jensen, Science, **277**, p. 1221 (1997).
2. S.J. Andersen et al., Acta Mater., **46**, p. 3283 (1998).
3. A. Thust and K. Urban, Ultramicroscopy, **45**, p. 23 (1992).
4. M.J. Hÿtch and W.M. Stobbs, Ultramicroscopy, **53**, p. 191 (1994).
5. W.E. King and G.H. Campbell, Ultramicroscopy, **56**, p. 46 (1994).
6. H. Zhang et al., Ultramicroscopy, **57**, p. 103 (1995).
7. R. Kilaas and V. Radmilovic, to be published in Ultramicroscopy, (2000).
8. V. Radmilovic, R. Kilaas, U. Dahmen, and G.J. Shiflet, Acta Mater., **47**, p. 3987 (1999).
9. P.J. Gregson, H.M. Flower, Acta Met., **33**, p. 527 (1985).
10. V. Radmilovic, G. Shiflet, G. Thomas, and E. Starke, Jr., Scripta Met., **23**, p. 1141 (1989).
11. Y.A. Bagaryatskii, Dokl. Akad. Nauk SSSR, **87**, p. 397(1952).
12. J. M. Silcock, J. Inst. Metals, **89**, p. 203 (1960-61).
13. G.C. Weatherly and R. B. Nicholson, Phil. Mag., **17**, p. 801 (1968).
14. H. Perlitz and A. Westgren, Arkiv för kemi, mineralogi och geologi, **16B**, p. 1 (1943).
15. L.F. Mondolfo, Aluminum alloys-Structure and properties, Butterworths, London, (1976) pp. 497-501.
16. F. Cuisiat, P. Duval and R. Graf, Scripta Met. **18**, p. 1051 (1984).
17. J. Yan, L. Chunzhi and Y. Minggao, J. Mater. Sci. Letters, **9**, p. 421 (1990).
18. J.I. Perez-Landazabal, M.L. No, G. Madariaga and J.S. Juan, J. Mater. Res., **12**, p. 577 (1997).
19. A.K. Gupta, P. Gaunt, and M.C. Chaturvedi, Phil. Mag. A, **55**, p. 375 (1987).
20. A. Thust, M. Lentzen and K. Urban, Ultramicroscopy **53**, p. 101 (1994).
21. S. Hovmöller, Ultramicroscopy, **41**, p. 121 (1992).

QUANTITATIVE MAPPING OF CONCENTRATIONS AND BONDING STATES BY ENERGY FILTERING TEM

J. MAYER*, J. M. PLITZKO[§]
Max-Planck-Institut für Metallforschung, D-70174 Stuttgart, Germany
*now at: Central Facility for Electron Microscopy, Aachen University of Technology,
D-52056 Aachen, Germany
[§]now at: Lawrence Livermore National Laboratory, Livermore, CA 94550

ABSTRACT

We have developed new methods to quantify the data acquired by electron spectroscopic imaging (ESI) in an energy filtering TEM. The analysis is based on recording series of energy filtered images across inner-shell loss edges or in the low-loss region. From the series of ESI images, electron energy loss (EEL) spectra can be extracted and subsequently analysed using standard EELS quantification techniques. From an ESI series one can measure the absolute amount (area density) of an element in the given sample area or the concentration ratios of one element with respect to other elements. Spectrum line-profiling has been shown to be an efficient way to acquire and present the information on the chemistry of an interface. The results obtained for different metallisation layer systems show that segregation in the monolayer range can still be analysed with high spatial resolution. For the study of the energy-loss near-edge structure (ELNES) a higher energy resolution is required. ESI series with narrow energy window width can be used to distinguish between different bonding states of a given element and is demonstrated for thin films of diamond and amorphous carbon.

INTRODUCTION

The increasing number of energy filtering transmission electron microscopes (EFTEMs) has given many microscopists the ability to apply the fast and very efficient tool of electron spectroscopic imaging (ESI) for analytical characterisation, rather than to record EELS spectra. Most commonly, ESI analysis is based on identifying the presence of a characteristic inner-shell loss edge of the element under investigation (for an overview see Reimer [1]). This can be accomplished by recording two or three ESI images at energy losses in the background region before the edge and the signal region just above the edge. The two methods mainly used are the three-window technique, which was first proposed by Jeanguillaume et al. [2], and the ratio map technique which was introduced by Krivanek et al. [3]. In this paper we will present results obtained with a new method for quantitative analysis which is based on a multi-window approach. Rather than acquiring only two or three ESI images, we acquire a series of ESI images around an inner-shell loss edge and, if required for quantification, in the low loss region [4]. The advantages of this method are: 1) information on the energy loss spectrum is obtained for each pixel in the image and the standard quantification methods developed for EEL spectra can be used for the analysis, 2) the background extrapolation and the signal integration regions can be extended over a large energy-loss range, and 3) the standard EELS methods for single scattering deconvolution and least squares fitting for overlapping edges can be used. From an ESI series, spectrum line profiles can be extracted and can be quantified e.g. in terms of the absolute thickness of thin segregation layers at interfaces. Furthermore, we show how the bonding information present in the near-edge fine structure can also be extracted from a series of ESI images acquired with higher energy resolution.

EXPERIMENTAL RESULTS AND DISCUSSION

The EFTEM investigations were performed on a Zeiss EM 912 Omega which was operated at 120 kV and is equipped with a LaB_6-cathode. ESI image series were recorded on a GATAN 1024x1024 slow scan CCD-camera using GATAN's DigitalMicrograph software. The required image processing routines were written in the script language within DigitalMicrograph and the quantitative analysis was performed using the GATAN EL/P program package.

Acquisition and Analysis of ESI series

In the experiments, we record series of energy filtered images over a predefined energy-loss region with an energy window of finite width δE and a stepwidth s between individual ESI images. ESI series have been obtained in the low-loss region and/or in the core-loss region of the energy-loss spectrum. Subsequently, the information on the EEL spectrum is extracted for each individual pixel or, by integrating over the corresponding pixels, for a given image area. The integration leads to an improvement of the signal to noise ratio (SNR) and can e.g. be carried out parallel to a given boundary [5].

A quantitative analysis can be performed on an individual spectrum, or a one- or two-dimensional array of spectra. In interface studies, the extraction of spectrum line profiles perpendicular to the interface delivers very valuable information, similar to what has been shown before in experiments performed on a dedicated STEM [6]. The line profile analysis can be done automatically by moving the selected area across the interface and extracting a spectrum for every position. The result of the line profile analysis can be illustrated in terms of so-called spectrum line profiles, which visualise the distribution of the elements across the interface, or in real concentration profiles, e.g. of segregation layers and diffusion gradients.

An experimental prerequisite for an accurate quantification is the knowledge of the exact slit width δE. Experimentally, δE can be measured by projecting the energy loss spectrum onto the CCD camera with the slit inserted. If the slit width δE deviates from the step size s, a correction factor $\kappa = s/\delta E$ has to be taken into account in the analysis. A possible drift of the slit position relative to the spectrum was never detected in our investigations.

The energy loss spectra extracted from an ESI series with n images can graphically be visualised in several different ways. In the following, we use linear interpolation between the individual data points. The resulting spectra resemble very closely the spectra which would be obtained with a parallel energy-loss (PEELS) detector with much higher sampling frequency. The background extrapolation and subtraction is performed via a power-law background fit. We have found that, using the spectra obtained by linear interpolation, very accurate background fits can be obtained. However, it should be kept in mind that the linear interpolation is only an approximation. To become more accurate, a modelling of the exact functional dependence of the intensity variation for windows with a finite width ∂E is required.

Analysis of Spectrum Line Profiles

The technique of acquiring spectrum line profiles by stepping a focused probe across an interface and recording a series of spectra on a parallel spectrometer is a well known technique. It is very useful for the one-dimensional analysis of elemental distribution and chemical bonding [6]. Elemental mapping in an EFTEM in the past was mainly used for the acquisition of qualitative elemental distribution images. However by recording series of energy filtered images in the core loss region it is now possible to obtain full spectral information of the distribution of one or several elements.

Fig. 1 shows an overview of an Al/TiN/Ti/SiO₂/Si sample [7,8]. A thin layer of Ti was sputtered between SiO₂ and TiN for better adhesion. In the experiments, we have recorded series

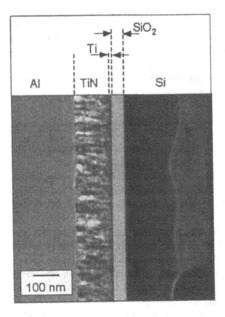

Fig. 1. Overview of an Al/TiN/Ti/SiO$_2$ metallisation layer system on a Si-substrate and schematic arrangement of the individual layers.

Fig. 2 (below). Spectrum line profile across the TiN/Ti/SiO$_2$-interface extracted from a series of ESI images which was acquired in the energy loss range from 360 - 600 eV.

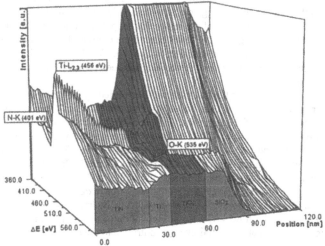

of energy-filtered images with energy losses from 360 eV to 600 eV with a step of $\Delta E = 10$ eV. This energy range covers the N K-edge (401 eV), the Ti $L_{2,3}$-losses (456 eV) and the O K-edge (535 eV). Spectra corresponding to a line scan perpendicular to the layer system can be extracted from the ESI series and the result is plotted in Fig. 2. The core loss edges corresponding to the different elements are marked in the figure. From such spectrum line profiles, one can distinguish between the TiN- and the Ti-layer, for which only the Ti peak is present. However, in the line scan no sharp transition can be seen between the Ti layer and the SiO_2. First the O-peak increases, then the Ti-peak decreases. This indicates, that Ti is oxidised at the interface to SiO_2. The extension of this oxidised layer is about 20 nm. Quantification reveals that the O/Ti ratio is about 2:1 directly at the SiO_2 interface and decreases monotonically down to pure Ti. Thus, we find a gradient in the O/Ti ratio and a composition similar to TiO_2 directly at the interface to SiO_2. We speculate that this oxidised interface layer causes the better adhesion of the TiN compared to a TiN layer without a Ti interlayer.

Determination of the area density of thin layers

As a second experimental example, we want to report on investigations performed in the materials system Al_2O_3–Ti–Cu [9]. A thin interlayer of titanium is introduced between an Al_2O_3 substrate and a Cu metallisation layer to enhance the adhesion of the copper on the sapphire substrate [10]. Both the titanium interlayer and the Cu film were grown by molecular beam epitaxy. The thickness of the titanium layer can be controlled with an accuracy of better than one monolayer during the deposition process in the MBE machine [10].

In the following, we want to discuss the quantification of the number of atoms per unit area of the Ti interface layer with nominally 1 nm thickness. In the experiment, a series of 30 ESI images in the energy range between 380 and 670 eV was acquired which includes the Ti-$L_{2,3}$-edge (456 eV) and the O-K-edge (535 eV). The slitwidth was calibrated to 10 eV and every image was acquired with 10 s exposure time and 2-fold binning of the CCD camera pixels. The collection semiangle β in this case was 12.5 mrad and the convergence angle 1.6 mrad. Fig. 3a shows one ESI image from the series around the Ti-L-edge which illustrates the arrangement of the materials in the sample. The Bragg contrast of individual grains in the polycrystalline copper is clearly visible and in the left part of the image the copper was removed during the ion milling of the sample. Three ESI images of the series around the titanium L-edge were used to calculate a titanium elemental map, which is depicted in Fig. 3b. The Ti distribution seems to be very homogeneous, except in the area where the copper has been removed. In this region an increase in intensity can be seen.

Line profile analysis across the interface was performed by integrating the signal parallel to the interface in areas of 1 x 50 pixels, which corresponds to 1.5 x 75 nm on the specimen [9]. Fig. 3c shows an example of these spectra line profiles. In the case of this profile the selected area was positioned in the left part of the image, where an intensity increase in the Ti layer can be seen. The titanium and the oxygen edges are clearly visible. The apparent width of the Ti-layer is larger than its nominal width because of the combined effect of a slight inclination of the film, the focus spread caused by chromatic aberration and the blurring caused by the point spread in the CCD-camera. Some overlap between the extension of the Ti-peak and the extension of the O-signal from the Al_2O_3 substrate can be seen. This may be indicative of a partial oxidation of the Ti; however, a slight inclination of the film and the blurring caused by the mechanisms discussed above certainly also contribute to the oxygen signal in the Ti layer.

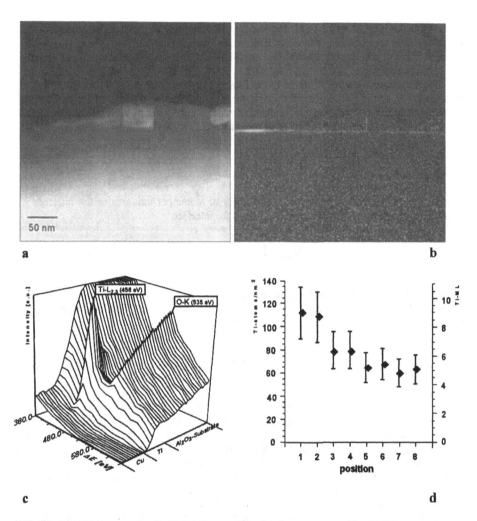

Fig. 3. **(a)** ESI image of a Cu/Ti/Al₂O₃-metallisation layer system. The ESI image was acquired at an energy loss of 380 eV and is the first out of an ESI series of 30 images with 10 eV increment. **(b)** Titanium elemental distribution computed from three images out of the series. The nominal thickness of the Ti layer is 1 nm, however a strong thickness variation is clearly visible. **(c)** Spectrum line profile across the left part of the interface shown in (a). **(d)** Absolute thickness of the titanium layer along the interface in Ti-atoms per nm^2 and in monolayers.

Furthermore, the lack of the overlying Cu layer may be an additional reason for the oxidation of the titanium found in this position. For the quantitative analysis of the Ti-signal, the low loss intensity $I_l(\Delta)$ has to be known. Thus, in a second experiment, information on the EEL spectrum was recorded in the low-loss region up to typically 150 eV, including the zero-loss peak. With the help of an EFTEM, the required information can be obtained from the whole imaged sample area in a single series of ESI images running from the zero-loss to energy losses of about 150 eV. The typical slitwidth and stepwidth is 10 eV and the exposure times are only a few seconds per image.

By integration over the Ti signal in the line profile and using the low loss intensities, the area density N_a of the titanium atoms forming the interlayer can be determined in a first step [4]. Using this number and the specimen thickness t which can be determined from the low loss, we can then compute the volume density n_a of titanium atoms within the layer:

$$n_a = \frac{N_a}{t}.$$

(1)

In a next step, this can be converted into number of atoms per unit area of the interface N_{int} which is obtained by integrating the signal across the interface

$$N_{int} = \int_{di} n_a \, dx$$

(2)

where d_i is the image width of the boundary layer and x is the co-ordinate perpendicular to the boundary. N_{int} is given in [atoms/nm^2] in the interface plane. This can finally be converted into the thickness d [in nm] of the layer which is given by:

$$d = N_{int} \, ? \, \frac{A}{\rho \, ? \, N_A}$$

(3)

where A is the molar weight of the element or compound and N_A is Avogadro's number.

After performing this in one location, the analysis is continued along the interface to determine the thickness variation, which is evident from the result in the elemental map (Fig. 3d). Spectra from eight different areas were analysed and the results are plotted in the diagram shown in Fig. 3d in terms of atoms per nm^2 in the interface plane as well as the equivalent in monolayers. In the larger area with homogeneous thickness of Ti, the value determined using equation (3) is (1.1±0.3) nm, which corresponds to (4.7±1.0) ML. These results are in good agreement with the expected values. The higher titanium concentration on the left side in Fig. 3d may be caused by variations during the MBE process, or by a possible accumulation of titanium during ion-beam thinning after the removal of the copper overlayer.

Analysis of near edge fine structure

The presence of an edge in the energy-loss spectrum not only reflects the occurrence of the corresponding element, the near-edge fine structure (ELNES) of the edge also contains information on the three-dimensional atomic co-ordination of this element in the sample. As a model system, we have studied CVD grown diamond films on Si substrates [11]. At the interface between the film and the substrate an amorphous layer is formed which mainly consists of amorphous carbon [12, 13]. An analysis of the ELNES makes it possible to distinguish between the two different phases of carbon, i.e. diamond and amorphous carbon.

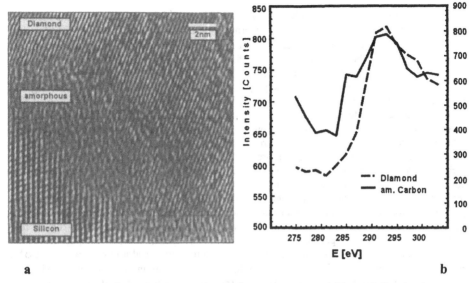

a   **b**

Fig. 4. **(a)** HREM image of the interface between the Diamond film and the Si substrate showing the presence of an amorphous layer. **(b)** ELNES of the Diamond film and the amorphous layer reconstructed from a series 20 ESI images. The result shows that the interfacial layer is formed by amorphous carbon.

Fig. 4 shows a high resolution micrograph of a typical interface area (acquired on a JEOL 4000 EX), which exhibits surface roughness and also contains an amorphous film between the Si substrate and the diamond film. This amorphous film with a thickness of 1 - 4 nm could also be identified in a tripod polished TEM specimen, excluding that it is only an artefact produced during ion beam thinning [14]. The amorphous layer is present almost everywhere along the interface and there exist only a few isolated islands where the diamond film is in direct contact with the substrate. The areas with no detectable amorphous layer can be found for example at elevated surface steps and are thought to be the nucleation centres for the diamond film [14]. The diamond film subsequently overgrows the areas onto which amorphous carbon has been deposited during the initial deposition steps prior to diamond nucleation. Since the two phases of interest, diamond and amorphous carbon, are formed by the same chemical element, they can not be distinguished from the presence of its edge alone. Rather, the near edge fine structure (ELNES) has to be used in order to separate the areas in which the two different phases occur. It should be kept in mind that EEL spectra could not be obtained from the 1 nm thick layer at the diamond/silicon interface on the Zeiss EM 912 Omega, because the beam on an analytical TEM equipped with a tungsten or LaB$_6$ filament can not be focused into a 1 nm probe. This would require the use of a FEG-STEM with its higher brightness for small probe diameters, as has been demonstrated for the same system by Muller et al. [15].

In order to reveal this difference in the ELNES by electron spectroscopic imaging, we have acquired a series of ESI images across the onset of the carbon K-edge. An energy-window width of 5 eV was chosen, which is a good compromise between the required energy resolution and maximising the signal in each individual ESI image [14]. The energy increment between the individual ESI images was set to 2 eV. We choose an energy increment which is smaller than the actual energy window width to make sure that one of the ESI images of the series is centred around the π^*-peak and another one around the σ^*-excitations. In total, the ESI series consisted of 20 images from $\Delta E = 265$ eV to $\Delta E = 303$ eV. The exposure time for each image was 10 seconds, i.e. the total acquisition time for the whole series was 200 seconds. From the whole series of ESI images, information on the ELNES of the carbon K-edge can be retrieved for any given area in the image. The energy-loss spectrum is obtained by simply extracting the intensity from the same area in the series of ESI images and plotting it as a function of the corresponding energy loss. Basically, this can be performed for each individual pixel in the images. However, the resulting spectra would be very noisy. In order to reduce the noise, the intensities were integrated over a certain area in the images. Prior to this, drift correction can be applied to the individual ESI images in order to align the corresponding areas properly in the series of images. The magnitude of the drift correction can either be determined by cross correlation or by visual inspection. From the drift-corrected series, we extracted the integrated intensities of a line profile with a length of 150 pixels and a width of one pixel which was placed in the centre of the amorphous layer. Thereafter, the line profile was shifted parallel into the diamond layer. The resulting intensity data are plotted in Fig. 4b. The carbon K-edge of the material forming the amorphous layer clearly shows a π^*-peak, whereas the edge from the diamond film shows an onset at about 4 eV higher energy losses and a more pronounced σ^*-peak. Qualitatively, the ELNES features reproduced in Fig. 4b are in good agreement with the shape of reference spectra [14], with an energy resolution which is lowered to about 5 eV, as defined by the slit width used for the ESI series. The carbon K-edge of the amorphous layer is superimposed on a much stronger background than the K-edge from the diamond film, which reflects both the increasing thickness towards the substrate and the amount of Si which is presumably dissolved into the amorphous layer.

CONCLUSIONS

We have demonstrated that ESI series can not only be quantified in terms of relative concentrations but also in terms of absolute numbers of atoms per nm^2. The present investigations on thin film systems clearly indicate that quantitative ESI analysis is applicable to segregation and precipitation in the monolayer range. In this case, drift determination and correction are very important to suppress artefacts and to maximise the signal-to-noise ratio. The main advantage of our ESI series technique compared to the well-known 'spectrum-imaging' technique [16, 17] is the short time in which two-dimensional analytical information can be obtained. The main advantage of the 'spectrum-image' method is the higher energy resolution in spectra obtained on a STEM with a field emission source. This illustrates that, if only the chemistry has to be analysed and time or number of pixels becomes a concern, then ESI will be the method of choice and results can be obtained even in the submonolayer range, which has also been demonstrated recently on an EFTEM with a FEG source [18].

Furthermore, we have shown that ESI series obtained in an energy-filtering TEM make it possible to obtain two-dimensional information on the variation of the ELNES on a nanometer scale. As in the former case, ESI makes it possible to obtain the information in much shorter time than in the traditional scanning approach. In comparison, the main advantages of a dedicated STEM are the higher energy resolution of ~ 0.5 eV and the better spatial resolution in the range of 0.5 - 1 nm. However, using an EFTEM with an FEG emitter, a similar spatial and energy resolution can be reached in ESI studies which means that the ESI approach is clearly advantageous if spatially resolved information is sought. Or in other words, FEG instruments equipped with an energy filter make it possible to select the most appropriate way to analyse the energy-loss space in each case - via PEELS acquisition in STEM mode or via ESI series in the TEM imaging mode.

ACKNOWLEDGEMENTS

We gratefully acknowledge support by the Volkswagen-Stiftung under contract I/69931. We would like to thank A. Strecker, U. Salzberger, U. Eigenthaler and K. Hahn for their assistance in the experiments, and R. Spolenak and J. Marien for supplying and preparing the FIB samples.

REFERENCES

[1] L. Reimer in "Energy-Filtering Transmission Electron Microscopy", ed. L. Reimer, Springer Series in Optical Sciences Vol. 71 (Springer, Berlin, 1995), p. 347.

[2] C. Jeanguillaume, P. Trebbia, and C. Colliex, Ultramicroscopy 3 (1978) 237.

[3] O.L. Krivanek, A.J. Gubbens, and N. Dellby, *Microsc. Microanal. Microstr.* 2, (1991) 315.

[4] J. Mayer, U. Eigenthaler, J.M. Plitzko, and F. Dettenwanger, Micron 28 (1997) 361.

[5] A. Berger, J. Mayer and H. Kohl, Ultramicroscopy 55 (1994) 101.

[6] C. Colliex, M. Tencé, E. Lefèvre, C. Mory, H. Gu, D. Bouchet, and C. Jeanguillaume, Mikrochim. Acta 114/115 (1994) 71.

[7] J. Marien, J.M. Plitzko, R. Spolenak, R.M. Keller, and J. Mayer, J. Microsc. 194 (1998) 71.

[8] R. Spolenak, O. Kraft, and E. Arzt, Microelectr. Reliabil. 38 (1998) 1015.

[9] J.M. Plitzko and J. Mayer, Ultramicroscopy 78 (1999) 207.

[10] G. Dehm, C. Scheu, G. Möbus, R. Brydson and M. Rühle, Ultramicroscopy 67 (1997) 207.

[11] M. Rösler, R. Zachai, H.-J. Füßer, X. Jiang, and C.-P. Klages, *Proc. 2nd Intern. Conf. on the Applications of Diamond Films and Related Materials.* Yoshikawa, M., Murakawa, M. (Eds), Tokyo (1993), p. 691 - 696.

[12] B.R. Stoner, G.-H.M. Ma, S.D. Wolter, and J.T. Glass, Phys. Rev. B 45 (1992) 11067 - 11084.

[13] Y. Tzou, J. Bruley, F. Ernst, M. Rühle, and R. Raj, J. Mater. Res. 9 (1994) 1566 - 1572.

[14] J. Mayer and J.M. Plitzko, J. Microsc. 183 (1996) 2.

[15] D.A. Muller, Y. Tzou, R. Raj, and J. Silcox, Nature 366 (1993) 725 - 727.

[16] C. Jeanguillaume and C. Colliex, Ultramicroscopy 28 (1989) 252.

[17] J.A. Hunt and D.B. Williams, Ultramicroscopy 38 (1991) 47.

[18] J. Mayer, S. Matsumura, and Y. Tomokiyo, Journal of Electron Microscopy 47 (1997) 283.

COMBINED HRTEM AND EFTEM STUDY OF PRECIPITATES IN TUNGSTEN AND CHROMIUM-CONTAINING TiB$_2$

W. MADER*, B. FREITAG*, K. KELM*, R. TELLE**, C. SCHMALZRIED**
*Institut für Anorganische Chemie, Universität Bonn, D-53117 Bonn, Germany, mader@uni-bonn.de
**Institut für Gesteinshüttenkunde der RWTH Aachen, D-52064 Aachen, Germany

ABSTRACT

The structure and chemical composition of two types of precipitates in the system TiB$_2$-WB$_2$-CrB$_2$ were studied by means of high-resolution TEM and energy filtering TEM. Type I particles (W$_2$B$_5$ structure) are precipitated at the basal plane of the hexagonal matrix whereas type II precipitates are thin platelets lying parallel to the $\{1\bar{1}00\}$ prism planes. Lattice imaging yields displacements of the metal positions with respect to the matrix. Information on the chemical composition at high lateral resolution is obtained from elemental maps of all chemical constituents using electron spectroscopic imaging (ESI). The type II precipitates show a decrease in the B and Ti concentration, whereas the tungsten concentration increases and the Cr is homogeneously distributed. The HRTEM results combined with the results of the elemental maps allow to develop a structural model based on the intergrowth of the β-WB structure in the TiB$_2$-rich matrix. The two deficient boron layers in W$_{0.5}$Ti$_{0.5}$B with a spacing of 0.38 nm can be used to examine the resolution limit of ESI.

INTRODUCTION

A typical and powerful domain of transmission electron microscopy is the study of precipitation, where the shape, crystal structure, orientation relationship and chemical composition of precipitates can be elucidated. When precipitates are very small, at least in one dimension, any information is extremely hard to achieve: Electron diffraction can not be used to obtain crystallographic information, and EDX or EELS may yield at best the overall chemical composition of the precipitate, however the positions of the elements in the crystal lattice remain unknown.

The present study is concerned with the structural and chemical characterization of thin precipitates in a transition metal boride applying high resolution TEM (HRTEM) and high resolution energy filtering TEM (EFTEM). The systems investigated are materials based on titanium diboride, TiB$_2$, which are candidates for cutting tools due to their high hardness and oxidation resistance at high temperatures. To increase the limited toughness in situ strengthening of the TiB$_2$-rich matrix can be realized by precipitation of borides of other transition metals. The phase diagram in the Ti-W-B system offers the possibility of precipitation from a supersaturated mixed crystal [1,2]. The mixed crystal regime can be increased by additions of chromium boride.

Among other phenomena such as decomposition two different types of plate-shaped precipitates in the TiB$_2$-rich matrix were studied in detail. The combination of HRTEM and EFTEM allows us to specify the structure and chemical composition of the particles. The highly resolved images of the distribution of all the elements is particularly valuable for understanding the segregation and precipitation tendencies of the chemical constituents. The orientation relationship and the precipitate shape will be discussed on the basis of lattice misfit between precipitate and matrix.

EXPERIMENTAL

Dense material with nominal molar composition $TiB_2:WB_2:CrB_2 = 50:40:10$ was prepared from the powders (grade A, H.C. Starck) by hot pressing and subsequent equilibration in the single phase field at 2000°C for 8h under argon atmosphere [1,2]. After annealing at 1600°C in the two-phase field the material was prepared for the TEM investigations by standard mechanical thinning and finally argon ion beam milling.

The microscopes used in this study were a Philips CM30ST/STEM and a Philips CM300UTFEG, both operated at 300 kV and equipped with a post-column imaging electron energy filter (Gatan Imaging Filter, GIF [3]). EEL spectra and images were recorded with the slow-scan CCD camera mounted in the imaging filter. The two microscopes are fitted with a Noran high-purity Ge detector to collect X-ray intensity for EDS analysis. The efficiency curve of the HPGe detector as a function of the X-ray energy was experimentally determined using a variety of standards.

RESULTS AND DISCUSSION

Electron diffraction and EDS analysis of the material after annealing at 1600°C shows the microstructure to consist of (i) plate-shaped grains of W_2B_5, (ii) regions with TiB_2 structure showing a spinoidal-like decomposition and (iii) TiB_2-rich grains with precipitates. Subject of the present study are the precipitates in the TiB_2-rich matrix. TiB_2 as well as the diborides of tungsten and chromium crystallize in the hexagonal structure of AlB_2 ($P6/mmm$), which consists of alternating hexagonal layers of boron and metal atoms (see fig. 1a). According to EDS analyses of the TiB_2-rich matrix Ti is substituted up to 1/3 by tungsten, however the crystal symmetry of TiB_2 remains unchanged.

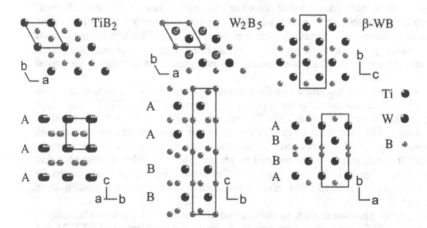

Fig. 1. Structural models of a) TiB_2, b) W_2B_5, and c) β-WB, each viewed along a and c.
B: light grey, Ti: grey, W: black. Note the different stacking sequence of metal layers.

Table 1. Crystallographic data of borides

	TiB_2 [4]	W_2B_5 [5]	β-WB [6]
space group	$P6/mmm$	$P6_3/mmc$	$Cmcm$
a [nm]	0.3028	0.2983	0.319
b [nm]			0.840
c [nm]	0.3228	1.388	0.306

W_2B_5 precipitates (type I)

Two types of plate-shaped particles can be identified. Type I particles precipitate parallel to the basal plane. Fig. 2 and fig. 3 show lattice images in the $[2\bar{1}\bar{1}0]$ direction of TiB_2 containing type I precipitates. The images were taken on thin regions under optimum focussing conditions with the CM30, where dark dots represent positions of metal columns. The spacing of the dots along c is $d_{0001} = 0.3228$ nm and perpendicular, i.e. along $[01\bar{1}0]$ $d_{01\bar{1}0} = a/2\sqrt{3} = 0.2622$ nm. While in perfect TiB_2 the dots are straightly aligned along c because of the AAA stacking sequence, displacements of the dots can be observed at some planes in the lattice images of figs. 2 and 3. The displacements can be measured to be $1/3\, d_{01\bar{1}0}$, and this can be explained by a change of the stacking of the metal layers e.g. from A to B as it is the case in the W_2B_5 structure. EDS analyses of regions with high density of such "stacking faults" reveal an increase in the W:Ti composition ratio with respect to defect-free regions. Hence the "stacking faults" represent intergrowth of thin layers of the W_2B_5 type-structure in TiB_2 along the basal planes.

A further proof for the intergrowth is provided by the somewhat larger spacing of the closed-packed layers in W_2B_5 compared to that in TiB_2. The increase in the layer spacing in W_2B_5 is caused by the corrugated boron layers where an A layer is followed by a B layer (cf. fig. 1b). In fig. 3 the lattice image of perfect TiB_2 is compared to an image of a region with two "stacking faults" taken from the same exposure. There TiB_2 with metal layers at position A changes to a metal layer at B, which turns over to a C layer at a small distance at the second "fault". The increase in layer spacing can be directly seen when comparing the lattice images in fig. 3.

The plate-like morphology of the precipitates as well as the crystallographic orientation relationship can be understood considering the close relation of both the crystal structures and the small misfit between the a-axes of TiB_2 and W_2B_5 (table 1).

Fig. 2. Lattice image in $[2\bar{1}\bar{1}0]_{TiB_2}$ Single AB stacking sequence in TiB_2-rich matrix is identical to precipitate of W_2B_5 with thickness of one unit cell. Displacement of $1/3\, d_{01\bar{1}0}$ in the metal positions at the fault is indicated.

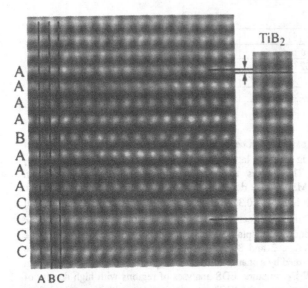

β-(W,Ti)B precipitates (type II)

Type II precipitates lie parallel to the $\{1\bar{1}00\}$ prism planes of the matrix and can be imaged best viewing along the c-axis of the TiB_2 matrix. However, the chemical composition and the structure of the particles could not be deduced from lattice images and EDS analysis. To obtain qualitative information on the chemical composition we applied energy filtering TEM by taking electron spectroscopic images produced by inelastically scattered electrons. ESI images were acquired in front of ionization edges (pre-edge images) and at ionization edges (post-edge images) of all the constituting elements. A background image is calculated from the pre-edge images using a power law model. To produce elemental maps the background contribution in the post-edge image must be removed after careful alignment of the images [7,8]. The exposure time as well as the energy window had to be increased from 4s/15eV for the boron images (B_K = 188 eV) to 60s/40 eV for the tungsten images (W_M = 1809 eV). Hence the critical parameter for such long exposure times is specimen drift [9], and the elemental maps of tungsten usually show somewhat wider contrast than e.g. the boron distribution.

The elemental maps of the four elements are shown in fig. 4. While the chromium distribution appears homgeneous, the plates show low intensity in the images of boron and titanium. Hence the concentration of these elements in the precipitates are decreased compared to the matrix. The elemental map of tungsten, however, reveals an enrichment at the precipitates. The B_K edge was used to obtain the distribution of boron at even higher resolution, and ESI images were taken with an energy window of 10 eV and 5 s exposure time. fig. 5 shows a boron map of a terminating precipitate which dissociates at the end forming a loop-like configuration. At the perfect, non splitted precipitate two dark lines are visible, and their distance can be determined to approx. 0.4 nm in the integrated profile across the lines shown in fig.5. Hence the precipitates contain two boron deficient planes at this distance [9].

Further information is obtained from lattice images of the precipitates viewed in [0001] of TiB$_2$. A typical unprocessed image taken at the CM30 is shown in fig. 6.

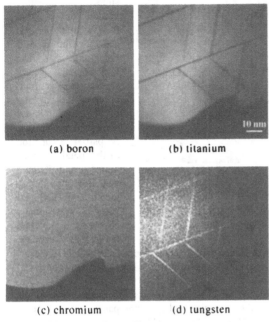

(a) boron   (b) titanium

(c) chromium   (d) tungsten

Fig.4: Elemental maps of (a) boron, (b) titanium, (c) chromium and (d) tungsten viewed along [0001]$_{TiB2}$.

the precipitate-matrix interface.

Based on this model image simulations using the EMS program package [10] were performed. However, the typical double row of black dots is not reproduced when occupying all metal sites in β-WB with tungsten. Therefore every second layer was occupied by titanium instead

Before modeling the precipitate structure it is helpful to realize that it must fit very well with that of the matrix, at least at the $\{1\bar{1}00\}_{TiB_2}$ prism planes. Among the different tungsten borides the β-WB is closest to the precipitate's structure and chemistry: The a-axis of TiB$_2$ and the c-axis of β-WB as well as the c-axis of TiB$_2$ and the a-axis of β-WB are very similar (see table 1). A model where β-WB with thickness equal to the length of the b-axis is fitting into the TiB$_2$-matrix is constructed. The orientation relationship is described by $(1\bar{1}00)_{TiB_2}$ ‖ $(010)_{β\text{-WB}}$ and $[0001]_{TiB_2}$ ‖ $[100]_{β\text{-WB}}$. The translation is chosen so that the metal positions in β-WB are aligned with those of TiB$_2$ in the direction perpendicular to

of tungsten in the unit cell. Based on this modified model the simulated images resemble the observed images, and the best matching image simulated for a crystal thickness of 4 nm and an under-focus of −30 nm is inserted in the lattice image in fig. 6.

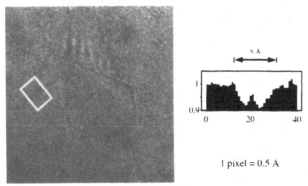

λ A

0   20   40

I pixel = 0.5 Å

Fig.5: Boron map of a type-II precipitate. The integrated profile on the right corresponds to the marked area on the map.

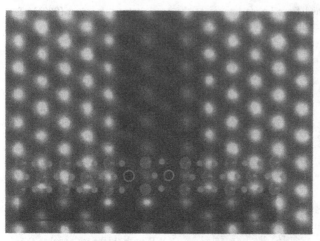

All the experimental findings support the structure model of a thin β -(W,Ti)B crystal slab for the type II precipitates . The lattice image clearly shows the continuation of the stacking of the metal layers across the precipitate. A partial occupation of the metal layers by tungsten and titanium was also observed by quantitative EDS analyses of occasionally appearing wider β-(W,Ti)B crystals,

Fig.6: Unprocessed image of a β-(W,Ti)B-precipitate in TiB₂ with image simulation and corresponding structure model.

where W:Ti ratios up to 3/2 was measured. Hence partial occupation with W and Ti of atomic rows might be possible. Finally, the boron-deficient layers is an inherent structural property of the monoboride.

Our future studies will concentrate on an even more precise characterization of the precipitates by exit-wave reconstruction, where imaging and specimen parameters are known much better than in single-shot microscopy. High resolution EFTEM at type I precipitates will be applied to directly show the boron distribution, which then should correlate with the stacking "faults" of the metal layers.

ACKNOWLEDGEMENTS

The authors gratefully acknowledge financial support by the Volkswagen-Stiftung under AZI/72911, and by the Ministerium für Schule und Weiterbildung, Wissenschaft und Forschung NRW through the project "Landeszentrum für Hochleistungselektronen-mikroskopie NRW" (AZ: IVA5-10301597).

REFERENCES

1. I. Mitra and R. Telle, J. Solid. State Chem. **133**, 25 (1997).
2. C. Schmalzried and R. Telle, 13th. International Symposion on Boron, Borides and Related Compounds, to be published (1999).
3. O.L. Krivanek et. al., Microsc. Microanal. Microstruct. **3**, 187 (1992).
4. J.T. Norton, H. Blumenthal and S.J. Sindeband, Trans. AIME **185**, 739 (1949).
5. R. Kiessling, Acta Chem. Scand. 1, **893** (1947).
6. B. Post and F.W. Glaser, J. Chem. Phys. **20**, 1050 (1952).
7. F. Hofer, P. Warbichler, and P. Grogger, Ultramicroscopy **59**, 15 (1995).
8. F. Hofer, P. Warbichler, B. Buchmayer, and S. Kleber, J. Microsc. **184**, 163 (1996).
9. B. Freitag, W. Mader, J. Microsc. **194**, 42 (1999).
10. P. Stadelman, Ultramicroscopy **35**, 43 (1987).

HIGH SPATIAL RESOLUTION X-RAY MICROANALYSIS OF RADIATION-INDUCED SEGREGATION IN PROTON-IRRADIATED STAINLESS STEELS

E.A. KENIK*, J.T. BUSBY**, G.S. WAS**
*Oak Ridge National Laboratory, Oak Ridge, TN 37831-6376, kenikea@ornl.gov
**University of Michigan, Ann Arbor, MI 48109-2104

ABSTRACT

The spatial redistribution of alloying elements and impurities near grain boundaries in several stainless steel alloys arising from non-equilibrium processes have been measured by analytical electron microscopy (AEM) in a field emission scanning transmission electron microscope. Radiation-induced segregation (RIS) has been shown to result in significant compositional changes at point defects sinks, such as grain boundaries. The influence of irradiation dose and temperature, alloy composition, prior heat treatment, and post-irradiation annealing on the grain boundary composition profiles have been investigated. Understanding the importance of these microchemical changes relative to the radiation-induced microstructural change in irradiation-assisted stress corrosion cracking (IASCC) of the irradiated materials is the primary goal of this study.

INTRODUCTION

Both microstructural (radiation hardening) and microchemical (radiation-induced segregation [RIS]) radiation effects contribute to irradiation-assisted stress corrosion cracking (IASCC) of stainless steels [1,2]. Similar to stress corrosion cracking of thermally sensitized stainless steels, the depletion of Cr and Mo at grain boundaries contributes to IASCC. Narrow (FWHM ~5 nm) RIS profiles form near grain boundaries for typical irradiation temperatures (~300°C) for light water nuclear reactors where IASCC is a critical concern. The evolution of the spatial extent and magnitude of the RIS profiles has been studied as a function of both material and irradiation conditions. X-ray microanalysis via spectrum lines and images is utilized to measure the RIS profiles of both major and minor alloying elements in several proton-irradiated stainless steels. The influence of pre-existing grain boundary enrichment and post-irradiation annealing of RIS profiles was also investigated.

EXPERIMENTAL

The compositions of a high purity heat (HP304L) and two commercial purity stainless steels (304SS and 316SS) are given in Table I. The HP304L was annealed at 850°C for 60 minutes plus water quenching; whereas the 304SS and 316SS were annealed at 1050°C and 1100°C, respectively, plus forced air cooling. AEM was performed on both as-received and 3.2 MeV proton-irradiated (0.1-3 displacements/atom (dpa) at 360-400°C) specimens. Details of the

Table I : Alloy Composition (wt%)

Alloy	Fe	Cr	Ni	P	Mo	Si	Mn	S	C	B	N
HP304L	69.67	19.65	9.55	0.001	0.01	0.01	1.1	0.003	0.006	<0.001	<0.001
304SS	70.64	18.30	8.50	0.03	0.37	0.65	1.38	0.03	0.04	<0.001	0.07
316SS	66.04	16.70	12.20	0.02	2.58	0.59	1.75	0.01	0.04	0.013	0.06

irradiations, material and specimen preparation and post-irradiation annealing are presented elsewhere [3,4]. Microanalysis was performed in a Philips CM200/FEG (200kV with an incident probe ~1.4 nm (FWTM) in diameter) using a room temperature Compustage double-tilt holder. Prior to analysis the specimen and the holder were plasma cleaned in argon and oxygen plasmas each for 5 minutes. EDS and PEELS spectrum lines and images were acquired and analyzed with an EMiSPEC (ES) Vision system. Quantification from EDS intensities to composition (wt%) was performed with experimental k-factors determined from wide-area matrix measurements. A rough estimate for the sulfur concentrations at the grain boundaries was performed by calculating the Mo L signal from the Mo K signal based on the measured Mo L/Mo K ratio in the matrix where no significant sulfur should be present.

RESULTS AND DISCUSSIONS

Typical RIS profiles for the HP304L material irradiated to 3.0 dpa at 400°C are presented in Fig. 1. From the RIS temperature dependence for this material [5], the RIS at 360°C would be greater in magnitude than that measured at 400°C. There is a ~8 wt% Cr depletion and ~7 and ~1 wt% enrichment of Ni and Fe, respectively, observed at the grain boundary. No segregation of minor alloying or impurity elements was observed. The average grain boundary composition is given as a function of dose in Table II (numbers of specimens characterized and of measurements made are both given). At a lower dose of 0.5 dpa, the magnitude of RIS was less and the average boundary composition was depleted to 17.8 wt% Cr and enriched to 11 wt% Ni. No grain boundary enrichment or depletion was observed for the unirradiated material.

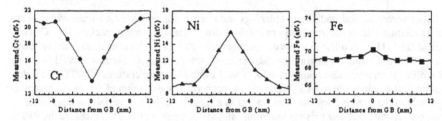

Fig. 1 Concentration profiles at grain boundary in HP304L irradiated to 1 dpa at 400°C.

Concentration profiles for the unirradiated and irradiated (1 dpa, 360°C) 316SS materials are given in Fig. 2 and average grain boundary concentrations are given in Table II. For the unirradiated material, enrichment of Cr, Mo and P (Δ=3, 6, 0.4 wt% relative to the adjacent matrix) as well as ~2 wt% Ni depletion is observed at the grain boundaries. No statistically significant enrichment or depletion of Mn, Si and S were observed. PEELS measurements did not indicate segregation of any interstitial impurities. However, atom probe analysis of the same material indicated significant enrichment of B (0.87 wt%) and C (0.14 wt%) to boundaries [6]. Irradiation to 1 dpa at 360°C resulted in "W-shaped" Cr profiles, decreases in Cr, Mn, Mo and P (3, 0.8, 0.5, 0.2 wt% relative to the unirradiated grain boundary composition), and increases in Ni and Si (4, 1 wt%). After irradiation to 3 dpa at 360°C, the "W-shaped" Cr profile was replaced by a more normal depletion profile (similar to Fig.1, although smaller in magnitude). The magnitude of the depletions or enrichments increased with dose.

The pre-existing enrichments in the unirradiated 304SS materials were similar, but less pronounced than those in the 316SS (except for 0.19 wt% C determined by atom probe). The effects of RIS on grain boundary composition were also similar (except for P which increased

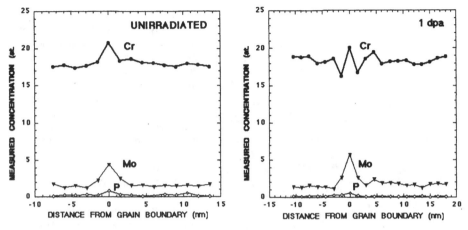

Fig. 2 Concentration profiles at grain boundaries in 316SS in the (a) unirradiated condition and irradiated to 1 dpa at 360°C.

significantly rather than decreasing with dose). In fact, the P and Mo levels were essentially constant from 1 to 3 dpa. It is possible that there is a tendency for co-segregation of these elements under irradiation similar to that observed in mill-annealed materials. Higher dose irradiations could test if this effect is the result of co-segregation or simply a transient. Such irradiations help explain the different P and Mo behavior between 304SS and 316SS.

Although the plasma cleaning generally minimized problems of contamination under the focused probe, some specimens did exhibit beam damage during spectrum line acquisition (Fig. 3). A slight localized thinning and a change of diffracting condition were observed in the upper and lower grains, respectively. The apparent thinning may be exaggerated by the high contrast

Table II: Grain Boundary Composition as Function of Dose (wt%)

Alloy	Dose (dpa)	# of Samples	# of Meas.	Fe	Cr	Ni	Mo	Mn	Si	P
HP304L @400°C	0.0	1	23	70.1	20.1	9.8	ND*	NM*	ND	ND
	0.5	1	29	71.2	17.8	11.0	ND	NM	ND	ND
	3.0	1	5	70.7	12.9	16.4	ND	NM	ND	ND
304SS @360°C	0.0	2	18	68.2	20.5	8.0	1.0	1.5	0.7	0.17
	1.0	3	50	68.2	18.7	9.3	0.7	1.1	1.1	0.87
	3.0	2	29	68.3	17.4	10.7	0.6	0.8	1.4	0.87
316SS @360°C	0.0	3	55	58.5	19.8	10.3	8.8	1.7	0.5	0.38
	1.0	4	48	57.8	17.1	14.0	8.3	0.9	1.6	0.27
	3.0	3	58	61.5	15.8	15.0	4.8	0.6	2.0	0.30

ND*: Element not detected; NM*: Element not measured

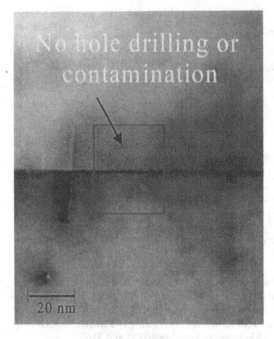

of the STEM image. For a stationary spot analysis, the worst case of thinning resulted in a ~40% decrease in the average count rate over a 100 second acquisition, indicating a 40% decrease in the local thickness. An alternative acquisition mode was used to judge the influence of the thinning on the measured profile. A spectrum image was collected with the same total acquisition time (420 seconds) and number of image rows as points in the spectrum line (21). As there were 20 points in each row of the spectrum image, the electron dose for each spectrum image point was 1/20 of that in the spectrum line. No thinning or contamination was observed after the spectrum image was acquired.

Fig. 3 STEM image after acquisition of both spectrum line (left) and spectrum image (rectangle).

For the unirradiated 316SS, the composition profiles acquired for the major elements (Fe, Cr, Ni, and Mo) by the two acquisition methods were very similar with the only differences arising from statistical variations. The concentrations measured at the grain boundary are given in Table III. The profiles for the minor elements (Si, P, and S) are given in Fig. 4. It must be noted that the sulfur levels both in this figure and in Table III are below the detection limit based on counting statistics, especially considering the peak overlap between S K and Mo L. The

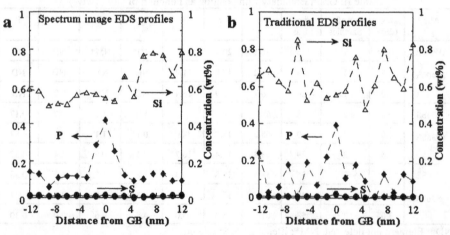

Fig. 4 Comparison of composition profiles for minor elements acquired by (a) spectrum image and (b) spectrum line approaches described in text.

Table III: Grain Boundary Concentrations (wt%) for Different Acquisition Modes

Acquisition Mode	Fe (+Mn)	Cr	Ni	Mo	P	Si	S
Spectrum Line	bal.	19.45	10.25	8.43	0.39	0.56	0.02
Spectrum Image	bal.	19.22	10.48	8.71	0.42	0.56	0.12

only obvious difference in these profiles is the larger variation in the matrix concentrations in the spectrum line profile. This difference was also seen in a second comparison of the two acquisition modes. It is not clear if the numerical or spatial averaging in the spectrum image approach is responsible for this difference. However, from the similarity of the measured profiles and concentrations, it appears that the localized thinning associated with a 20 second/point spectrum line does not modify the measured elemental profiles in these materials.

A series of in situ post-irradiation anneals was performed on the HP304L after irradiation to 1 dpa at 400°C. Both microstructural and microchemical changes were monitored as a function of annealing temperature and time. The measured RIS profiles for various conditions are given in Fig. 5. The RIS profiles for the 500°C anneals were not measured as no microstructural changes were observed. However, for the higher temperature anneals, significant coarsening of dislocation loops, as well as shrinkage and disappearance of smaller loops and defects, resulted in a decrease in the density of extended defects of between 53 and 68%. As seen in Fig. 5, little change in the RIS profiles occurred for the same anneals. Even for the highest temperature anneal (625°C, 40 minutes), the chromium level exhibited only a slight increase from the as-irradiated value of 15.4 to 15.8 wt%. Bulk anneals along with hardness testing to estimate the

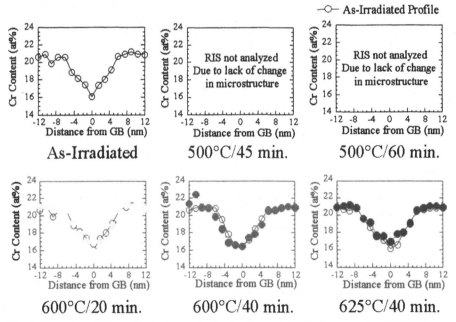

Fig. 5 Chromium RIS profiles before and after in situ, post-irradiation anneals of HP304L material irraddiated to 1 dpa at 400°C. As-irradiated profile shown for comparison.

change in mechanical properties are currently underway to confirm these results and check for the possible influence of thin foil effects on the annealing. The temperature separation in the recovery of radiation-induced microstructural versus microchemical changes should permit their relative importance in IASCC to be investigated.

CONCLUSIONS

Both traditional spectrum line profiles and profiles formed from spectrum images permitted the measurement of Cr, Mo, and P enrichment at grain boundaries in unirradiated stainless steels and radiation-induced segregation profiles in proton-irradiated stainless steels. Although some localized thinning was occasionally observed for spectrum lines, there was no significant impact on the profiles or the concentrations measured near grain boundaries. Summing of individual lines of a spectrum image yields line profiles without hole drilling or contamination. Radiation-induced segregation increases Ni and Si at grain boundaries and depletes Cr and Mo. The behavior of phosphorus under irradiation appears to depend on alloy composition or heat treat condition. Initially chromium depletion by RIS results in "W-shaped" profiles in materials exhibiting pre-irradiation enrichment of Cr, Mo, and P. In situ annealing of proton-irradiated stainless steel indicates that the recovery of irradiation-induced microstructural changes occurs prior to that of RIS profiles. The ability to separate the influence of these two irradiation effects will contribute significantly to the current understanding of irradiation-assisted stress corrosion cracking and possible mitigation of the problem in water cooled nuclear reactors.

ACKNOWLEDGEMENTS

Support was provided by the Division of Materials Sciences, U.S. Department of Energy under contracts DE-AC05-96OR22464 with Lockheed Martin Energy Research Corp., DE-AC06-76RLO1830 with Battelle Memorial Institute, and through the SHaRE Program under contract DE-AC05-76OR00033 with Oak Ridge Associated Universities. Research was performed in part at the Shared Research Equipment (SHaRE) User Facility at Oak Ridge National Laboratory. Grateful acknowledgements to P.L. Andresen, General Electric (alloys and bulk compositions), J. Gan and the Ion Beam Laboratory (irradiations), and the Electron Microscope Analysis Laboratory and staff at the University of Michigan.

REFERENCES

1. P.L. Andresen, in *Stress Corrosion Cracking,* ed. R.H. Jones (ASM International, Materials Park, OH 1992) pp.181-210.
2. E.A. Kenik, R.H. Jones, and G.E.C. Bell, J. Nucl. Mater. **212-215**, 52-59 (1994).
3. D.L. Damcott, J.M. Cookson, R.D. Carter, Jr., J.R. Martin, M. Atzmon, and G.S. Was, Nucl. Inst. and Meth. in Phys. Res. **99**, 780-783 (1995).
4. J.T. Busby, G.S. Was, and E.A. Kenik, in *Microstructural Processes in Irradiated Materials*, eds. S.J. Zinkle et al. (Mater. Res. Soc. Proc. **540**, Pittsburgh, PA 1999), pp. 495-500.
5. G.S. Was, et al., J. Nucl. Mater. **270**, 96-114 (1999).
6. E.A. Kenik, et al., in *Microstructural Processes in Irradiated Materials*, eds. S.J. Zinkle et al. (Mater. Res. Soc. Proc. **540**, Pittsburgh, PA 1999), pp. 445-450.

INVESTIGATION OF COPPER SEGREGATION TO THE Σ5(310)/[001] SYMMETRIC TILT GRAIN BOUNDARY IN ALUMINUM

Jürgen M. Plitzko, Geoffrey H. Campbell, Wayne E. King and Stephen M. Foiles*
Chemistry and Materials Science Directorate, University of California, Lawrence Livermore National Laboratory
P.O. Box 808, Livermore, CA 94550
*Science-Based Materials Modeling Department, Sandia National Laboratories, P.O. Box 869, Livermore, CA 94551

ABSTRACT

The Σ5 (310)/[001] symmetric tilt grain boundary (STGB) in the face centered cubic (FCC) metal aluminum with 1at% copper has been studied. The model grain boundary has been fabricated by ultra-high vacuum diffusion bonding of alloy single crystals. The segregation of the copper has been encouraged by annealing the sample after bonding at 200 °C. TEM samples of this FCC-material were prepared with a new low voltage ion mill under very low angles.
The atomic structure of the Σ5(310)/[001] STGB for this system was modeled with electronic structure calculations. These theoretical calculations of the interface structure indicate that the Cu atoms segregate to distinct sites at the interface. High resolution electron microscopy (HRTEM) and analytical electron microscopy including electron energy spectroscopic imaging and X-ray energy dispersive spectrometry have been used to explore the segregation to the grain boundary. The HRTEM images and the analytical measurements were performed using different kinds of microscopes, including a Philips CM300 FEG equipped with an imaging energy filter. The amount of the segregated species at the interface was quantified in a preliminary way. To determine the atomic positions of the segregated atoms at the interface, HRTEM coupled with image simulation and a first attempt of a holographic reconstruction from a through-focal series have been used.

INTRODUCTION

We have chosen a model grain boundary to investigate the segregation of an impurity to distinct sites in the boundary. Specifically, we use copper segregation in a aluminum Σ5(310)/[001] STGB. The phenomenon of segregation is of long standing scientific interest and has been studied extensively, both theoretically as well as experimentally [1, 2]. The comparison between theoretically predicted grain boundary structures and experimentally determined structures is important, especially for understanding the mechanical and physical properties of materials. Special grain boundaries such as the Σ5(310)/[001] STGB can give us first insights into the more general behaviour of grain boundaries. Studying a variety of bicrystals in different orientations may lead us to predict the properties of polycrystalline materials but certainly gives us a greater understanding of the complex behaviour of these real materials. The ability to choose and manipulate the occurance of special grain boundaries in a polycrystal can improve the material performance and can be used to design and engineer materials for optimum properties by using the knowledge of structure/properties relations for the grain boundaries.
With the controlled fabrication and preparation of bicrystals we are able to determine composition, structure and morphology of grain boundaries which depends on geometry, crystal orientation, impurity concentration and temperature. The limiting factor in this approach is the ability to fabricate well defined, precisely oriented interfaces, which is enabled here with the UHV Diffusion Bonding Machine [3].

Mat. Res. Soc. Symp. Proc. Vol. 589 © 2001 Materials Research Society

EXPERIMENTAL DETAILS

The single crystal of Al-1at%Cu was grown by the Bridgeman technique in a graphite mold with a growth axis of [001]. The crystal was oriented for cutting using Laue backscatter X-ray diffraction. The surfaces to be bonded were ground and polished to be co-planar with (310) to within 0.1°. This precision can be achieved by mounting the crystal slices in a specially designed goniometer equipped lapping fixture. The $\Sigma 5(310)/[001]$ STGB is then formed by a mutual misorientation of 180° of one crystal with respect to the other about an axis normal to the bonding faces. Details of this process are explained and discussed elsewhere [4].

The ultra-high vacuum (UHV) diffusion bonding process bonds the bicrystal under highly controlled environmental conditions. The two single crystals were bonded in the solid state at a temperature of 540 °C with a constant applied pressure of 1.0 MPa for 8 hours. Afterwards the bicrystal was annealed for 100 hours at 200 °C using a flowing Argon atmosphere of approximately 1 bar. For the TEM sample preparation, a 3 mm rod was cut by wire electric discharge machining (EDM) out of the bicrystal parallel to the interface. The rod was then cut with a diamond wire saw in 200 to 300 μm thick slices to create the 3 mm disks required for TEM. Standard grinding, polishing and dimpling were additionally performed. A new type of ion mill was used to gently thin the material. With this ion mill (Linda Technoorg IV3H/L), one can use in the beginning a high voltage gun operating between 2kV and 10kV and afterwards, for the finishing step, an ion gun operating at a low energy between 200 V and 2kV. Additionally one is able to choose a low incident ion beam angle of about 5 °. The material of our investigation was thinned for 3 hours with the high energy gun at 2kV from both sides and finally for 1 hour with the low energy gun at 200V and 5° incidence angle with liquid nitrogen cooling. The advantage and the improvement of the TEM samples by using double sided, low angle and low energy thinning has been shown by Strecker et al. [5]. The result and the improvement of ion beam thinning under these conditions can be clearly seen in the HRTEM image in Figure 4a where the surface topography is homogeneously flat over the whole imaged region. Additionally, samples were prepared using conventional jet polishing with solutions and conditions normally preferred for preparing TEM samples of metals and metal alloys.

The investigations were performed primarily using the recently installed Philips CM 300 FEG ST TEM at Lawrence Livermore National Laboratory (LLNL). This microscope operates with a field emission gun at 300 kV and is equipped with a SuperTwin objective lens (ST, C_s = 1.2 mm), a Gatan imaging filter (GIF) with a 2kx2k CCD camera and an EDX detector. Additionally, we performed measurements, particularly the acquisition of the HRTEM through focal series, at the NCEM (National Center for Electron Microscopy) at Lawrence Berkeley National Laboratory with a Philips CM 300 FEG UT TEM (UltraTwin objective lens, UT, C_s = 0.65 mm). The analytical results were partly obtained using a Zeiss EM912 Ω equipped with an Ω energy filter for electron spectroscopic imaging (ESI) and a dedicated VG STEM HB501 for energy dispersive X-ray spectroscopy (EDS), both are located at the Max-Planck-Institut fuer Metallforschung in Stuttgart, Germany.

RESULTS

The relation between grain boundary energy and impurity segregation to the interface have been theoretically calculated for the $\Sigma 5$ (310)/[001] interface in Al-1at%Cu within the Local Density Approximation (LDA). The calculations use a plane-wave basis and ultrasoft pseudopotentials [6]. Figure 1a shows the calculated structure of the interface with the (310) boundary plane in the

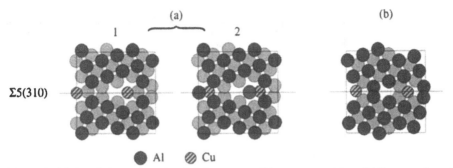

Figure 1: *(a) Atomistic structure (model) for two possible copper positions (b) Structure based on a coincidence site lattice construction.*

[001] viewing direction for the two positions of the copper atom considered. The overall structure is qualitatively similar to previous predictions for Al based on pair-potential calculations. Only substitutional sites in the boundary plane have been considered for the copper atoms at this time. The calculated energy for site (2) is lower by about 0.4 eV compared to the copper atom placed in the bulk. The copper atom is computed to be repelled from site 1 by 0.7 eV compared to the bulk. This site preference is consistent with a size effect interpretation since site (2) is locally compressed and copper is undersized in Al. Figure 1b represents the structure based on the coincidence site lattice construction (CSL) [7]. The difference between the two lattices is visible. Simulations of the HRTEM images were done using these two structural models.

To confirm the presence and determine the amount of the segregant, analytical electron microscopy was performed. Three electron spectroscopic images were acquired at the $L_{2,3}$-edge of copper around 931 eV with a 50 eV slit and 50 s exposure time to extract elemental maps (three-window-technique) of the copper distribution [8]. This specific investigation was performed on the Zeiss EM912 Ω, which operates at 120 kV and with a LaB_6 cathode. The overview of the investigated sample position and the resulting elemental map is shown in Figure 2a and b. Figure 2c shows the integrated line profile (integration was performed over 50 pixel lines).

Because of the small amount of the segregated species and the high energy loss of the investigated ionization edge the resulting distribution is heavily dominated by noise. Nevertheless, a bright line (peak in the integrated line profile) which represents the copper at the interface is visible. The copper signal at the edge of the hole is likely due to a preparation artifact. This sample was pre-

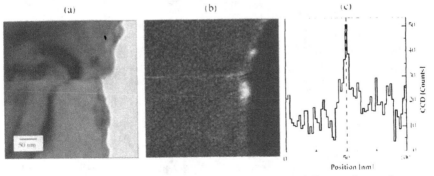

Figure 2: *(a) Bright field image of investigated sample position. (b) Copper elemental map and (c) integrated line profile (50 lines) over the grainboundary in the elemental map.*

303

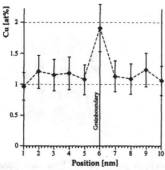

Figure 3: Illustration of an EDS line scan of Cu concentration across the investigated grain boundary.

pared by jet polishing and the copper might be a remnant of the chemical etching process. The possibility of sample drift during the acquisition, which might result in a false signal at the interface, was excluded by using several methods for the determination of the real drift between the three acquired images [9, 10]. But, due to the difficulty of investigating copper in such small amounts at the interface with EELS and ESI, the signal in this elemental distribution image (EDI) has to be further confirmed with multiple measurements at different sample positions. Also, the image quality and thus the quality of the result should be improved by using the field emission source at 300 kV in conjunction with the GIF. Precipitates which are rarely present and which might highlight the segregation could be observed, however the sample areas in these regions were too thick to be used for further investigation.

To reveal more precisely the segregation to the grain boundary, EDS line scans were performed and the resulting spectra were quantified. One of the linescans is shown in Figure 3. The amount of the copper in the analyzed volume is nearly doubled ($\approx$ 2 at%) in relation to the bulk concentration of $\approx$ 1 at% (average from different line scans). This illustrates clearly the significant segregation of copper to the grain boundary. High resolution images of the boundary structure were acquired using different consecutive defocus values. To acquire these through-focal series we have automated the acquisition process using the script language of the Gatan Digital Micrograph software. During the acquisition, the stability of the microscope and minimizing the drift of the

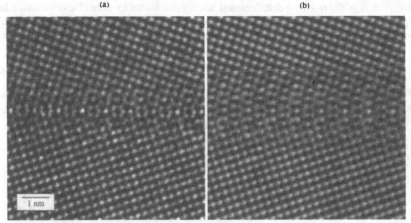

Figure 4: (a) HREM image of $\Sigma 5(310)/[001]$ STGB in Al-1at%Cu. (b) reconstruction of defocus series.

sample is crucial for allowing image processing of the series for holographic reconstruction. The through-focal series can now be acquired by starting the controlling script, without any further manipulation of the microscope. This acquisition procedure is now used routinely at the NCEM at LBL.

The main aim of recording the through-focal series is the possibility of exit wave reconstruction of the investigated boundary structure. The reconstruction is described in detail elsewhere [11, 12]. Shown in Figure 4a is one of the 30 images of the through-focal series which was obtained with the Philips CM300 FEG UT TEM. The most striking feature of this sample is the homogeneously flat surface of the sample over the field of view and the lack of distortions which are normally present in HRTEM images. This is directly due to the sample preparation with the low energy, low angle ion mill described above. The series was acquired in a defocus range starting at −130 nm. The defocus step was determined to be 2.2 nm. A first reconstruction has been performed and the resulting image can be seen in Figure 4b. The refinement of the reconstruction has to be done by using standard image simulation procedures to determine the exact defocus and imaging parameters at which the image was acquired. The starting structures for these simulations are shown in Figure 1a (site 2) and b. Image simulation was performed using the EMS software [13]. Figure 5a and b shows two simulated images for nearly the same defocus as in the original image but further simulations and refinements have to be done to improve the fit to the experimental image and the result of the exit wave reconstruction.

CONCLUSIONS AND OUTLOOK

These preliminary findings of segregation in this alloy for this specifically oriented grain boundary revealed the expected segregation behaviour. The structure of the grain boundary and the positions of the copper atom at the grain boundary are not yet determined exactly, but a tendency can be seen in these first results. To obtain more information about the atomic structure and the electronic structure, we are planning further experiments with 'Z-contrast' imaging and HREEL spectroscopy. Additionally investigations in the pure (copper-free) aluminum bicrystal will be done to compare the influence of the segregant on the grain boundary structure.

Finally we can say that the TEM sample preparation using a low energy and low angle ion mill can improve the image quality enormously despite the fact that point defects and defect agglomerates are normally produced in the crystals near the surface by ion beams in FCC-metals.

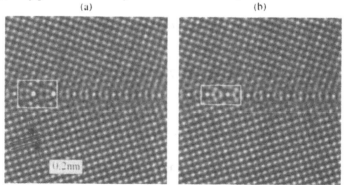

(a)   (b)

Figure 5: (a) EMS simulation (white framed boxes) for the CSL structure ($\Delta f \approx -149$ nm, t=4 nm) and (b) for the LDA (2) structure .

305

ACKNOWLEDGEMENTS

We would like to thank the NCEM and Christian Kisielowski for providing the insight into the reconstruction of defocus series and for his help at the Philips CM300 FEG UT TEM. The authors also wish to thank M. Kelsch, C. Song, A. Bliss and A. Strecker for sample preparation and useful discussions about the best preparation methods. We thank the Max-Planck-Institut fuer Metallforschung for access to the Zeiss EM912 Ω and Wilfried Sigle for his help at the VG STEM HB501. This work was performed under the auspices of the United States Department of Energy, Office of Basic Energy Sciences and Lawrence Livermore National Laboratory under Contract No. W-7405-Eng-48.

REFERENCES

1. P. Lejcek, S. Hofmann, *Critical Reviews in Solid State and Materials Sciences*, **20** [1] 1-85 (1995).

2. A. P. Sutton, R.W. Balluffi, *Interfaces in Crystalline Materials*, ed. (Oxford University Press, Oxford, 1995).

3. W. E. King, G. H. Campbell, A. W. Coombs, G. W. Johnson, B. E. Kelly, T. C. Reitz, S. L. Stoner, W. L. Wien and D. M. Wilson in *Joining and Adhesion of Advanced Inorganic Materials*, edited by A. H. Carim, D. S. Schwartz and R. S. Silberglitt (Mat. Res. Soc. Symp. Proc. **314**, Pittsburgh, PA 1993) pp. 61-67.

4. W. L. Wien, G. H. Campbell and W. E. King in *Microstructural Science*, edited by D. W. Stevens, E. A. Clark, D. C. Zipperian and E. D. Albrecht (Microstructural Science **23**, Materials Park, OH 1996) pp. 213-218 ;presented at 28th International Metallography Society Convention.

5. A. Strecker, J., Mayer, B. Baretzky, U., Eigenthaler, T. Gemming, R. Schweinfest, M. Ruehle in *Journal of Electron Micoscopy*, **48** [3] 235-244 (1999).

6. D. Vanderbilt in Physical Review **B41**, 7892 (1990).

7. W. Bollmann in *Crystal Defects and Crystalline Interfaces*, ed. (Springer-Verlag, Berlin, 1970).

8. C. Jeanguillaume, P. Trebbia, C. Colliex in *Ultramicroscopy*, **3** 237-242 (1978).

9. J.A. Hunt, private communication.

10. J.M. Plitzko, J. Mayer in *Ultramicroscopy*, **78** 207-219 (1999).

11. A. Thust, W.M.J. Coene, M. Op de Beeck, D. Van Dyck in *Ultramicroscopy*, **64** 211-230 (1996).

12. W.M.J. Coene, A. Thust, M. Op de Beeck, D.Van Dyck in *Ultramicroscopy*, **64** 109-135 (1996)

13. P. A. Stadelmann in *Ultramicroscopy*, **21** 131-146 (1987).

STEM ANALYSIS OF THE SEGREGATION OF Bi
TO Σ19a GRAIN BOUNDARIES IN Cu

W.SIGLE, L.-S. CHANG, W.GUST, M.RÜHLE
Max-Planck-Institut für Metallforschung und Institut für Metallkunde der Universität Stuttgart,
Seestraße 92, D-70174 Stuttgart, Germany. sigle@hrem.mpi-stuttgart.mpg.de

ABSTRACT
Cu bicrystals were doped with different amounts of Bi and annealed at various temperatures. The segregation at the grain boundary (GB) was analyzed by energy dispersive X-ray analysis (EDX) in the scanning transmission electron microscope (STEM). It is found that with both increasing temperature and increasing Bi content the amount of faceting of the GB increases, finally reaching complete faceting. This final state is distinguished by brittle behavior. By analyzing the shape of pores at the GB we found that the Cu surface energy anisotropy is probably increased by Bi surface segregation.

INTRODUCTION
It has been known for a long time that the addition of Bi to polycrystalline Cu leads to embrittlement of the material [1]. This is caused by the segregation of Bi atoms to the GBs which obviously weakens the adhesion of neighbouring grains and also leads to a faceting of most GBs [2]. Recently the Cu(Bi) phase diagram and the GB segregation have been studied in polycrystalline Cu [3]. It was found that below the solidus line a prewetting regime exists, where segregation is particularly strong. In the present work segregation was studied using a dedicated STEM which allows for much higher spatial resolution than Auger analysis. For a comprehensive overview of the present work we refer to [4].

EXPERIMENT
Cu Σ19a {331} GBs were produced by both Bridgman growth and diffusion bonding. Cylinders of 3mm diameter were doped with Bi and subsequently annealed. TEM specimens were prepared by electrolytic polishing. For details of specimen preparation see [5]. The analysis was performed using a VG 501HB STEM. During the acquisition of X-ray spectra the electron beam was scanned over a 3nm by 4nm area on the boundary. For the determination of the areal density of Bi on the GB the beam spreading within the specimen was considered [6]. In the following, areal densities will be given in number of monolayers (ML) where 1 ML is defined as a (331) plane which corresponds to 7 atoms·nm^{-1}. Errors are due to a variation among different measurements indicating some degree of inhomogeneity along the GB.

RESULTS
(i)   **Bridgman-grown crystals**
The Cu(Bi) phase diagram is shown in Fig.1. Filled circles mark the annealing/doping positions of analyzed specimens. Fig.2 shows the GB morphology together with measured segregation strengths for specimens doped with 25 at ppm Bi but annealed at different temperatures. A variation is mainly visible in the GB morphology. Whereas the GB is straight at 600°C, steplike facets (marked by arrows in Fig.2) appear after annealing at 700°C. Their length is about 20 nm. The facets show much stronger segregation than the unfaceted part of the GB. After

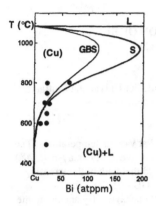

Fig.1: Cu(Bi) phase diagram [3]. L-liquidus, S- bulk solidus line, GBS-GB olidus line, (Cu)-Cu-rich solid solution. Only on the left side of the GB solidus line the Bi enrichment can be considered as areal GB segregation. In the region between the bulk and GB solidus lines a quasi-liquid phase exists along the GB. The doping/ annealing positions of the investigated specimens are shown by filled circles.

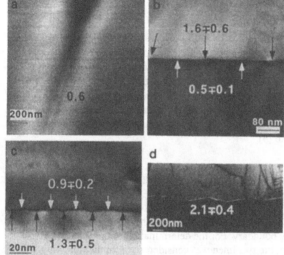

Fig.2: GB morphology as a function of annealing temperature and Bi volume concentration. (a) 600°C/25 atppm, (b) 700°C/25 atppm, (c) 800°C /25 atppm, (d) 800°C/65 atppm. Numbers denote ML of Bi.

annealing at 800°C a second type of facet becomes visible leading to V-shaped parts separated by regular GB segments. Again the segregation is stronger at the facets. No difference was detected between the two types of facets. An increase of the volume concentration of Bi to 65 at ppm (which is in the prewetting regime [3]) leads, at an annealing temperature of 800°C, to a completion of the faceting and to a segregation strength of about 2 ML. These specimens also exhibited very brittle behavior.

(ii)   **Diffusion bonded crystals**

Due to the residual surface roughness, pores are formed at the GB during diffusion bonding. All pores are faceted into identical shapes although the pores exhibit different sizes (1-5 µm) (Fig.3).

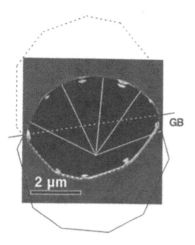

Fig.3: Faceted pore at the Σ19a GB of a Cu(25 at ppm Bi) bicrystal annealed at 600°C for 2 days. The GB is marked by a black line.

Therefore we assume that these shapes are equilibrium shapes which would be obtained from the Wulff construction [7]. An equilibrium polygon was constructed from the facets in one of the two grains. This polygon was rotated by 26° which corresponds to the tilt angle of the Σ19a GB. As would be expected for an equilibrium shape, it fits now excellently to the lower grain (dashed line). The mushroom-like objects located at the intersections of facets are highly Bi-rich. From this we assume that the facets were also covered by Bi atoms. This segregation is likely to change the shape of the γ-surface (γ-surface energy) as compared to pure Cu.

DISCUSSION

The temperature *and* Bi volume concentration dependence of the GB morphology shows that both parameters are involved in the variation of the GB shape. Since the whole system is trying to adopt a minimum in Gibbs energy, these results show that Bi reduces the GB energy for the observed facets (which are asymmetric Σ19a GBs) in a stronger way than it would for the symmetric Σ19a GB. Similar results were also observed for other high energy GBs like the Σ51 [6]. Presently it is not clear which prerequisites are necessary for a GB to become particularly low-energetic after Bi segregation. The temperature dependence (at constant volume concentration) clearly demonstrates that the observed GB phase transition is thermally activated. A possible scenario for this transition is given in Fig.4. At low amounts of segregation single Bi atoms are located at the GB and probably occupy particular GB sites. It could well be that the next nearest neighbors of each Bi atom rearrange so as to minimize energy [8]. This is indicated by rhombi in Fig.4. It is well known that, in the single phase regime of the phase diagram, Bi segregation can be described by the Fowler-Guggenheim isotherm [3] with an attractive interaction of Bi atoms. As a consequence, at an increasing level of segregation Bi atoms associate. We propose that in order to reach a low energetic configuration these "Bi dimers" undergo a thermally activated transformation (e.g. a shear) leading to the observed steplike facets at 700°C. This is shown in Fig.4. At even higher Bi concentrations these facets are completed. From the brittle behavior of these specimens we conclude that crack propagation along the GB becomes particularly easy if faceting is completed. This is reasonable since with isolated facets the crack has to propagate along (less brittle) non-faceted GB areas requiring the production of dislocations at the crack tip which is well known to increase the fracture toughness.

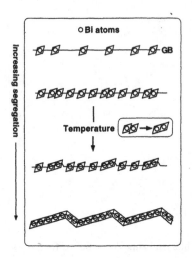

Fig.4: Possible scenario for faceting by Bi segregation.

From the Wulff construction we can directly calculate the surface energy anisotropy since the distance of the facets from the center (white lines in Fig.3) of the polygon is directly proportional to the surface energy. This leads to an anisotropy of about 8%. In pure Cu the surface energy anisotropy amounts to 3.5% at 830°C [9]. Values at lower temperatures are not available but the anisotropy will increase. Taking into account the tendency of GBs to facet, it would not be a surprise if the surface energy anisotropy were increased by the addition of Bi.

CONCLUSIONS
Systematic investigations of the Bi GB segregation for a particular GB in Cu have shown that the GB morphology is dependent on both the annealing temperature and on the segregation strength of Bi. A scenario for the GB phase transition was proposed. Apart from GB faceting, Bi also leads to a faceting of surfaces.

ACKNOWLEDGMENTS
The authors would like to thank Mrs. Sorger for the diffusion bonding, Mr. Bahle for the crystal growth, Mr. Stutz for spark erosion work, Prof. M. Finnis, Prof. V. Vitek and Dr. U. Alber for stimulating discussions. The organisational work of Dr. B. Baretzky is greatfully acknowledged.

REFERENCES
[1] W. Hampe, Z. f. Berg-, Hütten- und Salinenwesen 23,93 (1874).
[2] A. Donald, Phil. Mag. 34, 1185 (1976).
[3] L.-S. Chang, E. Rabkin, B.B. Straumal, B. Baretzky, and W. Gust, Acta Mater. 47,4041 (1999).
[4] W. Sigle et al., to be submitted to Acta Mat.
[5] U. Alber, H. Müllejans, and M. Rühle, to be published in Acta Mater. (1999).
[6] U. Alber, H. Müllejans, and M. Rühle, Ultramicroscopy 69, 105 (1997).
[7] G. Wulff, Z. Kristallogr. 34, 449 (1901).
[8] V. Vitek , private communication.
[9] M. McLean, Acta Met. 19, 387 (1971).

INVESTIGATING ATOMIC SCALE PHENOMENA AT MATERIALS INTERFACES WITH CORRELATED TECHNIQUES IN STEM/TEM

N. D. BROWNING, A. W. NICHOLLS$, E. M. JAMES, I. ARSLAN, Y. XIN*, K. KISHIDA**, S. STEMMER***
Department of Physics, University of Illinois, 845 W. Taylor St., Chicago, IL 60607-7059.
$RRC, University of Illinois, 845 W. Taylor St., Chicago, IL 60607-7058.
*now at National High Magnetic Field Laboratory, Florida State University, 1800 E. Paul Dirac Drive, Tallahassee, FL 32310
**now at NRIM, 1-2-1 Sengen, Tsukuba, Ibaraki 305-0047, JAPAN
***now at Department of Mech. Eng. & Materials Science, Rice University, 6100 Main St., Houston, TX 77005-1892.

ABSTRACT

A complete understanding of the complexities behind the structure-property relationships at materials interfaces requires the structure, composition and bonding to be characterized on the fundamental atomic scale. This level of characterization is beyond the scope of a single imaging or microanalysis technique and so to solve practical interface problems, correlation between multiple techniques must be achieved. Here we describe recent advances in the JEOL 2010F 200kV field-emission STEM/TEM that now allow atomic resolution imaging and analysis to be obtained in both TEM and STEM mode and discuss two applications of these techniques.

INTRODUCTION

Internal interfaces are known to play a dominant role in determining the overall bulk mechanical and electrical characteristics of many materials [1]. In characterizing the particular materials issues associated with interfaces it is useful to classify them into one of two groups; homophase (grain-boundaries in single phase materials) and heterophase (composite interfaces such as metal-ceramic interfaces). In addition, in each of these groups the interface can be clean, have segregated impurities or contain a different phase from the bulk (interphase). The full characterization of each of these different types of interfaces therefore presents a wide range of experimental problems. An obvious framework around which an understanding of interfaces can be constructed is a knowledge of the atomic structure. There are various techniques in transmission electron microscopy to study these atomic scale effects at interfaces. Both conventional transmission electron microscopes (TEM) [2,3] and scanning transmission electron microscopes (STEM) [4-6] provide the capability of imaging interface structures on the atomic scale. Each approach has its advantages. In the case of the TEM, the phase contrast imaging technique provides a very high-intensity image that, through extensive image simulation [7], can provide accurate atomic structures. The Z-contrast approach in STEM, on the other hand, produces a direct image of the atomic structure, i.e. no simulations are required, that has compositional sensitivity [5]. Ideally, we would like to be in a position to apply both techniques to the study of interfaces, using whichever one provides the simplest solution to the problem at hand.

Therefore, the phase contrast TEM technique and the Z-contrast STEM technique can both provide information on the atomic structure of interfaces. However, strain effects at the interface can make absolute quantification of composition difficult to achieve for both methods. To overcome this limitation, microanalysis can be performed using either energy dispersive X-ray spectroscopy (EDS) [8] or electron energy loss spectroscopy (EELS) [9] in either TEM or STEM mode. As with the imaging techniques, both have advantages and disadvantages. In the case of EDS, it is possible to get a very straightforward measurement of composition [10], while with EELS it is possible to get atomic resolution composition [11] and bonding [12] information (albeit in a manner that is not as straightforward to interpret). Again, with the large range of problems that exist, there will be times when one of the techniques will provide the most straightforward means to get the desired information, and ideally we would like to be in a position to choose which one to use during the experiment. For complete analysis there are also diffraction techniques that can provide significant information on interfaces [8,13].

To study materials interfaces, we therefore need to be able to employ a wide variety of techniques and ideally, switch between TEM and STEM mode depending on the information we require. The recent development of the JEOL 2010F 200kV Schottky field emission STEM/TEM has now made it possible to accomplish precisely this [14,15]. Depending on the materials system being investigated it is possible to perform all of the functions of a state-of-the-art TEM and all of the functions of the highest resolution dedicated STEM during the same experiment. In this paper, we discuss the practical aspects of tuning the instrument for this wide range of experimental techniques. As most of the recent developments have been concerned with the STEM mode of operation, we shall focus primarily on a description of those techniques and only briefly mention the TEM techniques where relevant (i.e. to demonstrate that the microscope is still a functioning TEM). Having defined the techniques, we go on to discuss two examples of where multiple TEM and STEM techniques were used to investigate materials interfaces; the analysis of grain boundaries in $(Ba,Sr)TiO_3$ thin films and the analysis of the effect of As at the film-substrate interface on the quality of CdTe (111) on Si (111) substrates.

STEM TECHNIQUES

The key to high-resolution STEM is the formation of an electron probe of atomic dimensions. Figure 1 shows the lens arrangement in the JEOL 2010F. Essentially, the electron optics of the microscope above the specimen are aligned in such a way as to make the probe as small as possible on the surface of the specimen. For a given acceleration voltage and emission type (e.g. Schottky field emission or cold field emission) the size of the probe is dependent only on the spherical aberration coefficient of the probe forming lens (0.14nm for the 2010F).

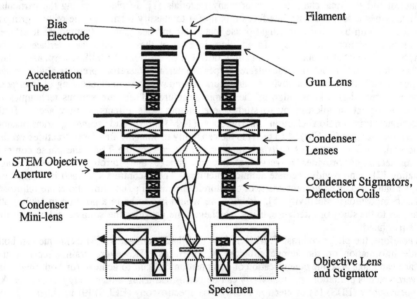

Figure 1. (a) The probe forming optics in the JEOL 2010F

For a given spherical aberration coefficient and acceleration voltage, the probe size is optimized most readily with the electron "Ronchigram", or "shadow image". This is because the intensity, formed at the microscope Fraunhofer diffraction plane, varies considerably with angle, and this variation is a very sensitive function of lens aberrations and defocus [17,18]. When the excitation of each illumination electron optical component (i.e. lens or stigmator) is slightly changed, very small misalignments become apparent by translations in the pattern that depart from circular symmetry. Furthermore, the presence or absence of interference fringes in the pattern indicates the amount of incoherent probe broadening due to instabilities and the effect of a finite source size. Figure 2 shows a schematic ray diagram for the formation of a Ronchigram and an example of the Ronchigram obtained from Si <111> after the lenses and stigmators have been optimized [16,17]. A key aspect of the alignment of the microscope using the Ronchigram is that the operating parameters of the microscope can be optimized (focus, stigmation, placing detectors on axis) and the tilt of the specimen can also be defined. The ability to tilt the specimen to a zone axis in this manner avoids many of the difficulties associated with the use of older dedicated STEM microscopes for atomic resolution imaging.

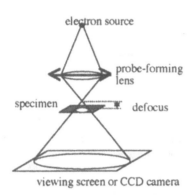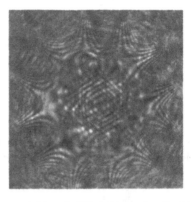

Figure 2. (a) Schematic ray diagram for a Ronchigram and (b) typical Ronchigram for Si <111> showing interference corresponding to the 0.192nm {220} periodicity .

Z-contrast Imaging

Z-contrast images [18] are formed by collecting the high-angle scattering on an annular detector (Figure 3). Detecting the scattered intensity at these high-angles and integrating over a large angular range effectively averages coherence effects between neighboring atomic columns in the specimen. Thermal vibrations reduce the coherence between atoms in the same column to residual correlations between near neighbors [18], a second order effect. This allows each atom to be considered as an independent scatterer. Scattering factors may be replaced by cross sections, and these approach a Z^2 dependence on atomic number. This cross section effectively forms an object function that is strongly peaked at the atom sites, so for very thin specimens where there is no dynamical diffraction, the detected intensity consists of a convolution of this object function with the probe *intensity* profile (Figure 3). The small width of the object function (~0.02 nm) means that the spatial resolution is limited only by the probe size of the microscope. For a crystalline material in a zone-axis orientation, where the atomic spacing is greater than the probe size, the atomic columns are illuminated sequentially as the probe is scanned over the specimen. An atomic resolution compositional map is thus generated, in which the intensity depends on the average atomic number of the atoms in the columns. This result also holds true for thicker specimens. In this case, the experimental parameters cause dynamical diffraction effects to be manifested as a columnar channeling effect, thus maintaining the thin specimen description of the image as a simple convolution of the probe intensity profile and an object function, strongly peaked at the atom sites [18]. The phase problem associated with the interpretation of conventional high-resolution TEM images is therefore eliminated. In thin specimens, the dominant contribution to the intensity of a column is always its composition, although due to the higher absorption of the heavy strings the contrast does decrease with increasing specimen thickness and in very thick crystals there is no longer a high resolution image. The effect of changing focus is also intuitively understandable as the focus control alters the probe intensity profile on the surface of the specimen. A typical Z-contrast image is shown in Figure 4.

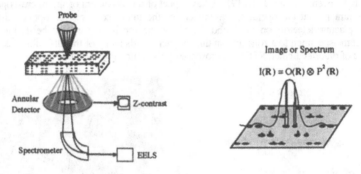

Figure 3: (a) schematic of the STEM. (b) The Z-contrast image can be interpreted as a simple convolution of the experimental probe and the object function.

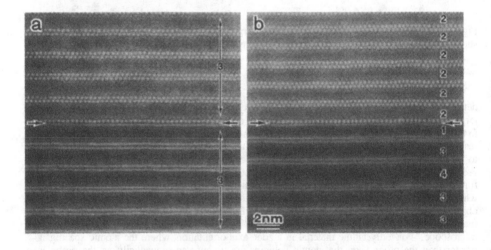

Figure 4. Z-contrast image of a c-axis twist boundaries in a $(Bi_2Sr_2Ca_{n-1}Cu_nO_y)$/Ag tape (indicated by arrows). The numbers (n) in the figures represent the superconducting phases.

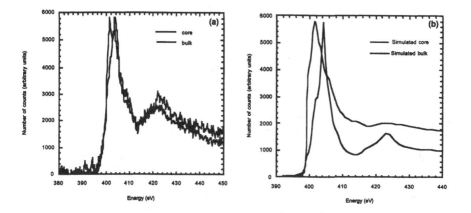

Figure 5. (a) Experimental EELS spectra obtained from a GaN dislocation and (b) multiple scattering simulations reproducing the changes in fine-structure with full clusters.

Electron Energy Loss Spectroscopy (EELS)

As can be seen from Figure 1, the annular detector used for Z-contrast imaging does not interfere with the low-angle scattering used for EELS [9]. This means that the Z-contrast image can be used to position the electron probe over a particular structural feature for acquisition of a spectrum [11,12]. The physical principle behind EELS relates to the interaction of the fast electron with the sample to cause either collective excitations of electrons in the conduction band, or discrete transitions between atomic energy levels, e. g. $1s \rightarrow 2p$ transitions [9]. The ability to observe discrete atomic transitions allows compositional analysis to be performed by EELS (the transitions occur at characteristic energy losses for a given element). Furthermore, the transitions to unoccupied states above the Fermi level allows the degree of hybridization between atomic orbitals to be determined, i.e. information on local electronic structure (bonding) changes can be ascertained. As an example of the type of localized bonding information that can be obtained, figure 5 shows experimental EELS spectra obtained from a dislocation core in GaN [20]. The spectrum from the core column shows a change in density of states (i.e. bonding) that is consistent with the loss of crystal symmetry at the core (as verified by the multiple scattering simulations [21] also shown in figure 5).

Energy Dispersive X-ray Spectroscopy (EDS)

The detectability of segregants at interfaces by EDX depends both on the spatial resolution and the sensitivity of the detector. With the Schottky field emitter used in the 2010F, beam currents of 0.5nA are possible in 1nm probes. This is comparable to the performance of dedicated STEMs [22]. A measure of the sensitivity of the EDS system can be made by measuring the peak-to-background (P/B) ratio from the NIST thin film Cr specimen [23]. Use of this specimen allows direct comparison with results from dedicated STEMs designed for EDS analysis [22]. Table 1 shows a comparison of the performance of the 2010F with two VG dedicated STEMs. The P/B value for the 2010F is much better than those published for 200kV thermionic source microscopes (<3000 (b)) and almost matches the performance of the specially modified VG HB603 at 200kV. Although EDS analysis at atomic resolution is not possible it is clear that detection of segregants at less than monolayer coverage is possible with the 2010F.

	300kV VG HB603	200kV VG HB603	200kV JEOL 2010F	100 kV VG HB 601
P/B	5770	4760	4010	3200

Table 1. Comparison of peak to background ratios in STEM instruments and the JEOL 2010F.

TEM TECHNIQUES

A full description of TEM techniques at this point is beyond the aim of this paper, and is something that has been covered in great detail in many publications (see for example [8,24]). It is our intention only to point out that in optimizing the JEOL 2010F for STEM, the electron optics for the TEM aspect of its use have not been changed. Figure 6 shows the lens and detector arrangement in the microscope below the specimen. As has been discussed before in the section on Z-contrast imaging and EELS, the STEM detectors do not interfere with each other or for that matter with any of the detectors used for TEM operation. Additionally, none of the lens parameters have been changed and so the entire range of TEM experiments for which the instrument was originally designed can be performed with the microscope. To change from TEM to STEM requires only a change in the lens settings (with the control of the deflection/scan coils being taken over by the scanning unit). For the purposes of the experiments described in the latter sections of this paper, we shall make use of high-resolution phase contrast imaging, conventional TEM imaging and CBED functions of the JEOL 2010F TEM.

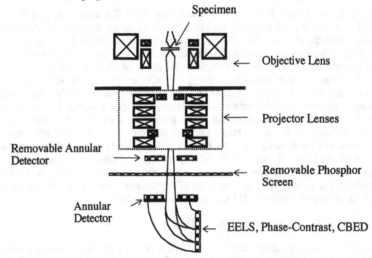

Figure 6. Schematic showing the lens arrangement in the JEOL 2010F below the specimen.

CORRELATED TECHNIQUES FOR INTERFACE ANALYSIS

Nonstoichiometry at grain boundaries in $(Ba_xSr_{1-x})Ti_{1+y}O_{3+z}$ thin films

As a first example [25] of the use of correlated techniques to study interfaces, we examine grain boundaries in $(Ba_xSr_{1-x})Ti_{1+y}O_{3+z}$ (BST) thin films that are being developed as potential high permittivity dielectrics for very large scale integrated capacitor applications (DRAMs) [26]. In these materials, the (Ba+Sr)/Ti ratio is the primary microstructural parameter used to control the dielectric and electrical behavior of the films. However, the physical mechanisms by which the (Ba+Sr)/Ti ratio controls the electric and dielectric properties in thin films are not yet understood. A necessary step in understanding the composition dependence of the dielectric and electrical properties is to determine the mechanisms by which the excess titanium is accommodated in BST thin films. Samples of two different (Ba+Sr)/Ti ratios, 46.5/53.5 (y=0.15) and 49/51 (y=0.04), respectively, were investigated. The films were grown by chemical vapor deposition and strongly {100} textured [27]. The BST grains in both samples were found to have typical lateral dimensions of 10 - 20 nm, and are known to be columnar from previous work [27]. Figure 7 shows a typical high-resolution TEM image of the grain structure in the y=0.15 sample. As can be seen from the image there does appear to be amorphous areas at the grain boundaries (which is not present for the y=0.04 sample).

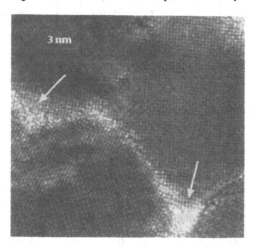

Figure 7. High-resolution image of the y=0.15 sample. The arrows indicate amorphous phase.

Ti $L_{2,3}$ and O K-edges (EELS) were recorded from the grains and the grain boundaries. Figure 8 shows the fine structure of the oxygen K-edges in both samples, recorded on and off the grain boundaries. In the y=0.04 sample, virtually no difference in the fine structure between the spectra recorded on and off the grain boundaries can be detected. However, in the y=0.15 sample, the fine structure of the spectra recorded on the grain boundaries shows significant differences compared to the grain interior. The most significant change occurs in the peak denoted "b", which is reduced in intensity in the spectra recorded on the grain boundaries. For comparison purposes, Figure 8 shows reference spectra of the oxygen K-edges obtained from TiO_2 and $SrTiO_3$, respectively. The principal features of the oxygen K-edge in BST are the same as those in $SrTiO_3$. The peak labelled "b" in the $SrTiO_3$ spectrum is shifted to higher energies in TiO_2 (it occurs there at approximately the same energy loss as the peak "c" in $SrTiO_3$). The reduced intensity of peak "b" in the spectra recorded on the grain boundaries in BST films with y=0.15 can be interpreted as the combined presence of energy-loss features characteristic of both TiO_2 and BST. Figure 9 shows the Ti $L_{2,3}$ edges from the two samples. In neither case can a pronounced change in the fine structure be observed. However, differences in the spectra between the two samples are observed. The edges in the y=0.15 sample are slightly broadened compared to those of the y=0.04 sample. This results in a smaller splitting of the L_3 and L_2 edges in the y=0.15 sample and is indicative of distortions in the Ti-O octahedra.

Using these EELS results we can calculate the Ti/O ratio at the grain boundaries in both samples. For the y=0.04 sample there is virtually no difference between the bulk and the boundaries. For the y=0.15 sample, there is significant increase in the Ti/O ratio. Based on the

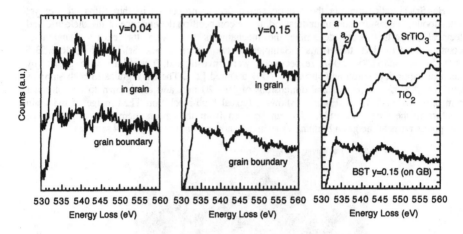

Figure 8. Oxygen K-edges from (a) the y=0.04 sample, (b) the y=0.15 sample and (c) shows a comparison between oxygen K-edges from $SrTiO_3$ and TiO_2 (rutile)

fine structure of the spectra and the observation of the amorphous phase at the grain boundaries in this sample we can conclude that at very high non-stoichiometry we have accommodation of Ti excess in the form of TiO_x at the grain boundaries. Additionally, the broadening of the Ti L-edge suggests that there are also distortions in the octahedra within the grains and there is at least some accommodation of Ti excess within the grains. These distorted octahedra could be a possible reason for the observed decrease in permittivity of the films with increasing Ti content.

The effect of As Passivation on CdTe(111) films grown Si (111) substrates

As a second example of the use of multiple techniques to investigate interface properties, we show an analysis of the effect of Arsenic (As) passivation of Silicon (Si) substrates on the subsequent growth of the II-VI semiconductor Cadmium Telluride (CdTe) [28,29]. In this work, two films were grown by MBE on nominal Si (111) substrates under similar growth conditions except that the substrates were cooled under a different initial flux, i.e. either As flux or Te flux. The microstructures of the two films grown on the Te and As treated Si (111) substrates are shown in the conventional TEM image in Figure 10. It is immediately obvious from the low magnification images that the two films have different structural features. The film grown on the Te treated substrate has very faulted microstructure, containing different twin grains, planar defects on two sets of {111} planes, and threading dislocations (Figure 10(a)). Furthermore, the polarity of the film is determined to be (111) A from the convergent beam electron diffraction (CBED) pattern. The film-substrate interface is also very rough, with 15nm modulations into the Si substrate due to the Te flux reacting with Si during cooling. A very different quality of film is observed for growth on the As-passivated substrate. Apart from the few wide lamellar microtwins confined to the 0.8 μm region close to the interface and a few threading dislocations, the epilayer is single crystal with excellent structural quality over large areas. Double crystal X-ray rocking curves from the CdTe (333) reflection using Cu-Kα₁ and a Si (133) crystal monochromater gives measurement of the full width at half maximum (FWHM) of 87 arcseconds for a 5μm thick layer [28]. The selected area diffraction pattern from the CdTe/Si interface shows the perfect alignment of CdTe (111) layer relative to the Si (111) substrate (Figure 10(e)). The polarity of the film is confirmed to be (111) B by CBED.

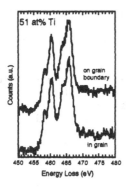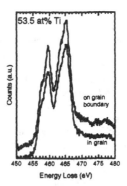

Figure 9. Ti $L_{2,3}$ edges from (a) the y=0.04 and (b) the y=0.15 sample.

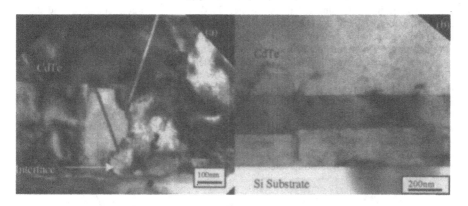

Figure 10: (a) Low magnification bright field TEM image showing a cross-sectional view (in the [1-10] direction) of the sample grown (a) on Te treated Si (111) substrate, (b) on As passivated Si (111) substrate.

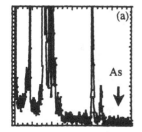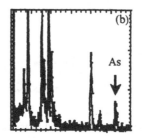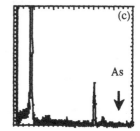

Figure 11: EDS spectra of (a) CdTe layer, (b) CdTe/Si interface, and (c) substrate, confirming the presence of a monolayer of As at the interface.

The difference in the quality of the films therefore appears to be related to the difference in the initial flux treatment of the Si (111) surface. To obtain further information on this effect we can go beyond the imaging process and analyze the CdTe/Si interface by energy dispersive X-ray spectroscopy (EDS). Figure 11 shows that the As that is initially deposited on the Si surface remains there for the subsequent growth of the CdTe film. The three spectra show that only when the 1nm probe is placed at the interface is there an As signal present (the two other spectra are obtained ~5nm from the interface). Furthermore, quantifying the spectra shown in figure XY gives an estimate of the As concentration close to one monolayer. These results are consistent with the in situ reflection high energy electron diffraction (RHEED) observations before initiation of the CdTe growth, and confirmed by x-ray photoemission spectroscopy.

Having generated a general picture of the film morphology and the composition of the film-substrate interface we can now move on to examine the interface in more detail using atomic resolution Z-contrast imaging. Figure 12 shows a raw atomic resolution Z-contrast image of the CdTe/Si interface, and the same image after Fourier filtering. The elongated white spots in the image are atomic columns separated by distances smaller than the 1.5Å resolution of the microscope. The first CdTe layer can be easily ascertained by the intensity difference at the interface, as CdTe has a higher atomic number than Si or As. The extra planes of the misfit dislocations are indicated, and the cores of the misfit dislocations are clearly seen located at the interface between the CdTe film and the substrate. By drawing a Burgers circuit around the cores, the edge component of the Burgers vector is 1/4[112], which is parallel to the interface, providing the explanation as to why there is no tilt of the epilayer with respect to the substrate. The average spacing between misfit dislocations is 1.9nm. These are the grown-in misfit dislocations that have relieved the majority of the mismatch strain. Z-contrast images (Figure 12) from a different area along the same interface, show that the first few layers of CdTe can also be deposited in a twinning position relative to the substrate. This indicates that CdTe can initially nucleate in two orientations with a twinning relationship. This means the film adopts either the substrate orientation with ABCABC‖ABC......, or in a twinning position to the substrate with ABCABC‖BACBAC.., where A, B, or C denotes the atomic double layer. It must be pointed out from the high-resolution observations that no islands or grains other than the few coherent twin boundaries are found in the region at the interface, the twinned islands are of small quantity, and the film/substrate interface is atomically flat.

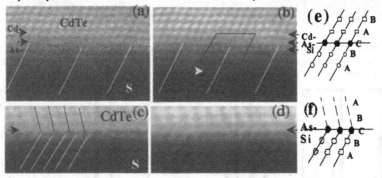

Figure 12: (a) A raw atomic resolution Z-contrast image of the CdTe/Si interface viewed along the [1-10] direction with (b) showing the Fourier filtered image of (a). (c) Atomic resolution Z-contrast image of a different region along the interface showing twin oriented layer growth with (d) showing the corresponding Fourier filtered image of (c). (e) Schematic showing the two possible orientations of the epilayer corresponding to (a) and (c) respectively (A,B and C denote a double atomic layer).

The above experimental observations show that As is crucial in obtaining the desired (111) B film with high crystalline quality. There are two main factors that result in the single domain high crystalline quality epitaxial CdTe (111) B growth on Si(111); the As-passivated Si (111) surface which results in the (111) B polarity and the nature of atomic steps and terraces on the nominally Si(111) surface which give rise to layer growth free of multi-domains [28]. It is well known that As replaces the outermost Si atoms of the double layer during passivation of the Si(111) surface in the MBE growth process [30], and it is known from the above investigations that As stays there after the growth. Both imply that As is present at the interface as a monolayer separating CdTe and Si during the growth. Based on the As-Cd, As-Te bond formation energies, Cd is favored as the first atomic layer forming a bond with the As at the interface, where the bonding configuration proceeds as Si-As-Cd-Te across the interface.

Even though it is known that Cd atoms have a very low sticking coefficient and would re-evaporate from the surface under only a Cd flux, the resultant CdTe (111) B layer growth implies that the presence of Te atoms supplied by stoichiometric flux is sufficient to stabilize the Cd on the surface. This occurs through a transition layer at the growth front. CdTe is initiated with Cd atoms forming one bond with As and three bonds with Te, producing a Te stabilized growth surface, which finally results in the (111) B growth. Because of the high interfacial energy associated with the large lattice mismatch between CdTe and Si and the requirement of low temperature initiation, layer deposition starts with island nucleation. CdTe nucleation could initiate both at step edges and on the terraces, and because the atomic structures of the Si (111) surface comprise double layer steps giving the same atomic configuration from terrace to terrace, all the nuclei would have only two possible positions, i.e. the twining positions. In this system, the associated lattice mismatch energy is the same for the two possible depositions. However, the twinning position nuclei might have a higher interfacial energy due to the "faulted" stacking sequence across the interface, and therefore the majority of the nuclei would adopt the orientation of the substrate (as is observed experimentally). Growth then proceeds in layer by layer fashion soon when the substrate temperature is raised to normal growth temperature after several tens of monolayer deposition. Because there is no tilt between the nucleating islands, the coherent double positioning boundaries are easily eliminated after a short nucleation stage.

CONCLUSIONS

The ability to perform many imaging and analytical techniques on the same microscope with atomic spatial resolution opens up a whole new range of experiments that can be performed to understand the structure property relationships at interfaces and defects. For the future, it may even be possible to go beyond the current resolution limitations and perform experiments on the atomic scale at elevated/ or lowered temperatures while the sample is being mechanically or electrically stressed.

ACKNOWLEDGMENTS

Aspects of this work were performed in collaboration with S. Sivananthan, S. Rujiwarat, R. Sporken, N. K. Dhar, S. J. Pennycook, S. K. Streiffer, and A. I. Kingon. The JEOL 2010F was purchased with support from the National Science Foundation under grant number NSF-DMR-9601792, and is operated by the Research Resources Center at the University of Illinois at Chicago. The research program on superconductors is sponsored by the National Science Foundation under grant number DMR-9803021; the research program on semiconductors is sponsored by the National Science Foundation under grant number DMR-9733895; and the research on oxides is sponsored by the Department of Energy under grant number DE-FG02-96ER45610.

REFERENCES

[1] A. P. Sutton and R. W. Balluffi, *Interfaces in Crystalline Materials* (Oxford University Press, 1995).

[2] M. M. McGibbon, N. D. Browning, M. F. Chisholm, A. J. McGibbon, S. J. Pennycook, V. Ravikumar and V. P. Dravid, *Science* **266**, 102 (1994).

[3] K. L. Merkle and D. J. Smith, *Phys. Rev. Letts.* **59** 2887 (1987).

[4] H. Ichinose and Y. Ishida, *Phil Mag A* **43**, 1253 (1981).

[5] M. F. Chisholm, A. Maiti, S. J. Pennycook, and S. T. Pantelides, *Phys Rev Lett* **81**, 132 (1998).

[6] P. D. Nellist and S. J. Pennycook, *Phys Rev Lett* **81**, 4156 (1998).

[7] F. Ernst, O. Keinzle and M. Ruhle, *J. Eur. Ceram. Soc* **19**, 665 (1999).

[8] D. B. Williams and C. B. Carter, *Transmission electron microscopy: a textbook for materials science* (Plenum, 1996)

[9] R. F. Egerton, *Electron Energy Loss Spectroscopy in the Electron Microscope* (Plenum, 1996).

[10] M. Watanabe and D. B. Williams, *Ultramicroscopy* **78**, 89 (1999)

[11] N. D. Browning, M. F. Chisholm and S. J. Pennycook, *Nature* **366**, 143 (1993).

[12] P. E. Batson, *Nature* **366**, 727 (1993).

[13] *Electron Diffraction Techniques*, edited by J. M. Cowley (Oxford University Press, 1992)

[14] E. M. James and N. D. Browning, *Ultramicroscopy* **78**, 125(1999).

[15] E. M. James, N. D. Browning, A. W. Nicholls, M. Kawasaki, Y. Xin and S. Stemmer, *J. Electron Microscopy* **47**, 561(1998).

[16] J. M. Cowley, *J. Elect. Microsc. Technique* **3**, 25 (1986).

[17] J. M. Cowley, *Appl. Phys. Lett.* **15**, 58 (1969).

[18] D. E. Jesson and S. J. Pennycook, *Proc. R. Soc. Lond. A* **449**, 273 (1995).

[19] P. D. Nellist and S. J. Pennycook, *Ultramicroscopy* **78**, 111(1999).

[20] Y. Xin, E. M. James, I. Arslan, S. Sivananthan, N. D. Browning, S. J. Pennycook, F. Omnès, B. Beaumont, J-P. Faurie and P. Gibart, in press *Applied Physics Letters*

[21] A. L. Ankudinov, B. Ravel, J. J. Rehr and S. D. Conradson, *Phys Rev B* **58**, 7565 (1998).

[22] C.E. Lyman, J.I. Goldstein, D.B. Williams, D.W. Ackland, S. Von Harrach, A.W. Nicholls and P.J. Statham, *J. Microscopy*, **176**, 85 (1994)

[23] D.B. Williams and E.B. Steel, *Analytical Electron Microscopy – 1987*, 228 (1987)

[24] J. C. H. Spence, *Experimental High Resolution Electron Microscopy*, (Oxford University Press, 1981)

[25] S. Stemmer, S. K. Streiffer, N. D. Browning and A. I. Kingon, *Applied Physics Letters* **74**, 2432 (1999).

[26] see A. I. Kingon, S. K. Streiffer, C. Basceri and S. R. Summerfelt, *MRS Bulletin* **21**, 46 (1996) and references therein

[27] C. Basceri, PhD dissertation, North Carolina State University, Raleigh 1997.

[28] S. Rujiwarat, Y. Xin, N. D. Browning, S. Sivananthan, D. J. Smith, Y. P. Chen and V. Nathan, *Applied Physics Letters* **74**, 2346 (1999).

[29] Y. Xin, S. Rujiwarat, R. Sporken, N. D. Browning, S. Sivananthan, S. J. Pennycook, and N. K. Dhar, *Applied Physics Letters* **75**, 349 (1999).

[30] R. D. Bringans, *CRC Crit. Rev. Solid State Mater. Sci.* **17**, 353 (1992).

ATOMIC SCALE ANALYSIS OF CUBIC ZIRCONIA GRAIN BOUNDARIES

E.C. DICKEY*, X. FAN*,†, M. YONG*, S.B. SINNOTT*, S.J. PENNYCOOK†
*Department of Chemical and Materials Engineering, University of Kentucky, Lexington, KY
40506-0046; ecdickey@engr.uky.edu
†Solid State Division, Oak Ridge National Laboratory, Oak Ridge, TN 37831

ABSTRACT

The core structures of two symmetric tilt [001] grain boundaries in yttria- stabilized cubic zirconia are determined by Z-contrast imaging microscopy. In particular, near-Σ=13 (510) and Σ=5 (310) boundaries are studied. Both grain boundaries are found to be composed of periodic arrays of basic grain-boundary structural units, whose atomic structures are determined from the Z-contrast images. While both grain boundaries maintain mirror symmetry across the boundary plane, the 36° boundary is found to have a more compact structural unit than the 24° boundary. Partially filled cation columns in the 24° boundary are believed to prevent cation crowding in the boundary core. The derived grain boundary structural models are the first developed for ionic crystals having the fluorite structure.

INTRODUCTION

Grain boundaries in yttria-stabilized cubic-zirconia (YSZ) have important ramifications for the macroscopic mechanical and electrical properties of the material [1]. While several studies have been conducted to understand the structure and chemistry of grain boundaries in YSZ [2], none have reported detailed grain boundary core structures (i.e. atomic positions). Here we provide a detailed experimental analysis of two high-symmetry YSZ grain boundaries, the near-Σ=13 (510) and the Σ=5 (310), corresponding to 24° and 36.8° symmetric tilt [001] grain boundaries, respectively. The grain boundaries are studied by Z-contrast scanning transmission electron microscopy (STEM), which reveals the projected cation sublattices parallel to the tilt axis. Based on these images, structural models of the grain boundaries are derived. Although the tilt angles between the two boundaries (24° and 36.8°) differ by only 13°, two completely different grain boundary cores are observed. The details of the grain boundary core structures will be discussed in detail below.

EXPERIMENTAL

The YSZ bicrystals were purchased from Shinkosh Co., Ltd., Japan. The near- Σ=13 (510) was a 24° [001] tilt boundary, only 1.4° away from the perfect Σ=13 orientation of 22.6°. The second bicrystal, within the growth tolerances (±0.5°), was a perfect Σ= 5 (310), or a 36.8° [001] tilt YSZ bicrystal. The tilt angles were confirmed by electron diffraction as discussed below. TEM samples were prepared from the bulk bicrystals by standard preparation techniques. First, a 2x1 mm rectangular plate containing the grain boundary was cut by a diamond wheel from the bulk specimen. The plate was mechanically polished to approximately 40 μm in thickness and then dimpled to about 15 μm. Finally, the specimens were thinned to electron transparency by ion milling with 3kV Ar ions while cooled with liquid N_2.

The Z-contrast STEM imaging was carried out on the 300kV VG Microscopes HB603U dedicated STEM at Oak Ridge National Laboratory equipped with a high-resolution objective

Mat. Res. Soc. Symp. Proc. Vol. 589 © 2001 Materials Research Society

lens capable of forming a probe 0.13 nm diameter. Transversely incoherent images were formed with a high-angle annular dark field detector, so that the intensities could be directly interpreted [3-5]. Because the scattered intensity scales with approximately the square of the atomic number, the projected cation column positions are revealed while almost no information regarding the oxygen positions is observed. To quantify atomic positions from the Z-contrast images, maximum entropy reconstructions were performed [6,7]. Using a Lorentzian profile for the probe function, the object function, corresponding to cation positions, was reconstructed by maximum-entropy image analysis [8,9]. The resulting object functions were used for quantitative analysis. Note that all object functions presented below are convoluted with a Gaussian for better visibility.

RESULTS

22° Symmetric Tilt Boundary

The tilt angle of the near-Σ=13 (510) bicrystal is confirmed to be 24° about the [001] by electron backscattered diffraction (EBSD) as shown in Fig. 1. Note the 24° rotation about the [001] sample normal between crystal A (Fig. 1a) and crystal B (Fig. 1b).

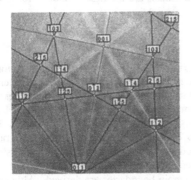

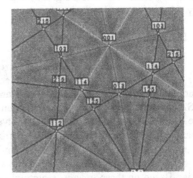

Fig. 1a EBSD pattern from crystal A of 24° YSZ bicrystal

Fig.1b EBSD pattern from crystal B of 24° YSZ bicrystal.

The Z-contrast image of this 24° symmetric tilt [001] ZrO_2 grain boundary is shown in Fig. 2, along with the corresponding object function obtained by maximum entropy reconstruction. The white dots in the image correspond to Zr columns, or the projected cation sublattice. Within the field of view, the 24° tilt boundary is composed of a periodic array of similar structural units (one of these units is indicated by the open circles on Fig. 2b). Small variations between the different units along the length of the grain boundary are currently being quantified. Since the image is only a representation of the projected structure, it is important to appreciate the cubic-ZrO_2 crystal structure (fluorite structure) in three dimensions. Fig. 3 shows a ZrO_2 unit cell in which the Zr atoms are distinguished by their position along the [001] direction, the beam direction. The Zr atoms denoted by filled symbols are one half of a unit cell below the Zr atoms denoted by the large open circles along the [001] projection. Although the projected distance between the Zr atoms in adjacent columns is 0.254 nm, the actual closest distance between Zr atoms is 0.359 nm.

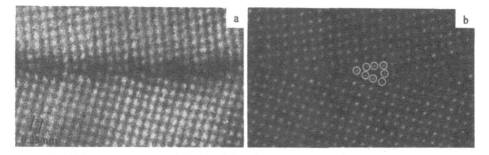

Fig. 2 (a) Z-contrast image of the 24° symmetric tilt [001] YSZ grain boundary and (b) the corresponding object function obtained by maximum entropy reconstruction.

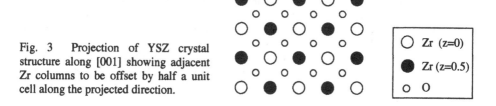

Fig. 3 Projection of YSZ crystal structure along [001] showing adjacent Zr columns to be offset by half a unit cell along the projected direction.

○ Zr (z=0)

● Zr (z=0.5)

o O

Based on the Z-contrast images, a structural model of the grain boundary is developed as shown in Fig. 4. Between each structural unit, there is a continuity of the (200) planes across the boundary indicating no in-plane rigid body translation normal to the tilt axis. Within the defect structural unit, two of the Zr columns (both having Zr atoms at the same depth) are very close to each other in the boundary core. The cation spacing in this region would be ~0.26 nm, over 25% smaller than the bulk spacing. Electrostatically, this situation would be energetically unfavorable because of the close proximity of the Zr cations. But, also note that the intensities of these two columns in Fig. 2 are lower that of the other Zr columns. The lower intensities can be interpreted as partially occupied Zr columns. Partial occupancy would allow the Zr atoms in the two adjacent columns to stagger along the beam direction and thus avoid close cation-cation positions. Similar observations have been made in other grain boundary studies including those of $SrTiO_3$ [10,11] and $YBa_2Cu_3O_{7-x}$ [12]. Partial occupancy to avoid cation crowding is emerging as a common relaxation mechanism for oxide grain boundaries.

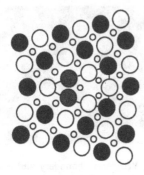

○   Zr (z=0)

●   Zr (z=0.5)

●   Zr (partially filled)

○   Oxygen

Fig. 4 Model of 24° YSZ grain boundary core showing one repeat unit. The Zr columns in gray are believed to be partially filled to prevent cation crowding.

The oxygen positions in the grain boundary model were chosen such that they were most similar to the positions in bulk ZrO_2, and some particularly ambiguous sites were selected on the basis of preliminary density functional theory (DFT) calculations [13]. In the fluorite structure, the oxygen ions are tetrahedrally coordinated by Zr ions while the Zr atoms fill cubic interstices in the oxygen sublattice. Note that most of the zirconium atoms in the 24° model (except those in the partially filled columns) remain cubically coordinated by oxygens, although the cube is severely distorted about some sites. Similarly, most of the oxygen sites near the boundary maintain a distorted tetrahedral coordination, except for the three columns in the very core of the boundary. Since the Z-contrast image does not give us information regarding the oxygen ions, it will be necessary to further explore the coordination of the oxygen ions in the boundary core with a combination of electron energy loss spectroscopy (EELS) and more complete atomistic simulations. Having accurate coordinates for the cation positions from the Z-contrast image, however, has provided an excellent starting model of the interface from which further refinements are being made.

36° Symmetric Tilt Boundary

Fig. 5 presents the maximum entropy refined Z-contrast image of the $\Sigma=5$ (310) YSZ grain boundary. Again, the boundary is composed of a periodic array of very similar basic structural units, one of which is highlighted with open circles in Fig. 5.

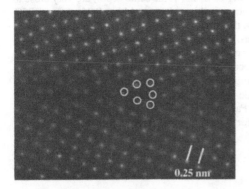

Fig. 5 Maximum entropy reconstructed Z-contrast image of the 36° [001] YSZ grain boundary. The open circles indicate the periodic structural unit.

The Σ=5 (310) grain boundary model, as derived from the experimental data, is presented in Fig. 6. As evidenced in the figure and by the coincident site lattice notation, this 36° has more translational symmetry leading to a smaller repeat unit than is found in the 24° boundary. Similar to the 24° boundary, this higher angle boundary maintains mirror symmetry across the boundary plane indicating no in-plane rigid body translation normal to the tilt axis. In the core of the structural unit, the two adjacent Zr columns that are at the same depth are sufficiently far apart from each other, at ~0.45 nm, so that no cation crowding occurs. Since none of the cation spacings in the 36° core is smaller than in the bulk, fully occupied cation columns are incorporated into the model. The grain boundary core, however, is much more compact laterally than the 24° boundary with fewer under- or over-coordinated sites as compared to the bulk environment.

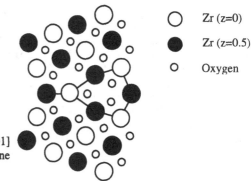

○ Zr (z=0)

● Zr (z=0.5)

○ Oxygen

Fig. 6 Model of 24° symmetric tilt [001] YSZ grain boundary core showing one repeat unit.

CONCLUSIONS

Z-contrast STEM was employed to image the atomic structure of 24° and 36° symmetric [001] tilt grain boundaries in YSZ. The images show that both grain boundaries are composed of periodic arrays of basic structural units unique to the particular boundary. From maximum entropy refined images, the atomic coordinates of the grain boundary units have been defined. In both boundaries no rigid body translations, which would break the mirror symmetry, are observed, which leads to a continuity of some (200) planes across the boundary plane. Although the experimental data could not reveal any rigid body translations along the beam direction, the models give no reason to suggest that such an offset would be favorable in either boundary. The atomic structures of both cores, however, are significantly different with an extended core in the 24° boundary. The experimental data also suggests that partially filled Zr columns may prevent cation crowding in the 24° boundary. Because almost no data regarding the oxygen positions is available from the Z-contrast images, the oxygen positions in the cores have thus far been only inferred. However, EELS and DFT calculations are underway to clarify the oxygen positions and further refine the boundary structures.

ACNOWLEDGEMENTS

The authors would like to thank the NSF Division of Materials Research who sponsors this research through contract #9976851 and the U.S. Department of Energy Division of Materials Sciences under contract no. DE-AC05-96OR22464 with Lockheed Martin Energy Research Corp.

REFERENCES

1. X. Guo, Solid State Ionics, **81** p. 235 (1995).
2. K.L. Merkel, G.-R. Bai, Z. Li, C.-Y. Song, *et al.*, Phys. Stat. Sol., **166** p. 73 (1998).
3. D.E. Jesson and S.J. Pennycook, Proc. R. Soc. Lond. A, **441** p.261 (1993).
4. D.E. Jesson and S.J. Pennycook, Proc. R. Soc. Lond. A, **449** p.273 (1995).
5. P.D. Nellist and S.J. Pennycook, Ultramicroscopy, **78** p.111 (1999).
6. S.F. Gull and G.J. Daniell, Nature, **272** p.686 (1978).
7. S.F. Gull and J. Skilling, IEEE Proc., **131F** p.646 (1984).
8. P.D. Nellist and S.J. Pennycook, J. Micros., **190** p.159 (1998).
9. A.J. McGibbon, S.J. Pennycook and D.E. Jesson, J. Micros., **915** p.44 (1996)
10. M.M. McGibbon, N.D. Browning, M.F. Chisholm, *et al.*, Science, **261** p.102 (1994).
11. M.M. McGibbon, N.D. Browning, A.J. McGibbon and S.J. Pennycook, Phil. Mag. A, **73** p.625 (1996) .
12. N.D. Browning, P.D. Nellist, D.P. Norton, *et al.*, Physica C, **294** p.183 (1998).
13. S.B. Sinnott, M. Yong and E.C. Dickey, (to be published).

SEM/EDX SPECTRUM IMAGING AND STATISTICAL ANALYSIS OF A METAL/CERAMIC BRAZE

PAUL G. KOTULA,[1] MICHAEL R. KEENAN[1] AND IAN M. ANDERSON[2]
[1]Sandia National Laboratories, PO Box 5800, Albuquerque, NM 87185-1405
[2]Oak Ridge National Laboratory, PO Box 2008, Oak Ridge, TN 37831-6376

ABSTRACT

Energy dispersive x-ray (EDX) spectrum imaging has been performed in a scanning electron microscope (SEM) on a metal/ceramic braze to characterize the elemental distribution near the interface. Statistical methods were utilized to extract the relevant information (i.e., chemical phases and their distributions) from the spectrum image data set in a robust and unbiased way. The raw spectrum image was over 15 Mbytes (7500 spectra) while the statistical analysis resulted in five spectra and five images which describe the phases resolved above the noise level and their distribution in the microstructure.

INTRODUCTION

Spectrum imaging, where a complete x-ray spectrum is collected at every pixel in a scanned image, is a powerful new tool for materials characterization. Spectrum imaging differs from mapping, where only windows around pre-selected energy ranges are acquired, in that the entire spectrum is collected at each pixel, as the beam is rastered across the specimen. This procedure allows even unanticipated elements to be mapped after the data has been collected. However, mapping by itself is potentially fraught with errors [1]. Additionally, the spectrum image cannot readily be visualized in its entirety. For these reasons, a robust and unbiased method is needed to extract all of the relevant information from these large raw data files. Ideally, one would like a spectrum describing the elemental distribution, and a corresponding image describing the distribution, for each phase in the microstructure. In the present work we describe the application of multivariate statistical analysis (MSA) to a spectrum image of a metal/ceramic braze.

EXPERIMENTAL

The braze characterized in this work joins polycrystalline alumina to Kovar (primarily Fe, Ni and Co) with a copper-silver eutectic alloy containing some titanium (an 'active' metal). The specimen geometry consisted of a sandwich of two pieces of alumina, two braze layers and a Kovar filler layer in the middle. The entire assembly was heat-treated to bond the interfaces and then a polished cross-section was prepared. The interface between the alumina and the copper-silver alloy was characterized in this work. The analysis was performed in a Philips XL30-FEG SEM operated at 10 kV, equipped with an Oxford super-ATW detector and XP3 pulse processor. The specimen was normal to the electron beam and the take-off angle was 35°. EDX spectrum images were acquired with an EMiSPEC Vision integrated acquisition system. A spectrum image of 100×75 pixels with 1000 channels per spectrum (10 eV/channel) was acquired from the specimen with a sampling density of 200 nm/pixel and a dwell time of 1 live second per pixel, for a total acquisition time of ~2 hours. The binary data file was imported into MATLAB where the statistical analysis was performed on a 500 MHz Pentium III with 512 Mbytes of RAM. The entire calculation took six minutes. The linear and orthogonal MSA calculation scheme followed that of Trebbia and Bonnet [2-5]. The MSA seeks to identify the independently varying sources of information in the raw data set and then orders these based upon the amount of variance they describe. The first several resultant components of information then describe the chemical composition and distribution of the phases in the microstructure. The rest of the components describe the noise in the data, which for the case of EDS is due to Poisson statistics.

Mat. Res. Soc. Symp. Proc. Vol. 589 © 2001 Materials Research Society

RESULTS

Figure 1 is a SEM secondary electron image of the analyzed area of the specimen. The microstructure consists of eutectic lamellae of Cu in Ag at the bottom of the image and polycrystalline alumina at the top of the image. The region of the spectrum image is delineated by a white rectangle. Figure 2 shows conventional maps for Ti, Fe, Co, and Ni, constructed after the spectrum image had been acquired. Figure 3 is the component spectrum and image describing contrast in the Cu/Ag eutectic phases. Figures 4 a and b are the component image and spectrum respectively from an interface intermetallic phase and Figures 4 c and d are from Ti segregated to the metal ceramic interface. Figure 5 is the sum of 25 raw spectra from the metal/ceramic interface.

DISCUSSION

Conventional mapping (Figure 2), utilizing the spectrum image data set, indicated Ti segregation to the metal/ceramic interface and segregation of Fe, Co, Ni, and Ti to an apparent intermetallic phase in pockets along the same interface. In microstructures containing many elements, both major and minor, with a potentially complex distribution, it is desirable to automate the procedure for identifying not just the elements present but the phases and their distribution. This requires MSA of the entire spectrum image.

The MSA indicated that there were five distinct information-bearing components (i.e., above the noise level) which describe the spectrum image. Each component of information consists of a component image, which is the amplitude or relative weighting of the associated component spectrum. The original spectrum from a given pixel can be reconstructed by summing the products of the individual component–image pixel intensities and the respective component spectra. Therefore, instead of thousands of spectra, five spectra and five images describe all of the information resolved above the noise level. Furthermore, no assumptions have been made as to the presence or absence of any of the phases identified or any elements associated with those phases. Since the particular form of MSA applied here is variance based, the resultant components of information contain the maximum variance in as few components as possible. Therefore, it is unavoidable to obtain component spectra containing both positive and negative peaks. The physical significance of this is that signals of opposite polarity in a given component spectrum are anti-correlated in the microstructure. This can most clearly be seen in Figure 3 which is the component spectrum and image from the Ag and Cu: Cu is positive in the component spectrum and correspondingly light in the component image; conversely, Ag is negative in the component spectrum and dark in the component image. Little contrast is observed in the other phases.

Figures 4a and b are the component image and spectrum from a (Ti, Fe, Ni, Co) intermetallic phase found at the metal/ceramic interface. The light pockets in Fig 4a are a (Ti, Fe, Ni, Co) intermetallic phase. The positive peaks in the component spectrum from this phase (Fig. 4b) correspond with the light regions of the component image. Figures 4c and d are the component image and spectrum corresponding to interfacial segregation of Ti. Again the light band at the metal/ceramic interface corresponds to the positive peaks in the component spectrum. Figure 5 is the sum of 25 raw spectra from the metal/ceramic interface. Even though only those spectra with the highest Ti signal were summed to form the spectrum in Figure 5, several other elements are found together with the Ti. It should be noted that there were no raw spectra from that interface showing only Ti. This is due to x-ray generation from the adjacent phases. In contrast, the MSA results of Fig. 4d indicate that the Ti at the interface is essentially by itself. In other words, the Ti signal from the interface varies independently from the elements with which it is spectrally overlapped in the raw data. Although both the intermetallic identified in Fig. 4a and the interface phase identified in Fig. 4c both contain Ti (as shown in Fig. 2, the Ti map), MSA has determined them to be two different phases. This solution was arrived at with no operator intervention. Maps of Fe, Co and Ni yielded essentially the same result but the power of MSA over conventional analyses is that no assumptions were made as to the presence of certain elements and thus the potential for missing important information is greatly reduced.

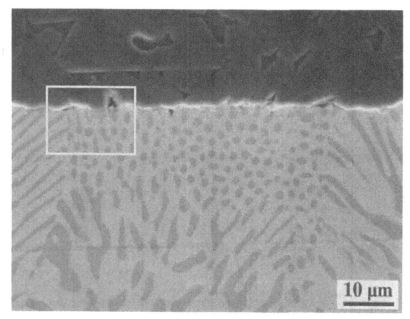

Figure 1. SEM secondary electron image of the analyzed region of the specimen.

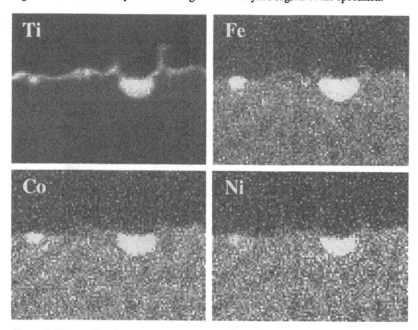

Figure 2. Conventional x-ray maps constructed from the spectrum image data set.

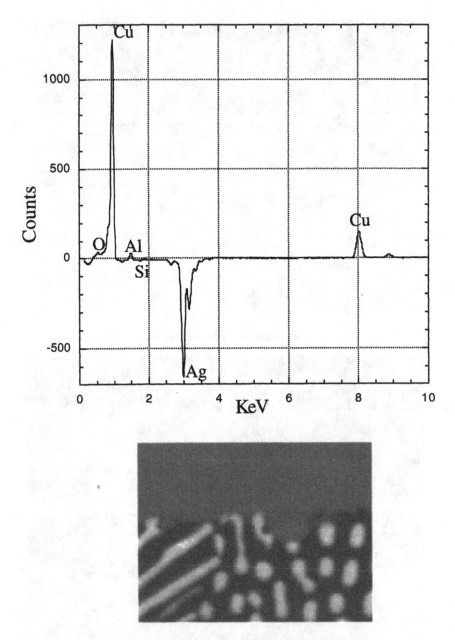

Figure 3. Component spectrum and image describing contrast in the Cu/Ag eutectic.

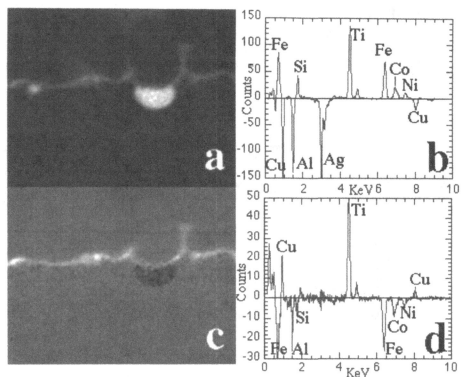

Figure 4. MSA component image and spectrum from (a, b) the Ti, Fe, Co, Ni intermetallic phase and (c, d) interfacial segregation of Ti.

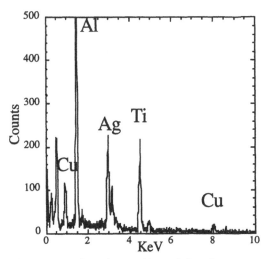

Figure 5. Sum of 25 raw spectra from the metal/ceramic interface.

CONCLUSIONS

MSA has been successfully applied to the characterization of a metal ceramic braze. The chemistry and distribution of six phases were identified in the microstructure without any operator intervention. Future work will focus on obtaining pure component spectra that can then be quantified.

ACKNOWLEDGMENTS

Sandia is a multi-program laboratory operated by Sandia Corporation, a Lockheed Martin Company, for the United States Department of Energy (DOE) under contract DE-AC04-94AL85000. Research at the ORNL SHaRE User Facility was sponsored by the Division of Materials Sciences, U.S. DOE, under contract DE-AC05-96OR22464 with Lockheed Martin Energy Research Corp., and through the SHaRE Program under contract DE-AC05-76OR00033 with Oak Ridge Associated Universities.

REFERENCES

1.   D. E. Newbury and D. S. Bright, Microscopy and Microanalysis **5** (1999) 333.
2.   P. Trebbia and N. Bonnet, Ultramicroscopy **34** (1990) 165.
3.   I. M. Anderson and J. Bentley, *Proc. Microscopy & Microanalysis 1997*, 931.
4.   I. M. Anderson and J. Bentley, *Mater. Res. Soc. Symp. Proc.* **458** (1997) 81.
5.   I. M. Anderson, *Proc. 14th ICEM: Electron Microscopy 1998* **1** (1998) 357.

CHARACTERIZATION OF THE INTERFACE BETWEEN LANTHANUM HEXA-ALUMINATE AND SAPPHIRE BY EXIT WAVE RECONSTRUCTION

B. WESSLER, A. STEINECKER, W. MADER
Institut für Anorganische Chemie, Universität Bonn, Römerstrasse 164, D-53117 Bonn, Germany, wessler@sg4.elmi.uni-bonn.de

ABSTRACT

Epitaxial thin films of rare-earth hexaaluminates on basal plane sapphire have been produced by chemical solution deposition. $LaAl_{11}O_{18}$ films with magnetoplumbite structure grow with $(0001)_{HA} \parallel (0001)_S$ and $[1\bar{1}00]_{HA} \parallel [2\bar{1}\bar{1}0]_S$ orientation relationship.

To investigate the stucture of the interface exit wave reconstruction of focus series was carried out using a field emission TEM. Due to the inversion of the imaging process major artefacts at the interface can be eliminated. Exit waves were simulated based on different interface models and were compared with the reconstructed waves to localize the positions of the atoms at the interface. Two different types of interfaces were observed in the samples. One of the types, in which the spinel block of hexaaluminate faces the sapphire with the mixed cation layer with occupied octahedra and tetrahedra, is characterized in detail. Face-sharing of coordination polyhedra is avoided to large extent.

INTRODUCTION

Rare-earth hexaaluminates have been attracting interest for laser [1], flourescence [2] and high temperature catalytic applications [3]. Moreover single crystalline thin films are discussed for their potential importance as buffer layers for the growth of hexagonal ferrites and spinel films. It is possible to produce single crystalline thin films of hexaaluminates on basal plane sapphire [4] or (111)-oriented yttrium aluminum garnet YAG [5] by aqueous solution routes followed by heat treatment. In our studies, precursors containing lanthanum and aluminum ions decompose to a polycrystalline film constisting of $LaAlO_3$, which converts into a single crystalline hexaaluminate film $LaAl_{11}O_{18}$ at temperatures $T \geq 1500°C$ by reaction with the underlying sapphire. The misfit between lanthanum hexaaluminate (HA) and sapphire (S) is 1.2%.

Rare-earth hexaaluminates, commonly referred to as $LnAl_{11}O_{18}$, are known to have the magnetoplumbite structure with space group $P6_3/mmc$ and were found to be nonstoichiometric with composition $Ln_{1-x}Al_{12-y}O_{19-z}$ [6] due to Ln and Al defects in the intermediate layer. The hexagonal structure consists of a spinel block which alternates with an intermediate layer, the mirror plane, containing aluminum, oxygen and the rare-earth cation. The spinel block is built by four oxygen layers in cubic stacking. In between those, two cation layers are occupied by octahedrally coordinated aluminum adjacent to a mixed layer, in which tetrahedral and octahedral sites are occupied.

Details of the terminating layers and the ion arrangement at the interface HA/S have not been studied so far. HRTEM is the optimum technique to study the atomistic structure of crystalline interfaces. Nevertheless, on applying conventional HRTEM the direct interpretation of electron image intensities bears difficulties with respect to the projected potential of the structure. This is a consequence of the nonlinear imaging process introducing artefacts and delocalization of the information due to lens aberrations.

First the electron wave interacts dynamically with the electrostatic potential of the object. By transmission through the specimen the amplitude and phase of the incident electron wave is modulated, yielding the complex object wave at the exit face of the object. Subsequently, the imaging process scrambles this amplitude and phase information, which can be described by a convolution of the object wave and the point-spread function of the microscope. The image wave is recorded by means of a detector in the image plane. Consequently, micrographs show only the square modulus of the amplitude of the image wave whereas its phase is lost.

Image analysis of focus series by restoring amplitude and phase of the exit wave function offers the advantage of eliminating those artefacts. Therefore the interpretation of the reconstructed phase and/or amplitude can be carried out straightforwardly and the obtained information is more reliable than the analysis of HRTEM 'single shots'.

EXPERIMENTAL
Preparation of films

A 0.1m aqueous solution of lanthanum nitrate hexahydrate (99,99%, Aldrich Chemical Company Inc., USA) and aluminum nitrate nonahydrate (>99%, Fluka Chemika, Switzerland) mixed with glycine (99+%, Aldrich) and poly(vinylalcohol) PVA (87-89% hydrolyzed; Aldrich) was used as precursor solution. Epi-polished basal plane sapphire substrates of size 5 x 5 x 0.5mm (Crystec, Germany) were cleaned with organic solvents (acetone, iso-propanol) and further annealed at 1300°C in air to minimize surface roughness. The precursor solution was deposited onto the substrates by spin-coating (KW-4a, Chemat Technology, USA) at 5,000 rpm for 30 s followed by pyrolysis on a hotplate at 300 °C for 10 minutes. This procedure was repeated once and then specimens were wrapped in platinum foil, placed in an alumina crucible and annealed at 1550°C for 5 hours in air using a heating/cooling rate of 15°C/min.

Characterization of films

After heat treatment the crystalline phases of the films were identified by x-ray diffraction XRD (X´Pert/PW1107/PW1140, Philips, The Netherlands). Film morphology was characterized by scanning electron microscopy (6300 FE-SEM, Jeol, Japan).

HRTEM investigations were performed using a Philips CM300 FEG electron microscope with Ultra-Twin lens (coefficient of spherical aberration C_s = 0.60 mm) fitted with a corrector for three-fold astigmatism and operated at 300 kV. Image intensities were recorded by a 1k x 1k slow-scan CCD camera (Gatan Inc., USA) at a magnification of 0.0193nm/pixel.

RESULTS AND DISCUSSION

First characterization of the films by XRD (θ-2θ scans; pole figure analysis) reveals the orientation relationship of the lanthanum hexaaluminate on sapphire to be $(0001)_{HA} \| (0001)_S$ and $[1\bar{1}00]_{HA} \| [2\bar{1}\bar{1}0]_S$ indicating the continuation of the orientation of the oxygen layers across the HA/S interface. The SEM micrograph (fig. 1) of the film surface shows large and smooth terraces and the characteristic morphology of the hexagonally shaped crystals. In the bright field cross-sectional TEM micrograph (fig. 2) a film thickness of approx. 50 nm is measured and a flat interface can be observed. No interfacial dislocations were detected, i.e. the interface appears to be fully coherent. To further check this phenomenon the distance between structural features in HA in HRTEM images was measured to be approx. 1% smaller at the interface than close to the film surface. Hence the film is in a strain state such that it matches the sapphire lattice at the interface.

A focus series of 20 images has been recorded in the zone axis $[1\bar{1}00]_{HA}$, one of which is shown in fig.3. The thickness and the tilt of the specimen as well as the starting defocus and the focus increment of the series were determined by means of conventional image simulation [7] of unit cells of the bulk material. The two bulk phases were studied independently yielding a common thickness $t = 11(\pm 1)$ nm and a crystal tilt $\tau \leq 0.2°$. Furthermore, the basic imaging parameters – starting defocus $z = 20(\pm 2)$ nm and focus increment $\delta z = -3.80(\pm 0.10)$ nm – could be fixed. The partial coherence may be described by a beam semi-convergence angle $\theta_c = 0.2$ mrad and a defocus spread with FWHM of $\Delta z = 5$nm.

Using the set of image intensities, the object wave has been restored by means of the Philips/Brite-Euram focal-series reconstruction package [8] up to a resolution of 8 nm^{-1}. The reconstructed phase with a section of the simulated phase and an interface model inserted is

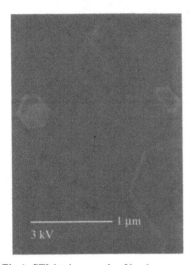

Fig.1 SEM micrograph of lanthanum
hexaaluminate on sapphire

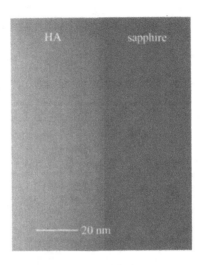

Fig.2 Cross-sectional TEM micrograph

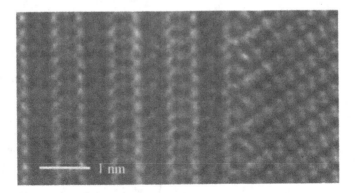

Fig. 3 One member of the focus series

337

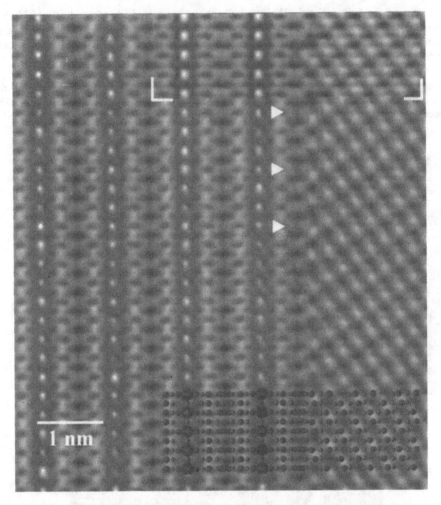

Fig.4 Reconstructed phase with simulated phase (in white edges) and interface model

shown in fig. 4. Compared to the intensity image in fig. 3 the interface can be seen to be clearly located.

Taking the structural data of lanthanum hexaaluminate [9] and sapphire [10] a three dimensional supercell of a HA-S bicrystal with orthogonal axes $A = 16.49$ Å, $B = 45.0$ Å and $C = 9.52$ Å was constructed by using the program CrystalKit [11]: (i) First, compression of the hexaaluminate by 1.2% was applied to achieve a common size of the oxygen sublattice of both phases ($a_{HA} = 5.561$ Å$\rightarrow a'_{HA} = 5.50$ Å). (ii) The lengths of the supercell along the interface (A) and along the beam direction (C) is determined by the smallest multiple of both lattices, i.e. $A = 6 \times 2.75$ Å$((O-O)_S) = 3 \times 5.50$ Å$(a_{HA}) = 16.49$ Å, and $C = 2 \times 4.76$ Å$(a_S; d(1\bar{1}00_{HA})) = 9.52$ Å. (iii) Finally the model of the interface was constructed by terminating the hexaaluminate with a closed packed oxygen layer followed by the mixed cation layer of the spinel block and ending the sapphire with a cation layer. The length of the simulation cell normal to the interface was fixed to $B = 45$ Å. Exit wave simulation was carried out using 4 subslices, i.e. $\Delta t = C/4$.

Simulations of the phase of the exit wave help to assign atom positions to bright spots in the phase pattern. As expected, in the lanthanum hexaaluminate bulk, the intense bright spots correspond exactly to the lanthanum positions in the $[1\bar{1}00]$ projection. The characteristic dumbbell-shaped contrast of the sapphire produced by aluminum columns in $[2\bar{1}\bar{1}0]$ is slightly shifted, which can be explained by the small specimen tilt. Within the spinel block dark channels corresponding to the unoccupied intersticies are visible.

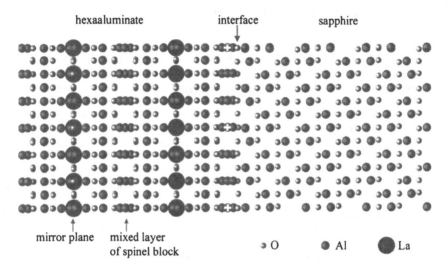

Fig. 5 Model of the interface between sapphire and hexaaluminate

Calculated electron waves based on different bicrystal models were then compared with the reconstructed wave. Best matching of simulated and reconstructed phases was obtained for the interface model, in which a cation layer of the sapphire is adjacent to the mixed cation layer within the spinel block of the HA. The supercell for this model is enlarged in fig. 5. A stacking sequence of ...$A_S B_S A_S B_{HA} C_{HA} A_{HA}$... leads to few face-sharing octahedra between both crystals. The second possible stacking – starting with C_{HA} – results in face-sharing between octahedra of S and tetrahedra of HA and is therefore not taken into consideration. Going into detail at the interface, it can be observed that the bright rods formed by the octahedra and tetrahedra in a row of HA are interrupted when an aluminum atom is very close to it (white triangles in fig. 4). We assume that the octahedral sites in the mixed layer of HA, marked with a white cross in fig. 5, are either unoccupied or show a considerably reduced occupation. In the case of fully occupied octahedral sites there would be three face-sharing AlO_6 octahedra – a structural motive not existing. Moreover the mixed HA layer is occupied by 3/4 with cations compared to 2/3 in the sapphire and thus a reduced number of cations would allow some charge balance. Reducing the occupation of those aluminum atoms in the HA by 2/3 in the simulation leads to a very good matching with the reconstruced phase. Nevertheless, an occupation factor could not be assigned with the applied methods.

Considering the epitactic system (111) spinel-(0001) sapphire, although higher mismatched, one may assume that the mixed cation layer will be favored as terminating layer as well. We did not observe any samples with the interface terminated by the layer occupied with exclusively octahedra. In the spinel the octahedra form the Kagomé motive in contrast to the sapphire with the honeycomb motive. These two layers do not match at all, because a high number of face-sharing octahedra would result.

A second type of interface was found, in which the mirror plane of HA faces a layer of sapphire which is occupied half. In the reconstructed phase every third bright dot corresponding to the lanthanum atoms is missing. Thus we may assume that there is either a vacancy or an oxygen ion replacing every third lanthanum atom at the interface.

CONCLUSIONS

Two types of interfaces between lanthanum hexaaluminate and sapphire appear atomically planar and are likely to be energetically favored. In all films the crystallographic orientation relationship is $(0001)_{HA} \| (0001)_S$ and $[1\bar{1}00]_{HA} \| [2\bar{1}\bar{1}0]_S$, and hence the orientation of the closed-packed oxygen layers is preserved across both types of interfaces.

The first type, in which the mixed cation layer within the spinel block of the HA is facing the sapphire, is characterized by a stacking sequence of $...A_S B_S A_S B_{HA} C_{HA} A_{HA}...$ of the oxygen sublattice across the boundary. Thus face-sharing between AlO_4 tetrahedra of the HA with AlO_6 octahedra of S is completely avoided. The face-sharing between 1/3 of the octahedra of HA with an adjacent pair of face-shared octahedra of S is found to be reduced by a lower cccupation of such octahedral sites in HA. This in turn is in accordance with arguments concerning charge balance which should occur between the positively charged spinel block of HA and sapphire.

Reconstruction of the exit wave from focus series is applied to achieve detailed information on *interphase* boundaries for the first time. The complex exit wave exhibits low noise; the reconstruction eliminates imaging artefacts and restores the highly localized information of the object structure, which is in particular valuable for interface studies. Thus image interpretation of the reconstructed data is very reliable.

ACKNOWLEDGEMENTS

The authors would like to thank F. F. Lange for helpful discussions.

REFERENCES

1. A. Kahn, A. M. Lejus, M. Madsac, J. Théry, D. Vivien, J. Appl. Phys. **52**, 6864 (1981)
2. J. M. P. J. Verstegen, A. L. N. Stevels, J. Lumin. **9**, 406 (1974)
3. M. Machida, A. Sato, M. Murakani, T. Kijima, H. Arai, J. Catalysis **157**, 713 (1995)
4. K. J. Vaidya, C. Y. Yang, M. DeGraef, F. F. Lange, J. Mater. Res. **9** (2), 410 (1994)
5. M. K. Cinibulk, J. Mater. Res. **14** (9), 3581 (1999)
6. N. Iyi, S. Tekenawa, S. Kimura, J. Solid State Chem. **54**, 70 (1984)
7. P. A. Stadelmann, Ultramicroscopy **21**, 131 (1987)
8. A. Thust, W.M.J. Coene, M. Op de Beeck, D. Van Dyck, Ultramicroscopy **64**, 211 (1996)
9. N. Iyi, S. Tekenawa, S. Kimura, J. Solid State Chem. **83**, 8 (1989)
10. L. Lutterotti, P. Scardi, J. Appl. Cryst. **23**, 246 (1990)
11. R. Kilaas, CrystalKit, Version 1.8.2., Berkeley, USA

ATOMIC AND ELECTRONIC STRUCTURE ANALYSIS OF Σ=3, 9 AND 27 BOUNDARY, AND MULTIPLE JUNCTION IN β-SiC

K. Tanaka, M. Kohyama
Department of Material Physics, Osaka National Research Institute, AIST, 1-8-31 Midorigaoka, Ikeda, Osaka, 563-8577, Japan, koji@onri.go.jp

ABSTRACT

The atomic structures of Σ=3, 9 and 27 boundaries, and multiple junctions in β-SiC were studied by high-resolution electron microscopy (HREM). Especially, the existence of the variety of structures of Σ=3 incoherent twin boundaries and Σ=27 boundary was shown by HREM. The structures of Σ=3, 9 and 27 boundary were explained by structural unit models. Electron energy-loss spectroscopy (EELS) was used to investigate the electronic structure of grain boundaries. The spectra recorded from bulk, {111}Σ=3 coherent twin boundary (CTB) and {211}Σ=3 incoherent twin boundary (ITB) did not show significant differences. Especially, the energy-loss corresponding to carbon 1s-to π* transition was not found. It indicates that C atoms exist at grain boundary on the similar condition of bulk.

INTRODUCTION

There has been increasing interest in SiC as high-temperature devices or high–performance ceramics. It is very crucial to investigate the structures and properties of grain boundaries (GB) of SiC, because grain boundaries in sintered ceramics or polycrystalline films have significant effects on the bulk properties. The study of GB by HREM is now one of the main tools for understanding materials. A rigid body translation of one grain relative to the other is an important part of the atomic relaxation of a grain boundary. The presence of a rigid body translation along the common <111> direction at {211}Σ=3 ITB was found in Si [1] and Ge [2], whereas no rigid body translation was found under different conditions in Si [1] or in diamond [3]. It was also pointed out that the translation state was sensitive to the environment of the boundary [4]. Finally, the inclination of the boundary plane can also be a possible factor in the decrease of the energy [4].

In the present study, {211}Σ=3 ITB and its junction with {111}Σ=3 CTBs, {221}Σ=9 boundary and its triple junction with {111}Σ=3 CTBs, and {552}Σ=27 boundary and its quadruple junction with {111}Σ=3 CTBs have been investigated by HREM and compared with a theoretical study.

EXPERIMENT

The transmission electron microscope (TEM) specimen of chemical vapor deposition (CVD) β-SiC was prepared by mechanical thinning and ion milling. CVD techniques can easily provide dense materials of high purity and interfaces that are well-defined even in covalent materials. Moreover, CVD specimens often exhibit preferred orientation during growth and this enhances the probability of coincident site lattice (CSL) grain boundary formation. HREM and EELS analysis was performed on CVD β-SiC with a preferred orientation of {220} and grain size of approximately 10μm using the field emission (FE) TEM (JEOL, JEM-3000F) equipped with Gatan Imaging Filter. Atomic models were given on the basis of structural unit model.

Mat. Res. Soc. Symp. Proc. Vol. 589 © 2001 Materials Research Society

RESULTS

In the case of Σ=3

Figure 1 (a) ~ (d) show the atomic models of the non-polar interfaces of {211}Σ=3 ITB in β-SiC. They include suitable reconstructions, in which arrows indicate the <011> bonds. These are symmetric and asymmetric models, respectively. Type A and B are constructed for the respective models. Si atoms are reconstructed in Type A, and C atoms in Type B. The relaxed atomic positions in fig. 1 are calculated by self-consistent tight-binding (SCTB) method.

The wrong bonds exist at the two types of positions. These are the intergranular bonds and the <011> bonds. The intergranular bonds exist on the {022} plane and cross the interface. The <011> bonds that are reconstructed connect the two atoms on the neighboring {022} planes, and double the periodicity along the <011> direction. The kinds and positions of the wrong bonds in Type A are inverted in Type B.

The grain boundary energies of fig. 1 (a) ~ (d) are 1.71, 2.37, 1.39, and 2.42 Jm^{-2}, respectively. Type A of asymmetric model is the most stable and Type B of symmetric model has the largest energy. The energy difference between Type A of the symmetric model and Type A of the asymmetric model is not so large, which indicates both structures can occur. The rigid-body translation in Type A of the symmetric model is a dilation of 0.001 nm and that in Type B is a dilation of 0.024 nm. The rigid-body translation in Type A of the asymmetric model is a shift of 0.072 nm along the <111> direction with a dilation of 0.016 nm. That in Type B is a shift of 0.068 nm along the <111> direction with a dilation of 0.032 nm.

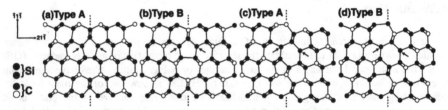

Fig. 1   Atomic models of {211}Σ=3 incoherent twin boundary in β-SiC. (a) and (b) are symmetric. (c) and (d) are asymmetric. Arrows indicate the <011> reconstruction bonds. Si atoms are reconstructed in Type A, and C atoms in Type B.

Figure 2 (a) and (c) are HREM images of the Σ=3 ITB in β-SiC. The length of incoherent twin is 6 {111} layers in figure 2 (a), 18 {111} layers in (c). [The lengths are indicated by numbers in fig. 2 (a) ~ (d).] As can be easily recognized, the {111} planes on different sides of ITB reveal no shift in fig. 2 (a), whereas no shift near the junctions with {111}Σ=3 CTB and a shift of about one fifth of the {111} plane distance near the center of {211}Σ=3 ITB in (c). This indicates that the rigid body translation along the <111> is zero in figure 2 (a), and zero near the junctions and about 0.05 nm near the center in (b).

Atomic models for Σ=3 ITB in fig. 2 (a) and (c) were given on the basis of the results of the theoretical calculations. Figure 2 (b) and (d) are structural unit models superimposed on HREM images corresponding to fig. 2 (a) and (c), respectively. White circles represent the reconstruction along <011> direction. It is obviously seen that structural unit which consists of symmetric 5-7-6 membered rings align along the boundary in fig. 2 (b), symmetric structural units are filled in near the junction between CTB and ITB and asymmetric 5-7-6 membered rings structural units are filled in at the middle of the ITB in (d). All structural unit models fit with the HREM images very well. It should be noted that the positions of all white circles

Fig. 2 (a) and (b) : HREM images of the Σ=3 incoherent twin boundary in β-SiC. (c) and (d) : Structural unit models superimposed on HREM images corresponding to (a) and (c), respectively. White circles represent the reconstruction along <011> direction.

correspond to the Si site and it was shown grain boundary energy was lower when Si was reconstructed than C by the theoretical calculation.

The obtained <111> translation was approximately 0.05 nm in experimental result and 0.072 nm in Type A in theoretical calculation, respectively. It should be noted that the structure of {111}Σ=3 CTB plays like an anchor because it is very stable and rigid, and a {211}Σ=3 ITB boundary always stay with a {111} coherent boundary in real materials, which is essentially different from the ideal bicrystal in the simulation. At the short boundary, it is difficult to translate along the common <111> direction because the top and bottom of the incoherent boundary are anchored by the {111} coherent boundaries. However, the <111> translation can occur at the relatively long boundary because the restriction from the coherent boundary is not so severe near the center of the incoherent boundary. The reason why the experimental value is smaller than the theoretical calculation one might be that atoms can not be fully relaxed because of anchor effect from {111} coherent boundaries.

Figure 3 shows C-K edge of EELS spectra from bulk, Σ=3 CTB and ITB in β-SiC. The π^* peak was found at Σ=3 ITB in diamond [3], but not found in β-SiC as seen in fig. 3. It indicates that C atoms exist at grain boundary on the similar condition of bulk.

As discussed above, the structures of Σ=3 ITB in β-SiC were explained very well by the prediction of SCTB calculation.

In the case of Σ=9 and triple junction

Figure 4 (a) shows the atomic model of the ideal triple junction (TJ) between one {221}Σ=9 and two {111}Σ=3 boundaries in diamond structure along [011] direction. It is seen from fig. 4 (a), atoms can not have proper bonds at {221}Σ=9 because of the height differences in {022} stacking. By introducing 1/4[011] translation, symmetric 5-6-7-6 membered rings structure can be achieved at Σ=9 boundary. And by introducing 1/9[41Ī] translation, zigzag 5-7-5-7 membered rings structure can be achieved. It was shown that the zigzag 5-7-5-7 structure has lower energy than symmetric 5-6-7-6 structure in Si [5]. However, introducing 1/4[011] translation at GB result in introducing screw dislocation at TJ and introducing 1/9[41Ī] translation result in introducing edge

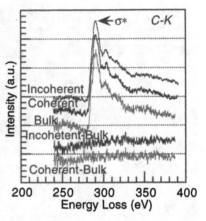

Fig. 3 EELS spectra from bulk, Σ=3 coherent and incoherent twin boundary in β-SiC.

dislocation. Therefore, both cases cause increasing total GB energy around TJ. On the other hand, zigzag 5-7-5-7 structure can be achieved by shifting GB plane by one atomic column as seen in fig. 4 (b). In this case, no extra dislocation is introduced, therefore total GB energy around TJ seems lower than the case of introducing translation.

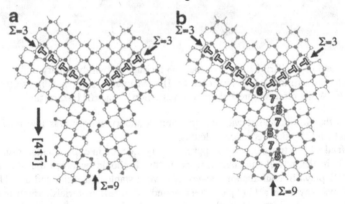

Fig. 4 Atomic model of the triple junction between one {221}Σ=9 and two {111}Σ=3 boundaries in diamond structure along [011] direction. Open and closed circles indicate the atoms with different height. (a) : Ideal position. (b) : One atomic column off.

Figure 5 (a) is a HREM image of TJ and (b) is a structural unit model superimposed on HREM images corresponding to fig. 5 (a). It is clear that Σ=9 boundary start from one atomic column away form the ideal TJ and stable zigzag 5-7-5-7 structure is achieved. It should be noted that choosing GB position has the same effect as translation.

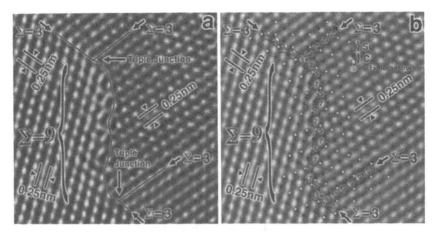

Fig. 5   HREM image of triple junction (a) and structural unit model superimposed on it (b). White circles represent the reconstruction along <011> direction.

In the case of Σ=27 and quadruple junction

Figure 6 (a) ia a HREM image of the quadruple junction (QJ) between one {552}Σ=27 and three {111}Σ=3 boundaries in β-SiC and (b) is a structural unit model superimposed on HREM images corresponding to figure 6 (a). It is seen that Σ=27 does not start from the ideal QJ and symmetric 5-7-6-5-6-7 membered rings structure is achieved near QJ but zigzag 5-7-5-7-6 membered rings structure is achieved far from QJ. GB energy of zigzag model is smaller than symmetric model according to SCTB calculation for β-SiC [6].

Figure 6 (c) and (d) are the atomic models of {552}Σ=27 boundary. Figure 6 (c) is a symmetric model and (d) is a co-existence model of symmetric and zigzag structure. As indicated fig. 6 (c), zigzag 5-7-5-7-6 structure like fig. 6 (d) is achieved if GB position shifts one atomic column from the symmetric structure position. To achieve zigzag structure, atomic relaxation is necessary. Also, the restriction from {111}Σ=3 boundary is so strong near QJ and there is no space for atomic relaxation. It might be the reason why symmetric structure is achieved near QJ and zigzag structure is achieved far from QJ.

CONCLUSION

HREM and theoretical calculation showed the atomic structures of Σ=3, 9 and 27 boundaries, and multiple junction in β-SiC were well explained by structure unit models. The translation state was sensitive to the environment of the boundary. A shift of GB position has the same effect as a rigid body translation. C atoms exist at grain boundary on the similar condition of bulk.

REFERENCES

1. H. Ichinose, Y. Tajima, and Y. Ishida in *Grain Boundary Structure and Related Phenomena*, (Supple. to Trans. of the Japan Institute of Metals, **27** Sendai, Japan 1986), p. 253-260.

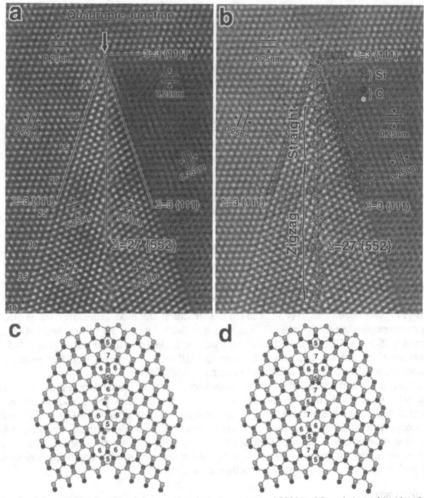

Fig. 6 (a) : HREM image of the quadruple between one {552}Σ=27 and three {111}Σ=3 boundaries in β-SiC. (b) : Structural unit model superimposed on (a). (c) : Symmetric model. (d) : Co-existence model of symmetric and zigzag structure.

2. A. Bourret, and J. J. Bacmann in *Grain Boundary Structure and Related Phenomena*, (Supple. to Trans. of the Japan Institute of Metals, **27** Sendai, Japan 1986), p. 125-134.
3. Y. Zhang, H. Ichinose, Y. Ishida, K. Ito, and M. Nakanose in *Diamond for Electronic Application*, edited by D. L. Dreifus, A. Collins, T. Humphreys, K. Das, and P. E. Pehrsson (Mater. Res. Soc. Proc. **416**, Pittsburgh, PA 1996), p. 355-360.
4. K. Tanaka, M. Kohyama, and M. Iwasa, Matl. Sci. Forum, **294-296**, p. 187 (1999).
5. M. Kohyama, Matl. Sci. Forum, **207-209**, p. 265 (1996).
6. M. Kohyama and K. Tanaka, to be published.

ELECTRONIC EFFECTS ON GRAIN BOUNDARY STRUCTURE IN BCC METALS

GEOFFREY H. CAMPBELL,* WAYNE E. KING,* JAMES BELAK,** JOHN A. MORIARTY,** AND STEPHEN M. FOILES§
* University of California, Lawrence Livermore National Laboratory, Chemistry and Materials Science Directorate, P.O. Box 808, Livermore, CA 94550
** University of California, Lawrence Livermore National Laboratory, Physics Directorate, P.O. Box 808, Livermore, CA 94550
§Sandia National Laboratories, Livermore, CA 94550

ABSTRACT

The dominant factor in determining the atomic structure of grain boundaries is the crystal structure of the material, e.g. FCC vs. BCC. However, for a given crystal structure, the structure of grain boundaries can be influenced by electronic effects unique to the element comprising the crystal. Understanding and modeling the influence of electronic structure on defect structures is a key ingredient for successful atomistic simulations of materials with more complicated crystal structures than FCC. We have found that grain boundary structure is a critical test for interatomic potentials. To that end, we have fabricated the nominally identical $\Sigma 5$ (310)/[001] symmetric tilt grain boundary in three different BCC metals (Nb, Mo, and Ta) by diffusion bonding precisely oriented single crystals. The structure of these boundaries have been determined by high resolution transmission electron microscopy. The boundaries have been found to have different atomic structures. The structures of these boundaries have been modeled with atomistic simulations using interatomic potentials incorporating angularly dependent d–state interactions, as obtained from Model Generalized Pseudopotential Theory. We report here new experimental and theoretical results for Ta.

INTRODUCTION

Atomistic simulations are an increasingly important means of understanding the behavior of materials under a variety of conditions. With this technique, an assembly of thousands, or up to billions, of atoms is defined in a computer simulation and allowed to interact according to certain rules and boundary conditions. The boundary conditions include temperature and states of stress, allowing the calculation of such properties as the equation of state or unstable stacking fault energy. The structure and properties of crystal defects can also be predicted, such as the configuration of a dislocation core or the formation and migration energies of interstitials and vacancies (see e.g. [1]). The rules of interaction are often very simple in order to speed computation. This simplification requires approximations to be made about the physics of the interacting atoms. Hence, in the development of models of interatomic interactions, an evaluation is necessary of whether the essential physics have been incorporated in the model. The models are validated by comparing their predictions with experimental observations.

Recently developed models of interatomic interactions incorporate angularly dependent contributions to model materials with directional bonding [2-7], such as the body centered cubic transition metals in which the d – bands participate in bonding. The strength of the directional component of the bonding has a major influence on the structure of crystal defects. The model of interatomic interactions with angular dependence that we use is the Model Generalized Psuedopotential Theory [5]. We have applied it to modeling the $\Sigma 5$ (310)/[001] symmetric tilt grain boundary (STGB) in niobium [8], molybdenum [9], and now tantalum as a critical test of its accuracy. We report here on the new results for Ta.

Grain boundaries are a particularly good test case for atomistic simulations. Perhaps the biggest reason is that high quality experimental data on the atomic structure of grain boundaries can be obtained through high resolution transmission electron microscopy (HREM). The limited resolution of the electron microscope, however, does place some stringent constraints on precisely which

Mat. Res. Soc. Symp. Proc. Vol. 589 © 2001 Materials Research Society

grain boundaries can be imaged at atomic resolution [10]. This limitation can be overcome by first choosing an amenable boundary to study both theoretically and experimentally, in this case the Σ5(310)/[001] STGB, and then fabricating that very boundary for experimental study. The capability that enables this approach to the problem is the ultra-high vacuum (UHV) Diffusion Bonding Machine [11].

ATOMISTIC SIMULATIONS

The Model Generalized Pseudopotential Theory (MGPT) [5, 12, 13] of interatomic bonding in BCC transition metals has been shown to represent both bulk and defect properties in the case of Mo [1, 9, 13]. Useful MGPT potentials have now also been determined for Ta over a wide volume range and preliminary results for selected mechanical properties have been calculated [14].

In a recent study [15] of bonding in BCC transition metals and its representation using semi–empirical and *ab–initio* methods, the Σ5 (310)/[001] STGB in Nb and Mo was used to compare the different representations, including MGPT. To facilitate total energy *ab–initio* calculations [16] a minimal 20 atom periodic cell containing two grain boundaries was constructed. Through calculations with larger simulation cells, this small cell was shown to give reliable atomic structures at the grain boundaries. This same minimal simulation cell, containing only one (310) repeat length between the grain boundaries was employed here in our present MGPT Ta studies. The system was brought to the position of minimum energy using a standard molecular dynamics code [17] and the method of simulated annealing along the following path: The ideal CSL constructed [18] grain boundary was created on the computer and the atoms were allowed to relax without any shift of the two grains. The effect of systematically shifting the two grains was then explored. The resulting energy, relative to the relaxed unshifted state, for a [001] shift is shown in Fig. 1.

The predicted structure for the Σ5(310)/[001] STGB in Ta is shown in Fig. 2(a) and (b). This boundary has the [001] crystal direction as the common tilt axis and Fig. 2(a) views the structure along this direction. A notable feature is seen in the view perpendicular to the tilt axis (Fig. 2(b)) in that the crystal planes are not perfectly aligned, rather there is a shift of one crystal with respect to the other.

EXPERIMENT

In order to experimentally investigate the structure of the grain boundary shown in Fig. 2, we fabricated it by diffusion bonding precisely oriented single crystals of high purity tantalum. The single crystal of Ta was grown at the Institute for Solid State Physics in Chernogolovka, Russia, from high purity stock material provided from Cabot Corp. The Ta was further purified by zone refining in high vacuum. The liquid zone was formed by electron beam heating from an annular electrode and five passes of the liquid zone were taken,

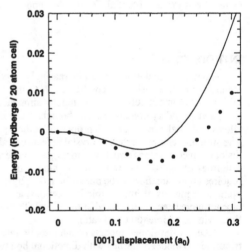

Figure 1: A plot of the change in grain boundary excess energy in the atomistic simulation as a function of the relative shift of the crystals along [001]. The line represents a rigid translation of one grain with respect to the other. The points represent calculations where the grain translations were held fixed and a relaxation allowed. The lowest point is the full 3D relaxation of the structure.

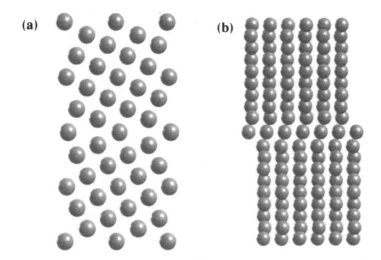

<u>*Figure 2:*</u> *The predicted structure of the* $\Sigma 5$ *(310)/[001] STGB in Ta from atomistic simulations using interatomic potentials derived from MGPT. In (a) the structure is viewed along the common [001] tilt axis and in (b) it is viewed perpendicular to the tilt axis along [130].*

with the last pass starting in a seed crystal oriented for growth along [310]. The crystal diameter was 18 mm with approximately 200 mm length.

Two 5 mm slices of the single crystal boule perpendicular to the growth axis were taken by electric discharge machining (EDM). The crystal slices were oriented with Laue backscatter x-ray diffraction and lapped in specially designed fixtures to orient the faces to within 0.1° of (310). The faces were polished with flat polishing techniques to achieve a flatness of less than 100 nm across the 15 mm diameter faces. Finally, the mutual orientation of the two slices was set by a 180° rotation of one crystal with respect to the other around an axis perpendicular to the slice face. The process is fully described elsewhere [19].

The UHV Diffusion Bonding Machine has been described in detail elsewhere [11]. The Ta crystals were sputter cleaned immediately prior to bonding to remove the oxide layer that forms upon exposure to the atmosphere. Auger spectra taken of the surface confirmed the removal of the oxide. The crystals were bonded at 1500°C for 8 h under an applied load of 1 MPa. The bicrystal was further heat treated in a separate UHV apparatus at 2600°C for 48 h in order to eliminate any residual porosity remaining at the grain boundary after the initial diffusion bonding step.

Two cylindrical cores were taken of the bicrystal by wire EDM cutting. The cores were 3 mm in diameter and contained the grain boundary along their axes. The two cylinders were cut with their axes along different directions: one along the common [001] direction and the other perpendicular to the first along the common [1 $\bar{3}$ 0] direction. The cylinders were sliced to give 3 mm disks of 300 μm thickness. The disks were gently lapped to 150 μm thickness and 50 μm deep dimples were ground centered over the grain boundary on both sides. The final 50 μm was thinned by electropolishing with an electrolyte of 1% hydrofluoric acid, 5% sulfuric acid, 34% butoxyethanol, and 60% methanol at −20°C. The dimpling assisted in locating the perforation to intersect the grain boundary. Nonetheless, perforations seldom intersected the grain boundary. In these cases, the specimens were immersed in a solution of 60% sulfuric acid, 25% nitric acid, and 15% hydrofluoric acid that enlarged the specimen hole while retaining thin area until it intersected the grain boundary.

The high resolution imaging was performed on a Philips CM300FEG–ST. The images of the

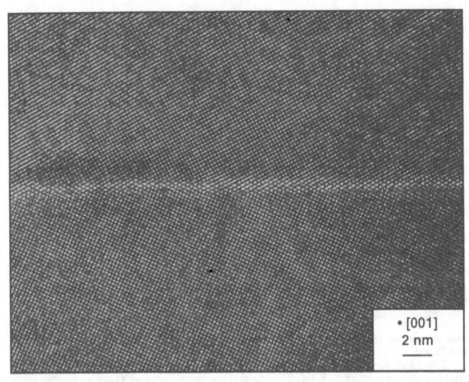

Figure 3: A high resolution electron micrograph of the Σ5 (310)/[001] STGB in Ta with the direction of view parallel to the common [001] tilt axis. The direction is the same as in Fig. 2a.

grain boundary structure along [001] were acquired on film. The images along [1 3̄ 0] were acquired on a 2k 5 2k CCD array after energy filtering for only zero loss electrons with a 20 eV window on a Gatan Imaging Filter.

RESULTS

The high resolution image acquired parallel to the tilt axis is shown in Figure 3. The grain boundary displays a plane of mirror symmetry in this projection. To reveal the shift predicted by the atomistic simulations, an image perpendicular to the tilt axis is required. A high resolution micrograph acquired in this direction is shown in Figure 4(a). Each crystal is imaged along [1 3̄ 0]. The two crystal planes with the largest interplanar spacing containing [1 3̄ 0] are (002) and (310) at 1.65Å and 1.04Å, respectively. The information limit for this microscope is approximately 1.2Å. Thus only the (002) set of atomic planes are imaged as fringes in this image. The shift of the crystals along [001] is revealed by sighting down the (002) fringes in the image and noting their alignment as they cross the grain boundary. The shift is most easily seen in a glancing angle perspective view, such as that shown in Figure 4(b). The disregistry of the (002) planes as they cross the grain boundary is evident.

SUMMARY

The high resolution images of the Σ5 (310)/[001] STGB in Ta show that the boundary contains a rigid body shift of the crystals along [001]. The shift is evident from a qualitative inspection of

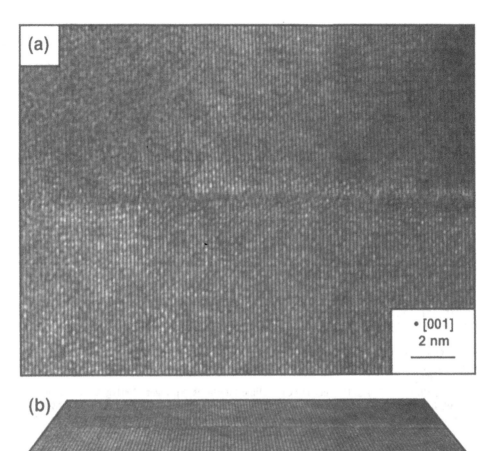

Figure 4: (a) A high resolution electron micrograph of the Σ5 (310)/[001] STGB in Ta with the direction of view perpendicular to the common tilt axis. The direction is the same as in Fig. 2b. (b) A glancing angle perspective view of the same image as shown in (a).

Fig. 4(a), assisted by the glancing angle perspective view shown in Fig. 4(b). The magnitude of the shift is approximately 1/4 a_0 which is comparable to the shift predicted by the atomistic simulations.

The results of the present study on Ta, when combined with previously reported results on Nb [8] and Mo [9], demonstrates the sensitivity of grain boundary structures to the non–central nature of bonding in these BCC transition metals. Potentials such as those obtained from MGPT, which is based on a physical understanding of electronic structure and bonding in metals, are required to model defect structures in BCC transition metals.

ACKNOWLEDGEMENTS

We would like to thank S.L. Weinland for assistance with specimen preparation and V. Vítek and C. Elsässer for many enlightening discussions on the calculation of grain boundary structure. This work was performed under the auspices of the United States Department of Energy and the Lawrence Livermore National Laboratory under contract number W–7405–Eng–48.

REFERENCES

1. W. Xu and J. A. Moriarty, *Phys. Rev. B*, **54** [10] 6941-6951 (1996).
2. M. I. Baskes, *Phys. Rev. B*, **46** [5] 2727-2742 (1992).
3. S. M. Foiles, *Phys. Rev. B*, **48** [7] 4287-4298 (1993).
4. A. G. Marinopoulos, V. Vítek and A. E. Carlsson, *Philos. Mag. A*, **72** [5] 1311-1330 (1995).
5. J. A. Moriarty, *Phys. Rev. B*, **42** [3] 1609-1628 (1990).
6. D. G. Pettifor, M. Aoki, P. Gumbsch, A. P. Horsfield, D. Nguyen Manh and V. Vítek, *Mat. Sci. Eng. A*, **192/193** 24-30 (1995).
7. A. E. Carlsson, *Phys. Rev. B*, **44** [13] 6590-6597 (1991).
8. G. H. Campbell, S. M. Foiles, P. Gumbsch, M. Rühle and W. E. King, *Phys. Rev. Lett.*, **70** [4] 449-452 (1993).
9. G. H. Campbell, J. Belak and J. A. Moriarty, *Acta Mater.*, in press.
10. P. Pirouz and F. Ernst in *Metal - Ceramic Interfaces*, edited by M. Rühle, A. G. Evans, J. P. Hirth and M. F. Ashby (Pergamon Press, New York, 1990) pp. 199-233.
11. W. E. King, G. H. Campbell, A. W. Coombs, G. W. Johnson, B. E. Kelly, T. C. Reitz, S. L. Stoner, W. L. Wien and D. M. Wilson in *Joining and Adhesion of Advanced Inorganic Materials*, edited by A. H. Carim, D. S. Schwartz and R. S. Silberglitt (Mat. Res. Soc. Symp. Proc. **314**, Materials Research Society, Pittsburgh, PA, 1993) pp. 61-67.
12. J. A. Moriarty, *Phys. Rev. B*, **38** [5] 3199-3231 (1988).
13. J. A. Moriarty, *Phys. Rev. B*, **49** [18] 12431-12445 (1994).
14. J. A. Moriarty, W. Xu, P. Soderlind, J. Belak, L. H. Yang and J. Zhu, *J. Eng. Mater. Tech., Trans. ASME*, **121** 120 - 125 (1999).
15. T. Ochs, C. Elsässer, M. Mrovec, V. Vítek, J. Belak and J. A. Moriarty, submitted to *Philos. Mag. A*.
16. T. Ochs, O. Beck, C. Elsässer and B. Meyer, *Philos. Mag. A*, in press.
17. M. P. Allen and D. J. Tildesley, Computer Simulation of Liquids (Oxford University Press, New York, 1987).
18. W. Bollmann, Crystal Defects and Crystalline Interfaces (Springer - Verlag, Berlin, 1970).
19. W. L. Wien, G. H. Campbell and W. E. King in *Microstructural Science*, edited by D. W. Stevens, E. A. Clark, D. C. Zipperian and E. D. Albrecht (Microstructural Science **23**, ASM International, Materials Park, OH, 1996) pp. 213-218; presented at 28th International Metallography Society Convention.

352

STRUCTURAL STUDY OF A [100] 45° TWIST PLUS 7.5° TILT GRAIN BOUNDARY IN ALUMINIUM BY HREM

M. SHAMSUZZOHA, P. A. DEYMIER* AND DAVID J. SMITH**
School of Mines and Energy Development, and Department of Metallurgical and Materials Engineering, The University of Alabama, Tuscaloosa, AL 35487
*Department of Materials Science and Engineering, University of Arizona, Tucson, AZ 85721
**Center for Solid State Science and Department of Physics and Astronomy, Arizona State University, Tempe, AZ 85827.

ABSTRACT

A [100] 45° twist plus 7.5° tilt grain boundary in aluminium prepared by cold rolling and annealing has been studied by high-resolution electron microscopy. The direct interpretability of the image features in terms of atomic column positions allows structural models of the grain boundary to be developed. The boundary exibits a high concentration of steps due to the 7.5° tilt from a perfect [100] 45° quasiperiodic misorientation. Occurrence of co-incidence and pseudo co-incidence of atomic planes across the interface appears to play an important role in the formation of steps along this boundary. Local relaxation of atoms resulting from the perturbation of the [100] 45° twist bi-crystal determines the boundary structure.

INTRODUCTION

The structural study of grain boundaries and associated interfacial defects can be used to gain a better understanding of the correlation between structure and properties of polycrystalline materials. Theoretical studies of grain boundary structure have so far led to the development of various geometrical models for grain boundaries based on coincidence site lattice (CSL) [1], displacement-shift-complete (DSC) lattice [2-4] and the O-lattice theories [2]. These models provide an adequate framework for structural characterization of atomic-resolution images of some grain boundaries and have been successfully used to characterize the experimental structure of many periodic grain boundaries in metals and semiconductors [5-10]. These structural models of grain boundaries, although providing a significant understanding of the structure of periodic grain boundaries, are inadequate to characterize the structure of more general boundaries such as quasiperiodic or aperiodic boundaries. Since most of the grain boundaries found in polycrystalline materials are of low symmetry, it is important to direct efforts towards gaining knowledge about the structure of such boundaries. The present study was undertaken with an aim of understanding the atomic structure of quasiperiodic and aperiodic grain boundaries. This paper focuses on an aperiodic boundary and presents a high-resolution electron microscopy (HREM) study of a [100] 45° twist plus 7.5° tilt grain boundary in aluminium.

EXPERIMENTAL PROCEDURE

The aluminum bi-crystal used in this study was prepared by the cross-rolling and annealing method described elsewhere [11]. In order to obtain a thin foil specimen suitable for transmission electron microscopy, a cylindrical disc of 3mm diameter containing the boundary was produced by ultrasonic disc cutting. The disc thus obtained was polished gently on 600 grade SiC paper to reduce its thickness to 0.18mm. Finally the disc was electropolished at a voltage of 15V at -25° C in a solution 25% nitric acid and 75% methanol. Polishing was halted just after perforation. The thin foil was initially examined with a 200 keV Hitachi H-8000 transmission electron microscope to determine the angle/axis orientation of the boundary. High resolution electron microscopy (HREM) was performed with a JEM 4000 EX electron microscope operated at 400 keV at a typical magnification of 500,000 times. Under the experimental condition used for imaging, atomic columns appeared black, so that the image could be interpretable intuitively in terms of atomic column positions to within 0.03nm [12].

RESULTS AND DISCUSSION

A high-resolution image with simultaneous lattice imaging conditions in the adjacent grains of the investigated boundary and its associated bi-crystal is shown in Fig. 1. The image exhibits atomic columns projected along [011] and [001] of crystal 1 and 2, respectively. It also reveals that the boundary is comprised of a series of facets separated by steps. Most of the facets run parallel to the (100) plane of crystal 1 and ($7\bar{1}0$) of crystal 2. Geometrical analyses of the boundary misorientation reveal that the boundary can be described as [100] $_{1,2}$ 45° twist followed by a 7.5° tilt rotation about the common [011] $_1$, [001] $_2$ direction. The existence of a small tilt in the bi-crystal misorientation indicates that the boundary assumes a slight deviation from that present in an ideal [100] 45° quasiperiodic boundary. This boundary feature suggests that the experimental boundary might be treated as a near [100] 45° twist quasiperiodic boundary. Accordingly, we have tried to characterize the boundary in terms of crystallography that is associated with perturbations to a perfect [100] 45° twist quasiperiodic grain boundary.

Figure 1. HREM micrograph showing a [100] 45° twist plus 7.5° tilt boundary in Al with steps marked by arrows.

The structure of quasiperiodic grain boundaries can be characterized within the framework of the crystallography of quasicrystals [13]. Representative dichromatic patterns (DCP) of a [100] 45° twist misoriented Al quasicrystal as viewed along [100] $_{1,2}$ and [011] $_1$ or [001] $_2$ axes are shown in figures 2a

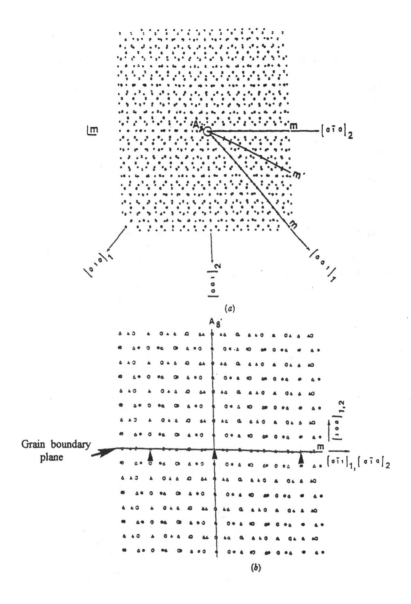

Figure 2.(a) [100] 45° dichromatic pattern as viewed along [100] of crystal 1 and 2. Circles and triangles represent lattices of crystals 1 and 2, respectively. (b) [100] 45° dichromatic pattern as viewed along [011]₁ and [001]₂. Filled and unfilled circles and triangles represent successive (011)₁ and (001)₂.

and 2b respectively. The DCP exhibits an 8'/mmm' symmetry, where the operation of symmetry indicated with a prime are colored operations relating atoms of two interpenetrating crystals and unprimed symmetry operators relate atoms within their respective crystal. The eight-fold colored axis is parallel to the rotation axis and perpendicular to a mirror plane that runs parallel to (100) of both crystals. The 8'/mmm' symmetry of the DCP is not consistent with any periodicity in the bi-crystal and therefore, eliminates the existence of any periodic structure in related [100] 45° twist grain boundary. However the symmetry operators present in the [100] 45° DCP provide some useful geometrical features. For example the symmetry operations allow the (02$\bar{2}$) of crystal 1 and (020) of crystal 2 to become parallel to each other. Across the (100) $_{1,2}$ grain boundary plane as in the DCP of figure 2b, such parallel planes of opposing crystals maintain a ratio of periods equals to √2, and perform pseudo-coincidence at the boundary when such a period is approached. In the (100) boundary of figure 2b the sites of such pseudo-coincidence between (02$\bar{2}$) of crystal 1 and (020) of crystal 2 are denoted by arrows. This boundary structure has been found to be closely resembling the structure of an experimentally observed [100] 45° twist boundary [14], which exhibits stepping at such sites of pseudo-coincidence.

Analyses of the DCP of [100] 45° twist bi-crystal after the introduction of varying amount of small tilt about the common [011] $_1$, [001] $_2$ direction revealed that a [100] 45°twist plus 8.13° tilt appears to produce a dichromatic pattern in which coincidence sites between (02$\bar{2}$) of crystal 1 and (020) of crystal 2 exist along a plane that closely resembles that observed in the investigated bi-crystal. Figure 3 shows the DCP of this bi-crystal misorientation. The pattern exhibits a 2/m symmetry, where two-fold axis is parallel to the tilt axis and perpendicular to a mirror plane that runs parallel to (011) of crystal 1 and (001) of crystal 2.

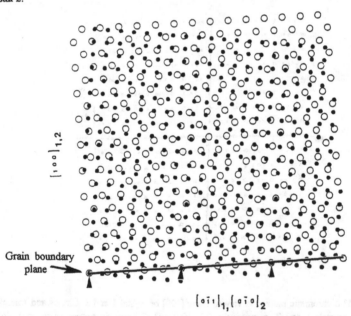

Figure 3 Dichromatic patterns of the [100] 45° twist plus 7.5° twist grain boundary as viewed along the [001]$_2$ or [011]$_1$ direction normal to twist axis. The symbols are the same as in figure 2. Coincident sites or near coincident sites present along the grain boundary plane of the DCP are designated by arrows.

This symmetry of the DCP, although lower than that existing prior to the introduction of tilt in the bi-crystal, also allows the formation of coincident or almost coincident sites along the $(7\bar{1}0)_1$ or $(100)_2$ grain boundary plane every 5 $(02\bar{2})$ planes of crystal 1 or 7 (020) planes of crystal 2. It is interesting to note that the boundary segment exists between the two neighboring coincident or almost coincident sites exhibiting a stacking ratio of 1.40 ($\approx\sqrt{2}$) for the $(02\bar{2})_1$ and $(020)_2$ planes. The atomic environment surrounding these special points of the boundary no doubt is favorable for strong atomic interactions. As a consequence, these boundary locations are likely to play critical roles in the overall stability of the boundary. Analyses of the experimental image as shown in figure 4 revealed that the grain boundary steps by only one half of the lattice spacing at locations separated either by 7 or 9 $(020)_2$ plane spacings, which appears to agree well with the spacing between two neighboring coincident or almost coincident points found in the DCP of figure 3. On a qualitative basis, it can be stated that the experimental grain boundary assumes stepping near the sites where pseudo-coincidence of atomic columns is expected. i.e where atomic environment is least distorted. The stepping allows the experimental boundary to achieve a stable structure by the coincidence of the appropriate lattices of opposing crystals.

Figure 4 Magnified image of grain boundary steps (shown by arrows) taken from segment of [100] 45° twist plus 7.5° tilt boundary shown in Figure 1

CONCLUSIONS

The boundary reveals certain sites at which the local atomic environment is least distorted. These sites arise when stacking of planes of opposing crystals existing normal to or close to the normal to the boundary plane assumes a ratio $\approx \sqrt{2}$. The boundary appears to perform stepping at such locations in order to achieve a stable structure in which many such coincidences of lattices exist.

ACKNOWLEDGMENTS

This research utilized the Central Analytical Facility of the School of Mines and Energy Development at The University of Alabama, and the Center for High Resolution Electron Microscopy in the Center for Solid State Science at Arizona State University.

REFERENCES

1. M. L. Kronberg and F. M. Wilson, Trans A.I.M.E., 85, 26 (1964).
2. M. Bollmann, Crystal Defects and Crystalline Interfaces, Springer, New York (1970).
3. R. W. Balluffi, in Interfacial Segregation, A.S.M., Metals Park, Ohio (1977).
4. D. A. Smith, and R. C. Ponds, Intern. Metals Reviews, 205, 61 (1976).
5. M. Shamsuzzoha, I. Vazquez, P. A. Deymier, and D. J. Smith, Interface Science, 3, 227 (1996).
6. M. Shamsuzzoha, D. J. Smith, and P. A. Deymier, Phil. Mag., 24, 1303 (1990).
7. J. M. Penisson, J. de Phys. CS 49, 87 (1988).
8. W. Skrotzki, H. Wendt, C. B. Carter, and D. Kohlstedt, Phil. Mag., A57, 383 (1988).
9. M. Elkajbaji, and J. Thibault-Desseaux, Phil. Mag., A58, 325 (1988).
10. C. D'Anterroches, and A. Bourret, Phil. Mag. A49, 783 (1985).
11. M. Shamsuzzoha, and P. A. Deymier, Scripta Metall. Mater., 24, 1303 (1990).
12. W. O. Saxton, and D. J. Smith, Ultramicroscopy, 18, 39 (1985).
13. A. P. Sutton, Progress in Materials Science, 36, 167 (1992).
14. M. Shamsuzzoha, P. A. Deymier, and D. J. Smith, Scripta Mater., 35, 327 (1996).

ATOMIC STRUCTURE OF GOLD AND COPPER BOUNDARIES

C.J.D. HETHERINGTON
University of Sheffield, Department of Electronic and Electrical Engineering, Mappin Street, Sheffield, S1 3JD, UK

ABSTRACT

A very thin bicrystal film, prepared by depositing gold onto a {111} germanium substrate, then annealing and removing it, has been studied by HREM. The crystals grow predominantly in the orientations (111) and ($\bar{1}\bar{1}\bar{1}$) which are related by a 60° rotation ($\Sigma 3$ CSL) and bounded by {112} facets. In contrast to the boundaries studied in thicker films, these {112} facets extend through the depth of the film. Another crystal within the film also had a {111} orientation, but rotated by around 20° from the other crystals. A short length of boundary was found that was edge-on, symmetrical, facetted on (213) planes, and corresponding to $\Sigma 7$ (or $\Sigma 21$). A copper {111} film did not show edge-on boundaries.

The rigid body translation across the facets were measured and found to vary depending on the proximity to facet junctions. An expansion at the boundary was the general case.

INTRODUCTION

Materials problem solving increasingly makes use of modelling of the atomic structure. The validity of the modelling methods relies to an extent on the availability of experimental data to compare with the simulated data. High resolution electron microscopy can reveal the atomic arrangement at suitable grain boundaries, suitable meaning parallel zone axes either side of an edge-on boundary. Modelling methods can now tackle the atomic structure of these boundaries.

Recently, developments in electron microscopy in this area have been made in the preparation of suitable boundaries and improvements in microscope resolution. These developments have increased the range of boundaries that can be observed to include metals as well as semiconductors, higher index zone axes as well as silicon (110) and boundaries in $\Sigma 3$ CSL gold other than the standard twin interface.

The technique for preparing the boundaries studied in this work involves depositing gold (or copper) on a single crystal semiconductor wafer [1]. More than one orientation may grow if the epitaxial alignment of the thin film on its substrate has multiple variants [2]. The resulting thin film will then form as a polycrystal and the maze of boundaries separating the crystals may become perpendicular to the plane of the thin film on annealing.

The imaging of these films is, on the one hand, routine in that it requires only straightforward HREM. On the other hand, the resolution required for these particular specimens is closer to 1Å than 2Å and that is possible only on machines built very recently.

This paper describes an extension to previous work on the films [3], including the observation of a new interface type in gold, and the first HREM observation of a copper bicrystal down the [111] direction.

EXPERIMENTAL

The preparation method of the {111} gold films and the TEM specimens was covered in detail in [3]. In brief, gold was deposited onto a germanium [111] substrate which was later removed chemically, leaving the film to be floated off onto a fine-mesh grid. The key element in this work was the preparation of a gold film thin enough to have the facets extend from the top surface to the bottom surface, without diverting into the in-plane {111} twin boundary.

Mat. Res. Soc. Symp. Proc. Vol. 589 © 2001 Materials Research Society

The nominal thickness of the film was 8nm, in contrast to a thickness of 30nm in the films studied in the earlier work. The film thickness was estimated by tilting a foil to 45° and measuring the projected width of the boundaries confirming a figure of around 8nm or less. The thickness, of course, may vary across the film. Estimates of film thickness may also be made from the [111] HREM images if we measure the projected width of inclined interfaces and make assumptions about the identity of the inclined plane.

The rotation from the <111> film normal also induces diffraction contrast between the (111) and ($\bar{1}\,\bar{1}\,\bar{1}$) grains. Any overlapping (111) and ($\bar{1}\,\bar{1}\,\bar{1}$) grains, brought about by an in-plane {111} twin boundary, would show up with an intermediate contrast level, but very few were seen.

{110} films of gold were formed by deposition onto a germanium [001] substrate.

The copper {111} film preparation was more complicated and required intermediate layers of cobalt and silver to be deposited first on the silicon [111] wafer [2], followed by ion-milling from the substrate side.

Microscopy was performed on the JEOL-ARMII in Stuttgart operating at 1250kV with the side-entry holder installed [4], giving a point-to-point resolution limit of 1.2Å, and the JEOL 2010F in Sheffield for the gold {110} film.

RESULTS

Gold {111} film, Σ3 bicrystal

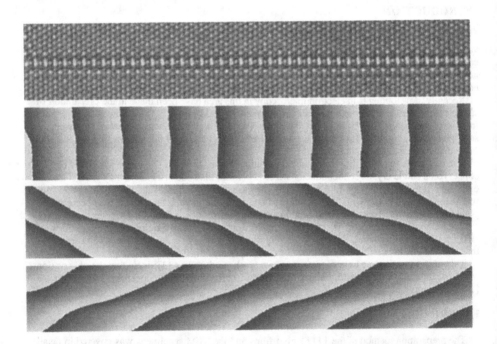

Fig. 1. HREM image of single (112) facet separating (111) and ($\bar{1}\,\bar{1}\,\bar{1}$) grains, spacing of{220} fringes is 1.44Å. The moiré fringes, generated digitally at 10x magnification, reveal the displacement between the two grains.

As in the earlier work, facetting on {112} planes was seen. Since the facets in this thinner film are believed to extend through the film thickness, measurement of any displacement at the boundary may now be performed with more confidence than in the earlier work [3]. In particular, expansion at the boundary is of interest in the modelling of the atomic structure of the facet and facet junctions [5]. Fig. 1 shows the displacements of the (220) fringes across a single (112) facet in the form of 10x magnification moiré fringes [6]. The facet is bounded by {121} facets (not shown) but, as it was about 150Å long, it is thought not to be affected by external constraints and therefore to have assumed the equilibrium structure. From the 60% shift of the moiré fringes inclined to the interface, the net displacement of the lattice across the facet is calculated to be an expansion of 1.0Å along the [112] direction with zero lateral translation. Other {112} facets examined have shown different displacements (e.g. a 0.55Å expansion in ref. [3]), possibly because they have been influenced by the displacements set by neighbouring facets.

If there is a case for an equilibrium configuration across a single (112) facet as argued above, then there may also be a case for an overall equilibrium configuration across a multifacetted length. The portion of boundary shown in fig. 2 contains several short (121) and (211) facets, but lies on average on a (110) plane. Also shown in fig. 2 is the moiré image from the (220) planes parallel to the overall boundary. The expansion is constant along the facets. The shift of the moiré fringes is measured at 65%, hence the expansion along the [110] direction is 65% of the d_{220} spacing, 0.94Å. The displacement of the other (220) fringes (shown as moirés in lower fig. 2) varies according to the particular (112) or (121) facet they are crossing.

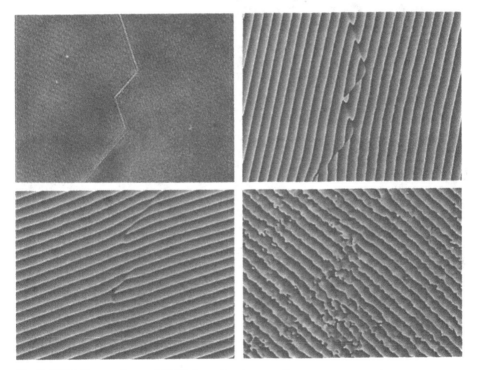

Fig. 2. HREM image of several {112} facets forming overall a (110) boundary orientation, with moiré fringe images revealing displacement of {220} fringes (see fig. 1 for details)

Gold {111} film, Σ7 bicrystal

One or two crystals were found in the {111} gold film that were rotated around <111> relative to the other crystals by approximately 22°, a diffraction pattern is shown in fig. 3. An explanation for this rotation in terms of nucleation on the (111) germanium surface is to be investigated. A consideration of the crystallography of the rotation leads to a suspicion that the Σ7 CSL may be involved. The Σ7 CSL may be described by a 38.2° rotation about the [111]. In this instance, if one bears in mind that a 60° rotation of the [111] projected lattice falls back onto itself, then a rotation of 60.0° – 38.2° = 21.8° is explained. If the 21.8° rotated crystal is of type [111] and the neighbouring crystal is also of type [111], then we have a Σ7 CSL. If the 21.8° rotated crystal is of type [111] but the neighbouring crystal is of type [1 1 1], then we must have a Σ21 CSL. And at the triple junction, the mathematics gives, as required, Σ21 -> Σ7 + Σ3.

In general, the boundaries between 0° and 21.8° crystals were inclined to the film normal. One section of boundary, however, was found that was edge-on (fig. 4) and it was, perhaps not unsurprisingly, at the symmetrical orientation, *i.e.* at the (123)/(123) plane.

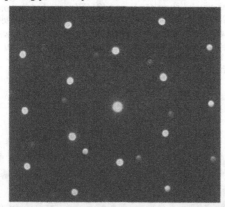

Fig 3 <111> diffraction patterns from neighbouring grains reveal a 21.8° rotation indicating a Σ7 CSL

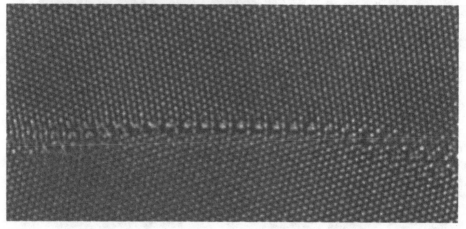

Fig. 4. HREM image of boundary between grains having Σ7 CSL, the straight facet is edge-on and lies on {123} type planes

Gold {110} film, Σ99 bicrystal

The {110} gold film was found to contain grains rotated with respect to each other by 90° about the film normal. They appeared in character somewhat similar to {110} aluminium films [7] with, for example, a facet located on (100)/(110) planes as shown at left in fig. 5. In contrast to the aluminium film, a region of interface was found that was not atomically sharp, but rather contained a slab of material that appears to be the 9R structure, fig. 5 right. The 9R structure has been observed also by Penisson and Medlin [8] and reported in copper by Ernst *et al.* [9]

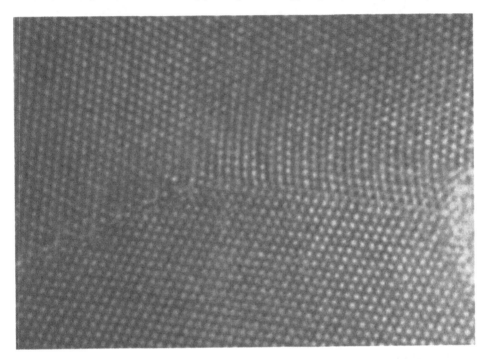

Fig. 5. HREM image of gold {110} film with boundary between grains containing 9R slab at right (JEOL 2010F)

Copper {111} film, Σ3 bicrystal

In the copper film observed in this work, the orientation was found to be <111> and HREM images show the 1.28Å {220} fringes. The thinner regions were assumed to be purely copper, and containing no intermediate layers. The copper film contained boundaries, indicating the existence of a Σ3 bicrystal. Work by Ernst *et al.* [9] on the copper Σ3 observed down [110] has shown that the interface may consist of a thin portion of 9R and the overall interface is broadened and inclined to the {112} plane. The interfaces in this work were found to be not precisely edge-on, but slightly inclined from the {112} planes. Interestingly, one portion of boundary was found that resembled the (112)/(121) facet junction in gold and is shown in fig. 6

Fig. 6. A copper bicrystal with boundary showing tendency to facet on {112} planes

CONCLUSIONS

Boundaries in $\Sigma 3$ gold bicrystals have now been observed on {112} and {110} facets as well as the standard {111} twin boundary. In a $\Sigma 7$ bicrystal, a symmetrical boundary was found on the {123} plane. In a {110} 90° bicrystal film of $\Sigma 99$ CSL, in which a symmetrical (557) twin boundary and an incoherent boundary lying on (100)/(110) have been previously reported, a region of boundary was observed to contain a thin slab of 9R structure. In a $\Sigma 3$ copper bicrystal, the orientations of the interfaces were not precisely edge-on.

ACKNOWLEDGMENTS

The author acknowledges contributions to this work from the following: Stefan Hinderberger and Ute Salzberger for preparing the thin films, Fritz Phillipp for assistance with the microscopy in Stuttgart and Roar Kilaas and Martin Hÿtch for help with image analysis software and Uli Dahmen and Doug Medlin for useful discussions. The work was supported by the Director, Office of Science, Office of Basic Energy Sciences, Materials Sciences Division of the U.S. Department of Energy under Contract No. DE-ACO3-76SF00098, and by the UK Engineering and Physical Sciences Research Council.

REFERENCES

1. U. Dahmen, J. Douin, C.J.D. Hetherington and K.H. Westmacott, Mat. Res. Soc. Symp. Proc. **139**, pp. 87-96 (1989).
2. K.H. Westmacott, S. Hinderberger and U. Dahmen, to be submitted (1999).
3. C.J.D. Hetherington, U. Dahmen and J-M. Penisson, Mat. Res. Soc. Symp. Proc. **466,** pp. 215-226 (1997)
4. F. Phillipp, R. Höschen, M. Osaki, G. Möbus and M. Rühle, Ultramicrosc. **56**, 1-10 (1994).
5. D.L. Medlin, private communication (1999).
6. M.J. Hÿtch, in Scanning Microscopy Supplement 10: <u>Signal and Image Processing in Microscopy and Microanalysis</u>, ed. P. Hawkes (Scanning Microscopy Int., Chicago, 1996).
7. U. Dahmen and K.H. Westamacott, Mat. Res. Soc. Symp. Proc. **229** pp. 167-172 (1991).
8. D Medlin, priv. comm. (1999) and J-M Penisson priv. comm. (1999).
9. F. Ernst, D. Hofmann, K. Nadarzinski, C. Schmidt, S. Stemmer and S.K. Streiffer, Mat. Sci. Forum Vols. 207-209, pp 23-34 (1996).

FORMATION OF AlN FILMS ON Ti/TiN ARC-LAYER INTERFACE WITH AL-0.5%Cu INTERCONECTS EVALUATED BY XPS AND ENERGY-FILTERED-TEM

J. GAZDA, J. ZHAO, P. SMITH, and R. A. WHITE
Advanced Micro Devices Corporation, Process Characterization Laboratory, 5204 E. Ben White Blvd., M/S 613, Austin TX 78741; jerzy.gazda@amd.com

ABSTRACT

Titanium/titanium nitride anti-reflective coatings (ARC) are widely used in the semiconductor industry during photolithography of aluminum metal interconnect lines. The quality and effectiveness of these coatings, however, depend strongly on the ability to control reaction products formed at film interfaces during processing. In the present study, formation of an Al-N compound at the interface between Ti/TiN ARC/BARC and Al-(0.5 wt. %)Cu interconnect was investigated. The effects of deposition temperatures for individual films and ensuing thermal cycling of the whole metal stack on formation of intermetallics were evaluated. The composition and chemical bonding state of an aluminum nitride interfacial layer was evaluated by x-ray photoelectron spectroscopy (XPS) of blanket wafers. These results were combined with measurements made by energy filtered transmission electron microscopy (EFTEM) of thickness and continuity of the film in specimens prepared by focused ion beam milling (FIB). Formation of AlN depends on the thermal cycling history of the metal stacks.

INTRODUCTION

The back-end interconnect lines in silicon integrated circuit technology are composed of stacked multilayer films. Typically, these films consist of two barriers that sandwich an Al-Cu alloy. The primary function of the barriers is to serve as a diffusion stop between Al and SiO_2. Barriers deposited on top of interconnect serve a second function, by providing anti-reflective coatings (ARC) for the subsequent photolithography process of the aluminum metal. Titanium and titanium nitride (TiN) films are widely used in the semiconductor industry because they can provide both of these functions. Primary reasons for their use are superior thermal stability and TiN's low reactivity with Al and Si $\geq 350°C$ [1-4]. However, the quality and effectiveness of these coatings depend strongly on the ability to control intermetallics formed at interfaces during processing [5]. The intermetallic phases play a major role in the final resistivity of interconnects as well as in their electromigration resistance. The most common phases formed during the fabrication of Ti-TiN/Al/Ti-TiN interconnects are AlN, $TiAl_3$ and Al_2Cu. The formation of intermetallics phases is diffusion limited and will continue until all available components are fully consumed [6]. These films are generally deposited in-situ in multi-chamber systems to insure controlled surfaces. In addition, careful control of deposition temperatures and subsequent thermal cycling reduces TiN/Al reactivity. In reality however, the ideal barrier does not exist - all films will react with a neighboring layer to some extent regardless of process parameters.

Our study evaluates Al-(0.5 wt. %)Cu interconnects with a Ti-TiN barrier and a Ti-TiN ARC metallization scheme. We will be focusing on the formation of AlN at the Al/ARC interface. In evaluation of this reaction product, we used EFTEM, FIB, and XPS.

Mat. Res. Soc. Symp. Proc. Vol. 589 © 2001 Materials Research Society

PROCEDURES

Thin films

The samples used in this study consisted of blanket and patterned films deposited on Si wafers. Blanket films were deposited on a thermal oxide (SiO_2) using a "short loop" operation procedure within the manufacturing environment. Patterned samples were pulled at predetermined operations during manufacturing and test structures were used for analysis. Both stacks, consisting of a Ti/TiN barrier (BARC), Al-(0.5 wt. %)Cu alloy, and Ti/TiN anti reflective coating (ARC), were deposited *in situ* using multi-chamber systems from Applied Materials Corp. using a Al-(0.5 wt. %)Cu target. During deposition, temperature was varied between 200°C to 400°C to induce different microstructure and grain size of Al film. To assess any post deposition effect, selected blanket samples were annealed at 400°C and 430°C.

XPS experimental procedures

Blanket samples were used for the XPS analysis. The only sample preparation step was the removal of ARC. This was done in order to minimize the ion sputtering induced chemical composition and chemical state variations. Fluorine based chemistry was used to selectively strip off the Ti/TiN ARC layer at the top of Al surface. XPS analysis was carried out in a Physical Electronic/Quantum-2000 X-ray photoelectron spectrometer using monochromatic Al-$K\alpha$ radiation (1486.6eV). Binding energies were calibrated by setting C1s at 285.0eV. The X-ray source was operated at 125W and 15kV. The analyzer is a hemispherical type with pass energy of 49.5eV for all the peaks. A PHI 11-066 ion gun was used to acquire an XPS depth profile using a beam voltage of 1kV, Ar^+ ion current of 1μA and 2x2mm raster size. The pressure of the test chamber was typically $8x10^{-9}$Torr during both data acquisition and ion sputtering. A photoelectron emission take-off angle of 45° was used. Zalar rotation was also used to improve the depth resolution. For quantification, XPS atomic sensitivity factors (ASF) relative to F1s=1.00 were used. For this work, the ASF applied are 0.499 for N1s, 0.733 for O1s, 0.256 for Al2p and 2.077 for Ti2p. The concentration of each element was acquired by dividing the area under XPS signal peak by the corresponding ASF.

TEM Sample Preparation

Blanket films were prepared for TEM evaluation using traditional metallographic grinding techniques, dimple polishing, and ion milling to perforation. Before TEM observation, specimen and sample holder were cleaned in a Fischione 1400 plasma cleaner for 15 minutes using 75%Ar-25%O_2 process gas.

The patterned specimens were prepared by focused ion beam (FIB) milling. The FEI Co. FIB-800THP workstation running *XP* software was used to prepare self-supporting specimens for these investigations. The 800THP is a single column instrument that uses Ga ions to remove material and excite secondary electrons used for image formation. Therefore, evaluation and milling steps were performed in succession. For the preparation, 30 pA - 15nA beams were used. Imaging was performed with 30pA beam to minimize Ga implantation and radiation damage to side and top surfaces of the specimens. The FIB preparation procedure reduces the preparation time from ~12 hrs required for a blanket film to ~4 hrs, including the grinding steps.

Energy Filtered TEM

Evaluation of the specimens was done with a Philips CM300FEG analytical transmission electron microscope equipped with a hot field emission electron source. Images were captured digitally with CCD cameras controlled by *DigitalMicrograph* running on a Macintosh computer.

The chemical analyses are facilitated by Noran Voyager X-ray energy dispersive spectrometer (EDS) system. Elemental distribution mapping and electron energy loss spectroscopy were performed using a post column Gatan imaging filter - GIF200.

Energy filtering increases image quality from thick foils and provides fast elemental distribution mapping capability. The zero-energy-loss electrons (ZLP) were used to acquire images of FIB prepared samples, typically <200nm thick. The energy slit width was set to 20 eV and the ZLP was centered in the slit upon each return to ZLP imaging conditions. Elemental maps were produced by an adaptation of the "three window method" [7]. For Cu maps, the L2,3 edge was used. Images were acquired with a 50eV slit centered at 958, 901, and 821 eV. For Ti maps, the L2,3 edge was mapped with a 30eV slit centered at 471, 436, and 471 eV. Finally, for nitrogen maps, images were acquired with a 20eV slit centered at 411, 385, and 353 eV. The elemental maps were then calculated correcting for sample drift. A power law equation was used to model energy loss background for the edges. Typically, the resulting elemental maps had a dynamic range of –500 to 1500 counts. The strength of the core-loss-edge signal depends on elemental concentration but is strongly attenuated by specimen thickness and diffraction contrast. To reduce plural scattering effects, specimen regions with a thickness of approximately 0.5 mean free path for the 300keV electrons were imaged. In FIB prepared sections, this low thickness was not attainable.

RESULTS and DISCUSSION

XPS data shows AlN formed at ARC interface

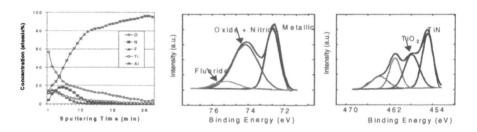

Figure 1   XPS results: (a) depth profile through blanket film with stripped ARC; and curve fitting of high resolution XPS spectra at the highest N concentration (b) Al2*p* peak and (c) Ti2*p* peak.

In order to minimize the sputtering ion induced chemical composition and chemical state variations, F based chemistry was used to selectively strip off the Ti/TiN ARC layer at the top of Al surface. The XPS depth profile was then carried out using low energy primary ions. Figure 1(a) shows the XPS depth profile of the as-stripped sample surface. With increasing sputtering time, Al concentration increased and O concentration decreased. It is also clearly shown that N concentration first increases and then decreases, reaching its maximum at about 2 minutes sputtering time. Small amounts of Ti residue and F (due to the strip chemistry) are also present at the sample surface. Figure 1(b) shows the curve fitting of the high resolution XPS spectrum of Al2*p* at the N peak position. The peak with binding energy close to 72.4eV was assigned as metallic Al. The peak at binding energy of 75.6eV was assigned to Al as a fluoride. The peak with binding energy of 74.4eV corresponds to Al in both Al_2O_3 and AlN chemical states.

Due to the binding energy overlap of Al chemical states from Al_2O_3 and AlN, the following data reduction procedure was carried out with the assumption that stoichiometric Al_2O_3 and TiO_2 are the only oxide forms present for Al and Ti. To evaluate the chemical form of aluminum compounds, a point was taken at the maximum of N concentration. The atomic concentrations there are 17% N, 25% O, 9% F, 10% Ti and 39% Al. From the curve fitting of Ti2p peak shown in Figure 1(c), it is found that half of the Ti is present as TiN and the half as TiO_2. Accordingly, 10% O is needed to form stoichiometric TiO_2. Assuming the remaining 15% O is in the chemical state of pure Al_2O_3, 10% Al is needed to form stoichiometric Al_2O_3. From curve fitting of the Al2p peak shown in Figure 1(b), 21% out of a total 39% Al is present as oxide and/or nitride. This leaves about 11% Al atoms that are chemically associated with N. In summary, there is aluminum nitride present at the as-striped sample surface.

EFTEM confirms the XPS results in blanket and patterned samples

Two primary effects were observed using EFTEM. An increase in the number density and size of the $TiAl_3$ intermetallics at the ARC interfaces with Al(0.5 wt. %)Cu resulted from longer cumulative annealing times. The second effect, was the formation of AlN film at the top and bottom interface of the Al(0.5Cu) layer in all the films regardless of the deposition and annealing temperature. An example of zero-loss electron images of blanket films is given in Figure 2 to illustrate annealing effects. The precipitates at the bottom and top of Al(0.5Cu) films were identified as $TiAl_3$ phase using selected area electron diffraction while their spatial distribution was revealed with Ti-$L_{2,3}$ EELS edge mapping.

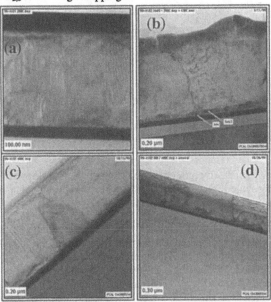

Figure 2   ZLP images of blanket wafer films showing increasing number density and size of $TiAl_3$ precipitates after post-deposition annealing. Films deposited at (a) 200°C, (b) 200°C and annealed at 430°C, (c) 400°C, and (d) 400°C and annealed at 400°C.

Figure 3 shows an example of the Ti $L_{2,3}$ map. The interesting feature, which led us to further investigation and final conclusion that an AlN film was present there, is the lack of Ti at the interface between Al(0.5Cu) and BARC.

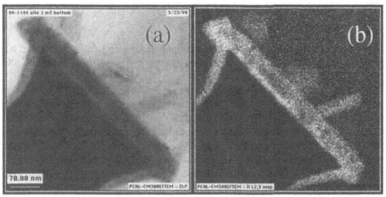

Figure 3 (a) ZLP image of bottom barrier-metal layer interface and (b) accompanying core-energy-loss-map for Ti $L_{2,3}$ EELS edge. Precipitates contain a high concentration of Ti. The interfacial layer without Ti is also visible as a break between TiN and TiAl$_3$.

An image of the ARC-Al interface built by overlapping N and Ti maps (N-K edge and Ti-$L_{2,3}$ edge) is shown in Figure 4(a). To construct the map, a ZLP image of the area was used as the background to help facilitate feature placement and reduce errors due to specimen drift. In this image, the AlN film is the continuous dark band at top of Al-Cu alloy. The N-edge signal also shows that the ARC, deposited as Ti and TiN films, transforms to a TiN film during post deposition processing. A typical interface after annealing of a blanket wafer is shown in Figure 4(b). The phase-contrast imaging shows the epitaxy of the interface.

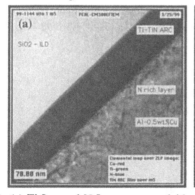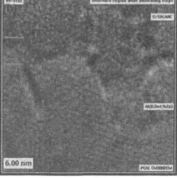

Figure 4 (a) Ti $L_{2,3}$ and N $L_{2,3}$ maps overlaid on ZLP image in pseudo-color image indicating presence of AlN film at top ARC-Al(0.5Cu) interface. (b) High resolution image of the interface showing epitaxial nature of the interface.

Formation of AlN at the interface of TiN with the Al alloy is favored from a thermodynamic point of view, although this effect has not been previously observed in VLSI interconnects. The reduction of TiN by Al can be described by the following two reactions:

$$1\,TiN + 4\,Al \rightarrow 1\,AlN + 1\,TiAl_3 \tag{1}$$

$$TiN + Al \rightarrow Ti + AlN \tag{2}$$

The excess Gibbs free energies of transformation are -21 Kcal/mol and -4.5 Kcal/mol, for reactions (1) and (2) respectively [6]. The formation of $TiAl_3$ is diffusion limited and continues until either Ti or Al is consumed [4].

The data collected in this study shows effectiveness of EF-TEM in evaluations of semiconductor devices. The imaging of AlN films would not be possible without this tool. The difficulty arises from lack of diffraction contrast, minimal mass contrast difference, and the epitaxy of the AlN film with both sides of the interface preventing high-resolution phase-contrast imaging. Because of the well known Ti $L_{2,3}$ – N K peak overlap and the small thickness of the AlN layer, EDS measurements also fail to provide unambiguous conclusions.

The largest source of error in the GIF mapping experiments results from specimen shift. Although automatic software algorithms can correct for misalignment while computing elemental maps, the pseudo-color images need to be aligned manually. In the present study, ZLP images of films were used as guides when composing the overlays. The individual maps were shifted pixel-by-pixel to align known landmarks.

CONCLUSIONS

1. Formation of AlN occurs at the interface between Ti/TiN ARC layers and Al(0.5 wt.% Cu) films that form typical interconnects in semiconductor devices. The presence of these films was confirmed by XPS and by EF-TEM.

2. The AlN films were formed during temperature excursions inherent to fabrication process of ULSI devices or intentionally induced during annealing of blanket wafers.

3. EFTEM allows direct evaluation of processing steps, close monitoring of product quality, and ultimately a reduction in cost of fabrication of fast ULSI circuits allowing transfer of these savings to customers.

REFERENCES

1. M. Wittmer, J. Vac. Sci. Technol. A **2** (1984) 273

2. A. Armigiliato and G. Valdre, J. Appl. Phys. **61** (1986) 390

3. C. Y. Ting, B. L. Crowder, J. Electrochem. Soc., **129** (1982) 2590

4. R. W. Bower, Appl. Phys. Lett., **23** (1973) 99

5. J. Sanchez, MRS Symp. Proc. **265** (1992) 131

6. K-H. Bather, H. Schreiber, Solid Films, **200** (1991) 93

7. R. F. Egerton, *Electron Energy-Loss Spectroscopy in the Electron Microscope*, 2nd ed., Plenum Press, New York, 1996

EFFECTS OF 'AS DEPOSITED' AND ALLOYING TEMPERATURES ON THE DISTRIBUTION OF Cu IN 0.5% Cu-Al FILMS.

P. L. SMITH, J. GAZDA
Advanced Micro Devices Corp., Process Characterization Analysis Laboratory, 5204 E. Ben White Blvd., Austin, TX 78741

ABSTRACT

The Al-0.5wt%Cu alloys in Ti-TiN/Al/Ti-TiN metal interconnects were investigated using blanket and patterned films deposited on SiO_2. Distribution of Cu and Ti within the Al films was determined using Resistivity (Rs), X-ray Diffraction (XRD), Rutherford Backscatter Spectroscopy (RBS), and Secondary Ion Mass Spectroscopy (SIMS). The location of Cu, Ti, Al and their intermetallics were studied in a CM300 field emission gun (FEG) TEM using Energy Filtered Transmission Electron Microscopy (EFTEM) with a Gatan Imaging Filter (GIF). No significant differences were observed with Rs or XRD textural analysis. Elemental distribution analysis from RBS and SIMS revealed relatively higher concentrations Cu at the Al/Ti-TiN barrier interface. This was attributed to sputtering bias during the start of the Al deposition and the subsequent snow plowing of Cu during the growth of the $TiAl_3$ intermetallic. TEM imaging and elemental analysis showed that Cu is completely dissolved into solution at deposition temperatures $\geq 300°C$. At temperatures $\leq 250°C$ some (Al_2Cu) precipitates were observed on grain boundaries. Continued thermal cycling does not effectively change the elemental distribution. Therefore, it can be concluded that the solid solubility of 0.5wt%Cu in Al was not reached at these low temperatures or during the subsequent thermal cycling.

INTRODUCTION

The reliability of Al (Cu) alloy interconnects in integrated circuit (IC) devices are of continued interest for all users. The presence of Cu, even at 0.25 at. %, has been found to be a major force in inhibiting the formation of highly resistive $TiAl_3$ intermetallics and in lowering the overall resistivity of interconnects [1,2]. Alloying Al with Cu also improves the reliability of thin metal lines by the reduction of electromigration [3]. It does this through formation of Θ-phase Al_2Cu precipitates, that block dislocation motion [4]. On the other hand, Al_2Cu precipitates are difficult to etch and can form micromasks and electrical shorts. The formation or dissolution of this phase is directly related to the thermal cycling during chip processing. However in most cases, manufacturing requirements, not microstructural considerations, drive the thermal cycling. Typically, Al(Cu) interlevel connections are built by sequentially depositing different films (AlCu, Ti, TiN, W) at elevated temperatures, patterning, and then encapsulating with SiO2. In these films, repeated annealing and quenching produces textured Al grain growth with Θ-phase (Al_2Cu) and intermetallic $TiAl_3$. The phase formation sequence and microstructure have been periodically observed and follow some general principles [5-7]. In as–deposited films the Θ precipitates are uniformly distributed in the grains and boundaries. When annealed below the solubility curve, the Θ precipitates coarsen at the grain boundaries at the expense of those at the grain interior. On the other hand, after annealing above the solvus temperature, almost all the Θ precipitates are found in the grain boundaries and triple points. These precipitates are coarser and fewer in number. Because of the need to scale down interconnect technologies today and future, the impact of Al(Cu) alloys, their residuals, and the associated intermetallics can only increase. This impact has generated a need to reduce the amount of Cu in the Al(Cu) alloys [8].

It is clear from the above that the improvement in performance of any interconnect scheme depends on understanding the Al/Cu/Ti interactions associated with the complex processes in today's fabs. In this work we characterize the Al 0.5wt%Cu full stack as a function of deposition and post deposition (alloying) temperatures. We focus on the distribution, location and phase of Cu as well as on the amount of TiAl$_3$ formed during processing.

EXPERIMENTAL

Process and Sample preparation

The experimental conditions of the present work were selected with a two-fold purpose: first to characterize our deposition systems and second to document the distribution and phases of Ti and Cu in the Aluminum films. The samples consisted of patterned and blanket depositions. Patterned samples were pulled at predetermined operations during manufacturing and test structures were used in analysis. Blanket films were deposited on thermal oxide. Both stacks consisting of a Ti/TiN barrier, Aluminum (0.5wt%Cu) and Ti/TiN anti reflective coating deposited *in-situ* using Applied Materials multi-chamber systems with a 0.5wt%Cu Al target. During deposition the temperatures were varied between 200°C to 400°C to induce different microstructure and grain size. To assess post deposition effects, select blanket samples were annealed at 400°C and 430°C in a two step alloying process. Blanket films were prepared for TEM using traditional metallography techniques. Patterned specimens were prepared using FEI Co. FIB-800THP-xP focused ion beam (FIB) milling.

Analytical Instrumentation

The principal characterization techniques used in this assessment were; Resistivity (Rs) which was preformed inline using a 4 point probe method. Rutherford Backscatter Spectroscopy (RBS). The RBS beam energy was 2.61 MeV (He$^+$ ions) with the detector set at 165 degrees to the beam. Samples were tilted 30° away from the detector. Secondary Ion Mass Spectroscopy (SIMS). A CAMECA ims6f double focussing magnetic sector mass spectrometer was configured so that the sputter ion beam column was at an angle of 30 degrees from normal to the sample. X-Ray Diffraction (XRD). A Bede D3 High-Resolution X-ray diffractometer with a rotating anode source operated at 50 Kilovolts and 250 mili-amperes was used to conduct nine-site analysis on unprepared full wafers. A Philips CM300FEG Transmission Electron Microscopy (TEM) equipped with SuperTwin™ objective lens and a hot field emission electron source. The chemical analyses were facilitated by a Noran Voyager X-ray energy dispersive spectrometer (EDS) system. Elemental distribution mapping and electron energy loss spectroscopy was performed using a post column Gatan imaging filter - GIF200. Energy filtering increased image quality from thick foils and provided fast elemental distribution mapping capability. Zero-energy-loss electrons (ZLP) were used to acquire images of FIB prepared samples, typically <200nm thick. The energy slit width was set to 20 eV and the beam was centered in the slit upon return to ZLP imaging conditions.

RESULTS AND DISCUSSION

XRD/Rs

The formation of intermetallics as a function of annealing time or temperature can be monitored by changes in the electrical resistivity of the films [10]. Because all transition barrier metals in this Al(Cu) interconnect scheme had the same deposition processes; any change in resistivity was attributed to the formation of Al$_2$Cu precipitates. Thus, Rs was treated as a reaction-rate parameter for the growth of Al₂Cu. The Rs data showed no significant change despite changes in the as deposition temperature. This fact indicated a uniform formation rate of Al$_2$Cu precipitation. Also, when the samples were subsequently alloyed, there was a parallel increase in Rs. This increase in Rs is related to the growth of TiAl$_3$, as it was documented with TEM. The grain size, texture, and the inline nine site Rs for the as-deposited and alloyed cases are summarized in Table 1A and B.

Table 1 – (A) are samples with different As deposited temperatures. (B) are As deposited with two step alloy at 430°C. Rs was taken at pre-alloy/,1st/ 2nd Alloy. Grain size was calculated from full width at half. Texture derived from rocking curve procedures.

A   **B**

As dep temp °C	Grain size Å	Texture <111>	Inline Rs Ω/sq	Two step Alloy °C	Grain size Å	Texture <111>	Inline Rs Ω/sq
200	3516	Medium	64.3	200	5623	Strong	64.7/69.3/74
350	4037	Strong	64.9	350	5623	Strong	64.6/68.6/73
350	5332	Strong	64.0	350	5623	Strong	64.6/69.5/74
400	4211	Strong	65.6	400	4944	Strong	63.7/70.1/75

*At the limit of maximum grain size which could accurately be resolved

RBS

The RBS Cu profiling was performed on blanket samples that were deposited at 200, 350 and 400C. A second set of samples from the same deposition series were analyzed after a two step (400/430C) alloying process. Prior to analysis a fluoride based chemistry was used to selectively strip off the Ti-TiN ARC. Data from these analysis provide information on the nature of the metallurgical interactions. Spectra (not shown) revealed increasingly elevated Cu concentrations with decreasing deposition temperature at the Cu-Al/Ti-TiN barrier. The Cu distribution in the 400C deposition was more uniform with much smaller peak at the interface. Upon alloying all the samples had similar profiles with uniform Cu concentration in the bulk and a small concentration bump at the interface. Some increase in Cu concentration at the interface was expected regardless of deposition temperature and is attributed to sputtering bias during the start of the deposition. While there is no direct evidence of Cu precipitation (Al$_2$Cu) it seem apparent that at lower deposition temperatures (< 200C) some is taking place. Also that subsequent alloying increases the solid solubility of Cu in the lower deposition temperature samples and in the end all Cu profiles look very similar regardless deposition temperature.

SIMS

The RBS samples were used by SIMS to confirm the as-deposited data. Figure 1 shows the overlaid spectra of the four deposition temperatures. The relative peak heights are indicative of Cu concentration at Al/Ti-TiN interface. This analysis confirmed the RBS results that at the lower deposition temperatures there is a small relative increase in Cu concentration at the Cu-Al /Ti-TiN interface.

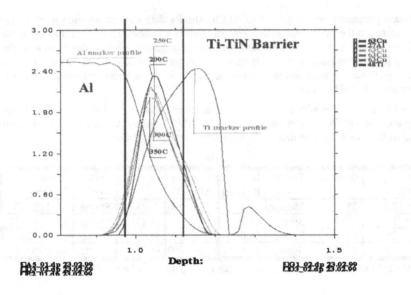

Figure 1   SIMS profile of Al/Ti-TiN stack with relative Cu concentration at interface

TEM

In order to establish an association between microstructure, resistivity data, and elemental depth profiling results we conducted cross-sectional TEM studies of the aluminum film and Al/Ti-TiN interface. The focus was on phase association and spatial distribution of Cu. The resulting images suffered reduction in intensity, but gained in sharpness of image features due to application of the GIF to remove inelastically scattered electron effects. Elemental maps were produced by an adaptation of the "three window method" [9].

Through these investigations we can make some general observations. The Aluminum had columnar grains with a grain size of about the thickness of the film, which increased with increasing deposition temperature. Thermal cycling increased the density of intermetallic formation at the Al/Ti-TiN interface. At depositions ≥250°C and regardless of post deposition thermal cycling, no Cu precipitation was seen in the grains or at the grain boundaries. However, at depositions ≤ 250°C some were observed in the grain boundaries.

Figure 2 shows a TEM bright field image of a representative patterned metal line deposited at 350°C and processed through several thermal cycles. Note grain size, lack of precipitation, and intermetallics at the interface. Figure 3 A is a bright field image of the barrier Al surface after ≥250°C deposition and alloying. The Ti elemental distribution is revealed in Figure 3B through EFTEM. A color composite-energy-loss-map (not shown) of the distributions of Ti, Cu and N revealed that Cu was being pushed away from the interface by the growth of $TiAl_3$. EDS measurements (not provided here) from these areas show an increase in

Cu concentration around the TiAl$_3$ grains. Literature values for TiAl$_3$ formation energy range from 1.6 to 1.24 eV, as 0 to 2% Cu is added to Al [10,11]. While literature values for CuAl$_2$ range from 1.2 to 1.59 eV [12]. Given these values and the apparent lack of Cu precipitation, it seems that TiAl$_3$ is a favored reaction.

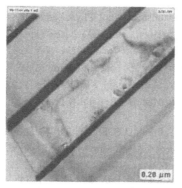

Figure 2   Cross-sectional TEM bright field image of a metal line thermally cycled several times.

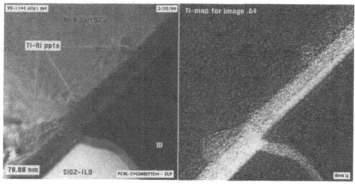

Figures 3   Cross-section of a thermally cycled metal line (A) in bright field TEM; (B), Ti EFTEM map.

CONCLUSIONS

In this work we characterized the Al-0.5wt%Cu films as a function of deposition and post deposition (alloying) temperatures. We focused on the distribution, location and phase of Cu, as well as on the amount of TiAl$_3$ formed during processing. In our metallization scheme at deposition temperatures of ≥250°C Cu does not precipitate out. Subsequent thermal cycling does not cause precipitation either. The reaction kinetics within this Cu concentration and temperature range favors the formation of TiAl$_2$. The density of these intermetallics increases with further thermal cycling. The TiAl$_3$ displaces Cu during its growth at the Al/TiTiN interface.

ACKNOWLEDGEMENT
The author, (P.L.S.), would like to express thanks to Clive Jones, Bill Brennan, Dr. Terry Stark, and Dr. Tim Hossain, for their help and encouragement.

REFERENCES
1. J. Tardy, K. N. Tu, APS, Physical Rev. B, **32** (1985), p. 2070
2. J. Krafcsik, J. Gyulai, C.J. Palmsrom, Appl. Phys. Lett. **43** (1983), p.1015 1
3. I. Ames, F.M. d'Heurle, R.E. Hoertmann, IBM J. Res. Dev. **14**,(1970) p.461
4. J. Sanchez, E. Arzt, MRS Symp. Proc. **265** (1992)
5. R. Venkatraman, J.C. Bravman, W.D. Nix, J. Electron Mater. **19**, (1990) p.1231
6. D. R. Frear, J.E. Sanchez, Jr., A.D. Romig, Metall. Trans. A **21** (1990) p.2445
7. J. E. Sanchez, Jr., E. Arzt, In Materials Reliability Issues in Microelectronics II, Vol. 265, p. 131 (1992)
8. Ramnath Venkatraman, J. C. Flinn, D.B. Fraser, J. of Electronic Mat. **19** p. 1231, 1990.
9. Egerton R. F. "Electron Energy-Loss Spectroscopy in Electron Microscope, 2nd ed., Plenum Press, New York (1996)
10. H. C. Huang, M. Witter, MRS, Symp. Proc. **25** p. 157 (1984)
11. R.W. Bower, Appl. Phys. Lett. **23**, p99 (1973)
12. R.A. Hamm, J. M. Vandenburg, MRS, Symp. Proc. **25** p.163 (1984)

INTERFACIAL INTERACTION BETWEEN CR THIN FILMS AND OXIDE GLASSES

N. JIANG, J. SILCOX
School of Applied and Engineering Physics and Cornell Center for Materials Research, Cornell University, Ithaca, NY 14853, njiang@ccmr.cornell.edu

ABSTRACT

The interfacial interaction between Cr thin films and multi-component oxide glasses has been observed by means of high spatial resolution electron energy loss spectroscopy. Besides a partially oxidized Cr thin layer, a ~5nm wide Cr diffusion layer is seen. Chromium oxidation at the interface results from the difference between the heats of oxide formation. Ion exchange between Cr^{2+} and alkaline earth ions then causes the formation of the diffusion layer. The electronic states of the Cr in this diffusion layer are different from that in the oxidized layer. Strong interaction between Cr and O in the diffusion layer suggests that such a layer could be responsible for the formation of a strong Cr/glass interface.

INTRODUCTION

Knowledge of interface properties is important in understanding the mechanisms of adhesion of metal films to glasses, which have been receiving wide industrial uses, e.g. in flat panel displays. Generally, Cr provides good adhesion to most oxides due to the formation of the oxide layers [1]. Serious efforts at determination of the reaction layer have been made [2]. However, none of these approaches have adequate spatial resolution to develop insights into the fundamental processes. Recently, high spatial resolution electron energy loss spectroscopy (SREELS) has been improved in scanning transmission electron microscopy (STEM) to probe electronic states and composition on a near-atomic spatial scale [3]. These advances provide an opportunity to re-visit the interfacial layer issue.

The work reported here provides a remarkably detailed description of Cr/glass interface by using SREELS. The distribution of both major and minor constituents at the interface has been found. Besides very thin oxidized Cr layers, a Cr diffusion layer is also seen. The formation of the diffusion layer is associated with the segregation of alkaline earth components, and will be discussed in detail.

EXPERIMENT

Thin Cr films were evaporated in high vacuum (10^{-6} ~ 10^{-7} Torr) on an alkaline earth boro-alumino-silicate glass (Corning Code 1737) at room temperature. Glass was polished by Corning Inc., and the surfaces were cleaned with a sequence of soapy water, acetone and methanol. Immediately before deposition, oxygen plasma cleaning was performed. The composition of glass is (in units of mole percent): SiO_2-B_2O_3(7.3)-Al_2O_3(11.5)-MgO(1.4)-CaO(5.0)-SrO(1.2)-BaO(4.4) [4]. Cross-sectional transmission electron microscope samples were prepared by the tripod polishing method. Low energy (~ 3kV) and low angle (~ 5°) ion milling was carried out as the final stage to clean the surfaces (less than 5min. on each side).

The specimens were observed in the Cornell VG HB501 STEM equipped with an Annular Dark Field (ADF) single electron sensitivity detector and a Parallel Electron Energy Loss Spectrometer (PEELS). The minimum attainable probe size is ≈ 0.22nm in diameter, the energy resolution is 0.7eV, and the spatial and energy drifts in this instrument are less than 0.3nm/min. and 0.03eV/min., respectively. A collector aperture with the semi-angle of 15 mrad was used in the experiments, which can exclude very high angle scattered electrons due to high angle Rutherford elastic scattering. In this study, a broadening of less than 1nm probe is expected.

A very important factor we must consider is irradiation damage of the glass. It is found that the glass can be decomposed spinodally under electron irradiation [5]. So very short acquiring times are necessary.

RESULTS AND DISCUSSION

The O K, Si L and Cr L core-edge EELS spectra are used to obtain the relative compositions of these elements across the interface. The background intensities preceding the edge have been fitted with an exponential form, $A \cdot \exp(r \cdot E)$, in which A and r are fitted parameters and E is the energy, which has been subtracted from the observed spectrum. As shown in Figure 1(a), the shaded region corresponds to a diffusion layer, in which the major composition gradually changes across the interface. It should be noted that the Si data indicated by arrows in Figure 1(a) might be overestimated. This arises from the overlap of the Si L-edge (100eV) with the Ba N_{45}-edge (93eV).

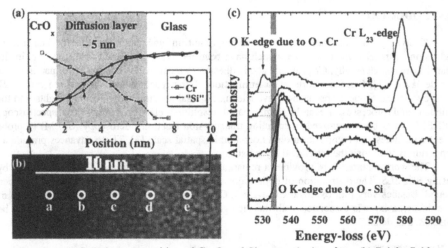

Figure 1. (a) Relative composition of Cr, O and Si across the interface. (b) Bright field image showing the Cr/glass interface. No sharp interface is seen. The white line shows the corresponding positions in (a). (c) EELS spectra showing the changes of O K and Cr L_{23}-edges across the interface. The alphabet, a, b, c, d, and e refer to the acquiring locations, marked as circles in (b).

Such a diffusion layer is also seen in the bright field image in Figure 1(b). The dark area on the left is from the high density of the Cr film, the light area on the right is from the low density of the oxide glass, while the darker than the bulk glass area in between reflects the presence of Cr in the glass.

The electronic structures across the interface were measured by the EELS spectra of the O K- and Cr L_{23}-edges, which provide information on the unoccupied O-p, and Cr-d electronic densities of states (DOS), respectively. The results are shown in Figure 1(c): spectrum a is recorded in the Cr film close to the interface, spectra b and c are from the diffusion layer, d is near the diffusion layer while e can be considered as the bulk glass.

For the Cr L_{23}-edge, not only does the intensity change across the interface, but the threshold also shifts to higher energy by about 0.6eV in the diffusion layer relative to in the oxidized Cr layer. The positions of L-edges in transition metals are often related to their oxidation states: the higher the threshold energy of L-edge, the higher the oxidation state [6]. This suggests that the strength of the Cr – O bond is stronger in the diffusion layer.

In multi-component oxide glasses, the O K-edge should be complicated. However, the measured O K energy loss near edge fine structure (ELNES) in the bulk glass shows only simple features, which are very similar to the O K ELNES in amorphous SiO_2. Using ELNES techniques, the local structure of Si, B and Al in this glass have been determined: Si tetrahedral SiO_4, as well as B trigonal BO_3, constitute the glass network, and Al tetrahedral $[AlO_4]^{1-}$ forms the network intermediate [5]. To compensate the excess negative charge of the Al tetrahedral, alkaline earth ion must bound with two $[AlO_4]^{1-}$ units. Since Si (also Al and B) has a weaker pseudopotential than O, the O K ELNES could be identified as originating from O – O scattering [7]. In other words, the first sharp peak results from scattering the excited electron by the second shell of atoms (most of them are oxygen atoms). So we can use the O K ELNES to monitor the glass network.

At the position c in the middle of the diffusion layer, a small bump appears at ~4eV before the major O K-edge, accompanying the presence of Cr L-edge. Meanwhile, a small increase of intensity in the shaded area in Figure 1(c) is also seen. As the probe moves to the interface (position b), the O K ELNES is very different from that in the bulk glass, although it remains a little characteristic of SiO_2. Both the first major peak (537eV) and the second bump (562eV) become wider and shorter. Such a feature has been found in a suboxide-SiO_x [8]. In addition, the peak characteristic appears at ~530eV, and the intensity in the shaded area increases significantly. On the Cr thin film side (position a), the Cr is partially oxidized near to the interface. The O K ELNES is characteristic of Cr oxide [9]. Due to unfilled d states in Cr, transitions to O 2p – Cr 3d hybridized states give the first 5eV of the O K ELNES. So we assert that the peak at about 530eV has Cr–O character, in distinction from the Si-O characterized major peak. The Cr–O peak increases as the Cr content increases.

Based on the crystalline structure of Cr_2SiO_4, the density functional calculations within the local density approximation (LDA) show that the 2p-projected local states on the O that connects the Si tetrahedron and distorted Cr octahedron also probe the shaded area [10]. This suggests that the increase in intensity in the shaded region (in Figure 1(c)) may reflect an increase of Si – O – Cr bonding characters.

However, the Cr – O and Si – O interactions cannot be the only contributions to the O K ELNES at the interface. The evidence shows the segregation of alkaline earth components to the interface region. In Figure 2, increases in the Ca L_{23}-edge and Ba M_{45}-edge intensities in the interface region are seen. In bulk glass, however, it is difficult to detect these elements by EELS in very short acquisition time due to the low concentration. Such an accumulation of alkaline earth elements at the interface could happen either before or after Cr deposition. There is no evidence, however, of the extra Ca and Ba at the interface of a-SiO_2 and alkaline earth boro-alumino-silicate glass. The same results are also expected for Mg and Sr. However; no efforts have been made to detect this effect due to the very low elemental content.

The segregation of alkaline earth ions at the interface may increase their contribution to the O K ELNES. We have recognized the O K ELNES in the alkaline earth rich region in the electron irradiation damaged glass [5]. As shown in Figure 3, the changes in the O K edge induced by the segregation of alkaline earth components is a little different from that induced by the Cr diffusion. The major difference is the change in the relative intensities between two major peaks. In addition, the sub-thresholds of the pre-edge have less then 1eV shift.

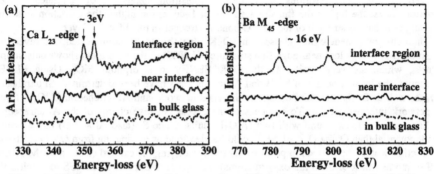

Figure 2. EELS spectra showing the segregation of Ca in (a) and B in (b) at the Cr/glass interface. "Interface region" means the location b in Figure 1(b), while "near interface" refers to the location d in Figure 1(b).

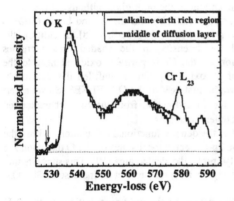

Figure 3. Comparison of O K edges in the middle of diffusion layer with that in the alkaline earth rich region induced by electron irradiation. The former is recorded in position c in Figure 1(b), and the latter is acquired in the bulk glass after irradiation damage [5]. Both spectra are normalized to the unit intensity under O K peaks. The dot horizon line is drawn to guide of the eye.

The driving force for Cr oxidation at the interface or Cr diffusion into glass is likely to be the difference between the heats of oxide formation ΔH^f. For example, ΔH^f for Cr_3O_4 (containing both Cr^{2+} and Cr^{3+}) is ~-1500kJ/mole, for SiO_2 is ~ -900kJ/mole, while ΔH^f for alkaline earth metal oxides are around -500kJ/mole [11]. (The negative represents the exothermic). So Cr oxidizes more strongly than either the alkaline earth metals or Si. It is noted that Cr has different valence states and the oxygen affinities decrease with the increasing valence [12]. This suggests that the initial oxidized Cr layer may consist of the lowest oxidized state Cr^{2+}. Comparing the Cr L edge thresholds between Cr film and this initial layer shows that the shift is much smaller than ~2eV, which is reported shift value between Cr and Cr^{3+} [13].

The formation of the diffusion layer may be due to the inter-diffusion between Cr^{2+} and alkaline earth ions in the bulk glass. A sketch of the inter-diffusion process is shown in Figure 4. Driven by the differences in the heats of oxide formation, the Cr^{2+} diffuses into glass, substitutes for alkaline earth ions in the glass, and forces them to segregate to the interface. In addition, the relatively high mobility of alkaline earth ions in glass networks can enhance such inter-diffusion [14].

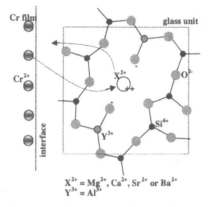

$X^{2+} = Mg^{2+}, Ca^{2+}, Sr^{2+}$ or Ba^{2+}
$Y^{3+} = Al^{3+}$

Figure 4. Schematic drawing showing the inter-diffusion Cr^{2+} and alkaline earth ions. The glass unit is based on the results that the SiO_4 and BO_3 are random network formers, $[AlO_4]^{1-}$ are network intermediate, and alkaline earth components are modifiers. The excess negative charge of $[AlO_4]^{1-}$ unit is compensated by alkaline earth ion.

The accumulation of alkaline earth ions at the interface may form a barrier layer for Cr further diffusion. Based on the "macroscopic atomic picture" model [15], the formation of any Cr-alkaline earth metal alloy is energetically unfavorable. The calculated heats of formation are all positive (e.g. ΔH_{Cr2Ba3} = +54 kJ/mole, where the + sign denotes an endothermic reaction) [16]. In addition, a positive charge layer also prevents Cr ions from diffusing into the glass substrates. So the Cr diffusion is transitory and should cease.

CONCLUSIONS

At the near-atomic resolution scale, the interaction at the Cr/oxide glass interface has been observed by EELS. Partially oxidized layers are seen, which is due to the high oxygen affinity of Cr. Cr^{2+} ions in the partially oxide layer can interact with alkaline earth ions in bulk glasses. As a result, Cr ions diffuse into glasses while alkaline earth ion segregate to the interfaces. A ~5nm wide Cr diffusion layer and accumulation of Ba and Ca at the interface are seen. At high enough concentrations, segregated alkaline earth metals serve as a barrier for further Cr diffusion. The electronic states of the Cr in this diffusion layer are different from that in the oxidized layer. Although there is no direct evidence to connect the diffusion layer with adhesion, it is quite plausible that the medium wide diffusion layer can provide better adhesion than those abrupt interfaces.

ACKNOWLEDGMENTS

This work is supported by NSF (Grant # DMR9632275) through the Cornell Center for Materials Research (CCMR), which also supports operation and maintenance of the STEM. We thank Dr. F. Fehlner of Corning Inc. for supply glass substrates, and J. B. Neaton and Professor N. W. Ashcroft of Cornell University for calculations. Acknowledgements are also due to M. Thomas and Dr. E. Kirkland for technical support. One of the authors (N. J.) would like to thank Professor G. H. Frischat for helpful discussion about Cr diffusion in glass.

REFERENCES

1. P. Benjamin and C. Weaver, Proc. Roy. Soc. London A **261**, 516 (1961).

2. E. C. Onyiriuka, L. D. Kinney and N. J. Minkowski, J. Adhesion Sci. Technol. **11**, 929 (1997).

3. P. E. Batson, Nature (London) **366**, 727 (1993); N. D. Browning, M.F. Chisholm, and S. Pennycook, *ibid*. **366**, 143 (1993); D. A. Muller, Y. Tzou, R. Raj, and J. Silcox, *ibid*. **366**, 725 (1993).

4. L. Tian and R. Dieckmann, J. Non-Cryst., in press.

5. N. Jiang and J. Silcox, in preparation.

6. J. Taftø and O. L. Krivanek, Phys. Rev. Lett. **48**, 560 (1982).

7. D. J. Wallis, P. H. Gaskell and R. Brydson, J. Microsc. **180**, 307 (1995).

8. D. A. Muller, T. Sorsch, S. Moccio, F. H. Baumann, K. Evans-Lutterodt and G. Timp, Nature (London) **399**, 758 (1999).

9. L. A. Grunes, R. D. Leapman, C. N. Wilker, R. Hoffmann and A. B. Kunz, Phys. Rev. B **25**, 7157 (1982).

10. J. B. Neaton and N. W. Ashcroft, private communication.

11. CODATA Key Values for Thermodynamics, *in CRC Handbook of Chemistry and Physics*, 79th, 5-1 (1998-1999).

12. T. B. Reed, *Free Energy of Formation of Binary Compounds*, (MIT press, Cambridge, 1971).

13. R. D. Leapman, L. A. Grunes and P. L. Fejes, Phys. Rev. B **26**, 614 (1982).

14. G. H. Frischat, *Ionic Diffusion in Oxide Glasses*, Diffusion Monograph Series No3/4, edited by Y. Adda, et al., (Trans. Tech. Publications, 1975).

15. A. R. Miedema, R. Boom, and F. R. de Boer, J Less-common metals **41**, 283 (1975).

16. F. R. de Boer, R. Boom, W. C. M. Mattens, A. R. Miedema and A. K. Niessen, *Cohesion in Metals: Transition Metal Alloys*, (North-Holland, Amsterdam, 1988).

CHARACTERIZATION OF INTERGRANULAR PHASES
IN DOPED ZIRCONIA POLYCRYSTALS

N. D. EVANS,*† P. H. IMAMURA,‡ J. BENTLEY† AND M. L. MECARTNEY‡

* Oak Ridge Institute for Science and Education, Oak Ridge, TN 37831-0117
† Metals and Ceramics Division, Oak Ridge National Laboratory, Oak Ridge, TN 37831-6376
‡ University of California, Irvine, CA 92697

ABSTRACT

Analytical electron microscopy at high spatial resolution in a scanning-transmission mode has been used to investigate the effects of glassy or crystalline material additions on grain boundary chemistry in yttria-stabilized zirconia polycrystals. Powders of additive phase were mixed into 3-mol% yttria-stabilized tetragonal zirconia polycrystals ('3Y-TZP') or 8-mol% yttria-stabilized cubic zirconia polycrystals ('8Y-CSZ'). Zirconias processed without additive phases were also examined.

Without additives, grain boundaries were depleted in zirconium and enriched in yttrium. In 3Y-TZP with 1 wt% borosilicate glass, silicon was observed only at triple points, but not in grain boundaries. In 3Y-TZP with 1 wt% barium silicate glass, barium was observed both along grain boundaries and at triple points, whereas silicon was detected only within the triple points. This suggests either the composition of the additive phase at the grain boundary is different from that at the triple points, or that barium ions segregate to grain boundaries during processing. In 8Y-CSZ with 1 wt% silica, silicon was observed in grain boundaries by an EDS spatial differencing technique. In 8Y-CSZ with 10 wt% alumina, EDS revealed aluminum at all grain boundaries examined.

INTRODUCTION

Superplastic effects in fine-grained ceramics, which offer potential to lower manufacturing costs via net shape forming, are dependent upon grain size, temperature, and stress. Two methods used to achieve superplastic deformation behavior in ceramics are either maintaining a submicron grain size during deformation, or adding an intergranular phase to limit grain growth and promote grain boundary sliding during sintering and isostatic pressing. Both glassy and crystalline additives which form such grain boundary phases have been successfully used to improve superplasticity in yttria-stabilized zirconia [1-4].

As part of a larger study, the scope of the work described here was to characterize the distributions, composition, and extent of grain boundary phases in 3Y-TZP and 8Y-CSZ, processed with and without grain boundary additives, in order to correlate these findings with the mechanical properties (described elsewhere).

EXPERIMENT

Commercially available (Tosoh Co., Ltd.) 8Y-CSZ and 3Y-TZP powders were mixed with additives to produce 8Y-CSZ with 1 wt% or 5 wt% silica, 8Y-CSZ with 10 wt% alumina, 3Y-TZP with 1 wt% borosilicate glass (designated 3YBS), and 3Y-TZP with 1 wt% barium silicate glass (designated 3YBaS). The composition of the borosilicate glass (mol%) was 83.3

383

SiO$_2$ 1.5 Al$_2$O$_3$ 11.2 B$_2$O$_3$ 3.6 Na$_2$O$_3$ 0.4 K$_2$O, and that of the barium silicate glass (mol%) was 45.8 SiO$_2$ 2.1 Al$_2$O$_3$ 21.6 B$_2$O$_3$ 29.3 BaO 0.5 As$_2$O$_3$. Zirconia powders processed without additives are designated 'Pure' 3Y-TZP or 8Y-CSZ. Details regarding specimen processing are reported elsewhere [4-6]. Following sintering and isostatic pressing, specimens were prepared for transmission electron microscopy by conventional grinding, polishing, and ion-milling techniques. Specimens were mounted on copper slotted washers and carbon-coated prior to examination.

The analytical electron microscopy (AEM) described here was performed at the SHaRE User Facility, Oak Ridge National Laboratory. An EMiSPEC Vision integrated acquisition system interfaced to a Philips CM200FEG with an Oxford Super-ATW detector and XP3 pulse processor was used to acquire EDS spectrum lines (typically 20 points, 0.5 to 2.0 nm spacing, 10 s dwell/point) across edge-on grain boundaries and across grain-boundary triple-points. Probes were selected to maximize the EDS signal while considering beam damage effects and best spatial resolution; probes were typically of 0.5 nA (1.2 nm FWHM) to 1.5 nA (2 nm FWHM). Spectra were "post processed" interactively with the Vision software to yield profiles of background-subtracted integrated intensities. Additionally, some spectra were acquired for 100 s live-time with probes which were either stationary or rastered (to reduce beam damage) within a region of interest.

RESULTS

Shown in Figure 1 are results from the 'Pure' zirconias. The submicron grain size is apparent in the image from a 3Y-TZP specimen (Fig. 1a). The location where an EDS spectrum line profile was acquired across the edge-on grain boundary is marked. From this, background subtracted FWHM intensities of the Zr Kα and Y Kα reveal zirconium depletion and yttrium enrichment has occurred at the boundary (Fig. 1b). The intensity ratios of the lines in Figure 1b (3Y-TZP) and corresponding results from a grain boundary in an 8Y-CSZ specimen are presented in Figure 1c; these indicate the chemical width of both boundaries is approximately 4 nm FWHM. Away from the grain boundary, the observed higher Y Kα/Zr Kα ratio in the 8Y-CSZ compared to the 3Y-TZP is consistent with the higher overall yttria content (8 versus 3 mol%).

EDS spectrum lines were acquired across grain boundaries of 8Y-CSZ with 1 wt% SiO$_2$ added. Zirconium depletion and yttrium enrichment at grain boundaries was observed (Figure 2a). Additionally, background fitting for Si K was obtained using a third-order polynomial fit to windows set below Si K and above Zr L peaks. However, this conventional profiling was complicated by the presence of Zr L and Y L peaks overlapping the Si K signal, resulting in a possibly unreliable Si K profile. (Note that the location of the maximum silicon signal at 8 nm does not correspond to the local extremes in the zirconium and yttrium signals at 9 nm.) However a spatial differencing technique was used to ascertain the presence or absence of silicon at the grain boundaries of these specimens. In this technique, a spectrum from a location away from the grain boundary (typically at or near an extreme of the spectrum line) was subtracted from that spectrum acquired at the grain boundary to obtain a spectrum of "excess" signal at the grain boundary. Prior to subtraction, the matrix spectrum was normalized to the grain boundary spectrum by a scalar selected so that the resulting integrated intensity (FWTM) of the two background subtracted Zr Kα peaks was identical. Correct normalization of the spectra was verified after calculation of the difference spectrum by ensuring the residual intensity (noise) of the Zr Kα signal oscillated evenly about zero (Figure 2b). The resulting

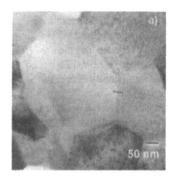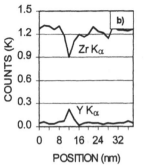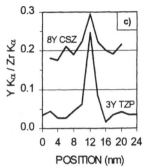

Figure 1. a) Grain boundary within 'Pure' 3Y TZP used to acquire EDS spectrum line; b) EDS profiles reveal Zr depletion and Y enrichment at the grain boundary; c) profiles acquired from grain boundary in 'Pure' 8Y CSZ specimens reveal similar segregation at boundary.

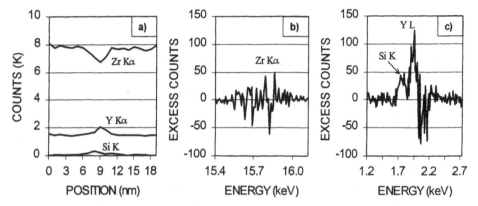

Figure 2. From EDS spectrum line from grain boundary of 8Y CSZ with 1 wt% silica added shows a) zirconium enrichment and yttrium depletion whereas presence of silicon, based on profile from SiKα, is inconclusive; b) residual of ZrKα from EDS spatial differencing of spectra; c) EDS spatial difference resolves Si at grain boundary.

difference spectrum, shown in Figure 2c, clearly shows both silicon (Kα at 1.74 keV) and yttrium (Y Lα at 1.92 keV) co-segregated to the grain boundary. The negative intensity associated with the Zr L peak at approximately 2 keV cannot be due to the depletion of zirconium at the grain boundary because the normalization was based upon the Zr signal. Rather, it is more likely that absorption of Zr L is occurring. Using calculated mass absorption coefficients of Zaluzec[7], an absorption correction for Zr L in 3Y TZP is needed for specimen thicknesses which exceed approximately 50 nm. The specimens examined in this study are thought to exceed this thickness.

Typical results from spectrum lines taken across grain boundaries of 8Y-CSZ processed with 10 wt% alumina are shown in Figure 3. The microstructure can be described as consisting of zirconia grains (~1 μm diameter) with alumina grains (~0.3 μm) at some multigrain junctions. Following background subtraction, conventional EDS profiles reveal the grain boundary is enriched with yttrium and aluminum, and depleted in zirconium, Figure 3a; the presence of the additive at the grain boundary is readily observed in the difference spectrum, 3b. Aluminum was found in all grain boundaries examined. It is interesting to note that, compared to 'Pure' 8Y-CSZ (Fig. 1c), the degree of yttrium enrichment at the boundary is reduced, but the total (Al + Y) amount of segregation is similar. A similar effect has been observed in 2.5Y-TZP doped with 5 wt% silica.[8] The presence of such segregated cations may significantly alter the chemical bonding states at the grain boundaries [9].

The results from 3Y TZP processed with 1 wt% borosilicate are presented in Figure 4. The STEM image, Figure 4a, shows the fine grain size of the superplastic material; the location across the grain boundary used for obtaining the EDS spectrum line is marked. While the zirconium depletion and yttrium enrichment were measured at the grain boundary (Figure 4b), silicon was not detected, even using the differencing technique (Figure 4c). Spectra acquired using probes rastered in the triple point regions indicate silicon is a major component of the material there.

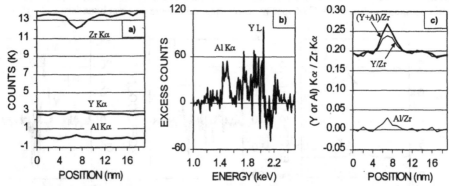

Figure 3. a) Spectrum line acquired from 8Y-CSZ with 10 wt% alumina indicates zirconium depletion, yttrium enrichment, and aluminum at the grain boundary; b) difference spectrum confirms presence of aluminum (Al Kα 1.48 keV) in grain boundary; c) Al may be substituting for Y as excess-cation/Zr ratio in 8Y-CSZ.

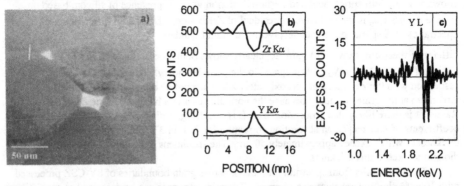

Figure 4. a) Location of spectrum line acquired from 3Y TZP with 1 wt% borosilicate; b) spectrum line reveals zirconium depletion and yttrium enrichment at grain boundary; c) spatial differencing of spectra from the grain boundary and from the matrix 15 nm away indicates silicon is not detectable in the grain boundary.

The location of a spectrum line acquired from 3Y TZP processed with 1 wt% barium silicate is denoted in the STEM image of Figure 5a. Zirconium depletion and both yttrium and barium enrichment (4 nm FWHM) were observed at the grain boundary (Figure 5b); silicon was not clearly detected, even using the differencing technique (Figure 5c). Both barium and silicon were observed in EDS spectra acquired from grain boundary triple points.

CONCLUSIONS

While overlapping Zr L and Y L peaks made the measurement by EDS of silicon at grain boundaries difficult, spatial differencing techniques provided a way to discern the presence (or

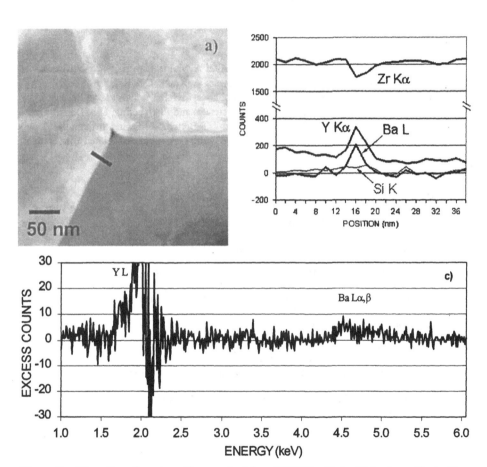

Figure 5. a.) Location of spectrum line acquired from 3Y TZP with 1 wt% barium silicate; b) EDS spectrum lines indicate zirconium depletion, yttrium and barium enrichment, and an absence of silicon at the grain boundary; c) barium, but not silicon, was observed in difference spectrum.

absence) of silicon at grain boundaries. In 3Y-TZP with 1 wt% borosilicate, silicon was observed only at triple points, but not within grain boundaries. It is possible that the borosilicate glass either did not wet the zirconia grains at the sintering temperature, or the materials processing did not produce full coverage of the zirconia by this additive. In 3Y-TZP with 1 wt% barium silicate, silicon was observed at triple points, but not in grain boundaries, whereas barium was detected in grain boundaries as well as triple points. This suggests the composition of the additive phase at the grain boundary may be different from that at the triple point. In 8Y-CSZ with 1 wt% silica, silicon was observed in grain boundaries by spatial differencing. In 8Y-CSZ with 10 wt% alumina, EDS revealed aluminum may be substituting for excess yttrium at grain boundaries. Future electron energy-loss spectrometry studies will help correlate changes in chemical bonding in the vicinity of grain boundaries to the superplastic properties of these ceramics as sintering phases are added.

ACKNOWLEDGMENTS

The authors are grateful for discussions with Adel A. Sharif, University of California - Irvine, Taketo Sakuma, University of Tokyo, and Edward A. Kenik, Oak Ridge National Laboratory. Research at the Oak Ridge National Laboratory SHaRE Collaborative Research Center was sponsored by the Division of Materials Sciences and Engineering, U.S. Department of Energy, under contract DE-AC05-00OR22725 with UT-Battelle, LLC, and through the SHaRE Program under contract DE-AC05-76OR00033 with Oak Ridge Associated Universities.

REFERENCES

1. Y. Yoshizawa and T. Sakuma, J. Am. Ceram. Soc. **73**, 3069 (1990).

2. K. Kajihara, Y. Yoshizawa, and T. Sakuma, Acta Metall. Mater. **43**, 1235 (1995).

3. P. Thavorniti and T. Sakuma, Mat. Sci. Eng. A **202**, 249 (1995).

4. A. A. Sharif, P. H. Imamura and M. L. Mecartney, Materials Science Forum **304-306**, 443 (1999).

5. A. A. Sharif, P. H. Imamura, T. E. Mitchell, and M. L. Mecartney, Acta mater. **46**, 3863 (1998).

6. P. H. Imamura, N. D. Evans, T. Sakuma, and M. L. Mecartney, J. Am. Ceram. Soc. **83**, 3095 (2000).

7. N. J. Zaluzec, EMSA Bulletin, **17**, 93 (1987).

8. P. Thavorniti, Y. Ikuhara, and T. Sakuma, J. Am. Ceram. Soc. **81**, 2927 (1998).

9. Y. Ikuhara, P. Thavorniti, and T. Sakuma, Acta mater. **45**, 5275 (1997).

THE EFFECT OF DIFFERENT OXIDIZING ATMOSPHERES ON THE INITIAL KINETICS OF COPPER OXIDATION AS STUDIED BY *IN SITU* UHV-TEM

MRIDULA D. BHARADWAJ, ANU GUPTA, J. MURRAY GIBSON* AND JUDITH C.YANG
Department of Material Science and Engineering, University of Pittsburgh, Pittsburgh, PA 15261.
*Materials Science Division, Argonne National Laboratories, Argonne, IL 60439

ABSTRACT

Effect of moisture on the oxidation of copper was studied using *in situ* UHV-TEM. The ultra high vacuum condition is required for minimum contamination effects. The initial observations show that the water vapor reduces the oxide as well as reduces the rate of oxidation if both oxygen gas and water vapor are simultaneously used. Based on these observations, we have speculated on the role of moisture in the solid state reactions involved in copper oxidation.

INTRODUCTION

Metal oxidation is an important chemical reaction that has great impact on technology and environment. In particular, the solid state reactions involved in oxidation processes have implications to metal corrosion. Since about 5% of the GNP of USA is spent on dealing with corrosion process, there is fundamental and practical research interest in oxidation mechanism. Further, copper has played a significant role in the development of oxidation theories, ranging from the classic oxidation studies on the epitaxial growth of the thermodynamically stable oxide scale to the surface science investigations of the dynamics of oxygen interaction with the bare metal surface[1]. Cu is a simple face centered cubic metal with lattice parameter a = 3.61Å and a melting temperature of 1083°C [2]. The low resistivity and good electromigration properties make it an attractive interconnect material for ULSI (ultra large scale integrated circuits) [3]. Two types of oxides, CuO and Cu_2O have been reported to form on Cu (001) due to oxidation [4]. CuO has monoclinic structure (space group C2/6, a = 4.662 Å, b = 3.416 Å, c= 5.118 Å, β = 99.49 Å) and Cu_2O is cubic, with space group Pn-3m (a = 4.217 Å) [5]. For the temperature and oxidation partial pressures used in these experiments, only Cu_2O is expected to form.

We had previously reported [6,7] our investigations of the kinetics of initial stages of Cu (001) oxidation using *in situ* UHV-TEM. It was shown that oxygen surface diffusion is the dominant mechanism for the initial oxide formation during oxidation in dry oxygen atmosphere. The present work focuses on the role of moisture in the solid state reactions in Cu oxidation. This has direct relevance to real world conditions as moisture is a significant component of air. Moreover, steam or hydrogen within oxide scale is known to prevent loss of contact between the scale and the metal [8]. *In situ* TEM technique used here permits the visualization of structural changes during the oxidation process in real time and provides information on buried surfaces. There have been previous *in situ* studies by TEM, where the specimen was oxidized in another chamber and transferred in vacuum to the microscope[9-12]. However, the initial stages of Cu oxidation have not yet been studied by *in situ* TEM with ultra high vacuum conditions, which is essential for minimum contamination effects.

EXPERIMENTAL

Single crystal 99.999% pure 1000 Å Cu films were grown on irradiated   NaCl which was cleaved along the (001) plane in an UHV e-beam evaporator system, where the base pressure was 10^{-10} torr. 1000 Å thickness was chosen so that the film was thin enough to be examined by TEM, but thick enough for the initial oxidation behavior to be similar to that of bulk metal. The Cu film was removed from the substrate and mounted on a specially prepared Si mount. The Si mount and the modified microscope specimen holder allow for resistive heating of the specimen up to 1000°C.

The microscope used for this experiment is a modified JEOL 200CX with a spatial resolution of 2.5 Å. An attached leak valve to the column of the microscope permits the introduction of gases directly into the microscope. In order to minimize the contamination, a UHV chamber was attached to the middle of the column, where the base pressure is less than 10^{-8} torr without the use of cryoshroud. The cryoshroud inside the microscope column can reduce the base pressure to approximately 10^{-9} torr when filled with liquid helium. A video camera mounted at the base of the column recorded real time pictures of the experiments. A 5μm objective aperture was used in order to enhance the contrast of the dark field images. For more details about the experimental apparatus, see McDonald et al [13]. The JEOL 200CX was operated at 100 keV.

The Cu film forms a native oxide on the surface due to air exposure. To remove this oxide, the Cu film was annealed at 350 °C for 15 minutes [14]. To remove the copper oxide formed due to *in situ* oxidation, the specimen was annealed at 350°C and methanol gas was leaked into the TEM column at 5×10^{-5} torr column pressure. Methanol reacts with the oxygen atoms to form a methoxy species, which is bound to the surface by oxygen. At 350°C, the methoxy species decomposes to give gaseous formaldehyde and water vapor (eq.1) [15].

$$CH_3OH_g + [O]_{ad} = HCHO_g + H_2O_g \qquad (1)$$

To oxidize the copper film, scientific grade oxygen gas (99.999% purity) was leaked into the TEM chamber at 4×10^{-4} torr pressure. Later, the oxygen supply was discontinued and water vapor was leaked into the chamber with column pressure at 4×10^{-4} torr. The water vapor came from a glass tube, containing de-ionized water, attached to the side of the column. In another experiment, methanol cleaning was directly followed by water vapor exposure. This was to study the effect of water vapor on pure Cu film. We also studied the effect of simultaneous flow of oxygen and water vapor on the oxidation rate.

The digitization of the   micrographs was done with UMAX-Astra1220U scanner. The software packages Adobe Photoshop 5.0 and Scion image 1.62c were used to determine the copper oxide island density as a function of time.

RESULTS

When water vapor is leaked into the microscope after methanol cleaning, surprisingly no reaction was seen. Figure 1(a) is a dark field image taken from Cu_2O (110) reflection after oxygen is leaked into the TEM column at 4×10^{-4} torr pressure at $350^{\circ}C$. Figure 1(b) is one of the subsequent dark field images taken from Cu_2O (110) diffraction spot after water vapor was leaked into the chamber substituting oxygen gas. A marked reduction in the Cu_2O island density is observed on exposure to the water vapor. Figure 1(c) is the Cu_2O dark field image showing complete reduction of Cu_2O islands.

Cu[100]   Cu[100]

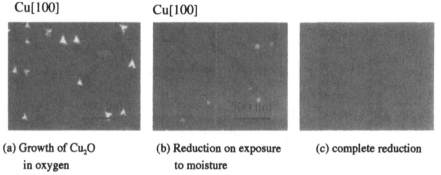

(a) Growth of Cu_2O (b) Reduction on exposure   (c) complete reduction
    in oxygen   to moisture

Figure 1. Dark field Cu_2O (110) TEM images on exposures to O_2 (a) and initial (b) and later
(c) stages of water vapor introduction.

We also observed from the Cu(200) dark field images that the indentations at the Cu_2O island sites fill in, demonstrating that Cu_2O reduction, as opposed to evaporation, occurs during exposure to water vapor.

In Figure 2(a-c) we show the selected area diffraction patterns of the Cu film after methanol cleaning(a) followed by 30 minute oxidation in dry O_2 atmosphere at 5×10^{-4} torr pressure which can be indexed as $(001)Cu_2O$, where the relative orientation between the copper oxide and Cu film is $[001]Cu//[001]$ Cu_2O and $(100)Cu//(100)$ Cu_2O (b) and subsequent reduction of Cu_2O by water vapor(c).

Figure 2. Selected area diffraction patterns of Cu film after methanol cleaning (a) followed by exposures to O_2 (b) and water vapor (c) at 5×10^{-4} torr and 350^0C.

When O_2 and H_2O vapor are leaked in simultaneously, we observed the formation of Cu_2O, but at a considerably slower rate. A plot of density of Cu_2O islands with respect to time is shown in figure 3, comparing the oxidation of the Cu film in dry O_2 and in O_2 mixed with water vapor at a total pressure of 4×10^{-4} torr and 350^0C.

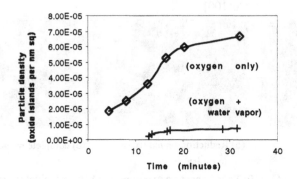

Figure 3. Plot of Cu_2O island density as a function of time

DISCUSSION

Figure 4 is the   thermochemical data for the oxidation reactions of several metals, including Cu [16]. For 350^0C, the dissociation pressure for Cu_2O is 10^{-18} torr. We calculated that the $P(O_2)$ inside the microscope column when only $H_2O(g)$ is leaked into the microscope is 10^{-14} torr (eq.2).

$$H_2O = H_2 + 1/2 O_2 \qquad\qquad (2)$$

Ellingham Plots

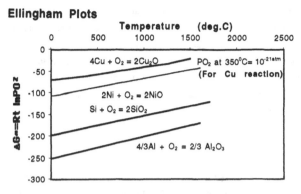

Figure 4. Ellingham plots for metal oxidation reactions.

Hence oxidation of Cu is expected when exposed to water vapor. But no oxide nucleation was observed despite the availability of oxygen.

This could be explained by considering the reaction of gaseous formaldehyde, produced by methanol cleaning, with the oxygen formed by dissociation of water vapor (eq.3).

$$HCHO \quad + \quad O_2 \quad = \quad H_2O \quad + \quad CO_2 \quad \ldots\ldots\ldots\ldots (3)$$

The calculated mole fraction of oxygen available for the above reaction ($\sim 10^{-11}$) was found to be negligible as compared to that of the available formaldehyde ($\sim 10^{-5}$). The entire oxygen could be consumed in the above reaction and hence we do not observe nucleation of Cu_2O.

The nucleation and growth of islands in figure 1(a) is as expected since the partial pressure of oxygen, when introduced into the column is 10^{-6} torr. This oxygen reacts with the formaldehyde as in equation (3) above. The mole fraction of oxygen available for reaction (3) is ~ 100 times more than that of formaldehyde. As a result, almost entire oxygen is left unused causing rapid nucleation of Cu_2O.

To explain the subsequent reduction of Cu_2O islands, we considered the possibility of carbon impurities on the Cu film. The reaction of this carbon with oxygen may lead to substantial reduction in the oxygen partial pressures to be below the threshold value of 10^{-18} torr (ref. figure 4). The minimum amount of carbon needed to accomplish the above was calculated as 9.6×10^2 atoms per μm^3 of Cu surface. Thus the possibility of carbon playing a role in Cu_2O reduction reaction is being investigated.

CONCLUSIONS

We have examined the effects of water vapor on the solid state reactions of copper oxidation under ultra high vacuum conditions. Our observations show the surprising effects that presence of moisture retards the rate of oxidation and water vapor causes reduction of Cu_2O. This contradicts the report by Pilling et al [17] in their classic paper on metal oxidation. Also, it is speculated that the surface contamination by carbon could play a crucial role in the reduction of copper oxide. The future experiments in this research project include investigations of the effect of electron beam on the oxidation rate in presence of water vapor as well as the effect of carbon and surface conditions. The quantification of these data should provide key insights into the structural effects of water vapor on oxidation reaction.

ACKNOWLEDGEMENTS

This research project is funded by the National Science Foundation (DMR#9902863) and Department of Energy (DEFG02-96ER45439).The collaborators are J.M.Gibson, B.Kolasa, M.Yeadon, Zhou G.Wen and Lori Tropia. The experiments were performed at the Materials Research Laboratory, University of Illinois at Urbana – Champaign.

REFERENCES

1. K.R. Lawless, Rep. Prog. Phys., **37**, 231-316, (1974).
2. N.W. Ashcroft and N.D. Mermin, *Solid State Physics*, Philadelphia, PA: Saunders College, 1976.
3. T. Ohba, Applied Surface Science, **91**, 1, (1995).
4. A. Roennquist and H. Fischmeister, Journal of Institute of Metals, **89**, 65, (1960-1961).
5. W.B. Pearson, *Pearson's Handbook of Crystallographic Data for Intermetallic Phases*, Ohio : American Society for Metals, Menlo Park, 1985.
6. J.C. Yang, M. Yeadon, B. Kolasa and J.M. Gibson, Scripta Materialia, **38**,1237 (1998).
7. J.C. Yang, M. Yeadon, B. Kolasa and J.M. Gibson, Applied Physics Letters, **70**, 3522-3524, (1997).
8. C.W. Tuck, M. Odgers and K. Sachs, Corros.Sci., **9**, 271-295, (1969).
9. R.H. Milne and A. Howie, Philosophical Magazine A, **49**, 665-682 (1984).
10. D.A. Goulden, Philosophical Magazine, **33**, 393-408, (1976).
11. K. Lawless and D. Mitchell, Memoires Scientifiques Rev. Metallur., LXII, 17-45, (1965).
12. K. Heinemann, D.L. Douglas, Oxidation of Metals, **9**, 379-459, (1975).
13. M.L. McDonald, J.M. Gibson and F.C. Unterwald, Rev. Sci. Instrum., **60**, 700, (1989).
14. J.C. Yang, M. Yeadon, D. Olynick and J.M. Gibson, Microscopy and Microanalysis, **3** (2), 121, (1997).
15. S. Frances, F. Leibsle, S. Haq, N. Xiang and M. Bowker, Surf. Sci., **315**, 284, (1994).
16. Richard A. Swalin, *Thermodynamics of Solids*, 2nd ed. (Wiley-Interscience Publishers, 1972), p.116.
17. N.B. Pilling and M.S. Bedworth, Journal of Inst. Metals, xxix, 529-82 (1923).

BONDING IN ION-IMPLANTED CARBON FILMS CHARACTERIZED BY TEM SPECTRUM LINES AND ENERGY-FILTERED IMAGING

J. BENTLEY,* K.C. WALTER,† and N.D. EVANS*‡
*Metals and Ceramics Division, Oak Ridge National Laboratory, PO Box 2008, Oak Ridge, TN 37831-6136, bentleyj@ornl.gov
†Los Alamos National Laboratory, Los Alamos, NM; now at Southwest Research Institute, San Antonio, TX
‡Oak Ridge Institute for Science and Education, PO Box 117, Oak Ridge, TN 37830

ABSTRACT

Ion-implanted diamond-like carbon (DLC) films have been characterized by techniques based on electron energy-loss spectrometry using an imaging energy filter on a 300kV TEM. Nitrogen implantation results in increased sp^2 bonding and a 1.3 eV shift to higher binding energies for carbon-K. Argon implantation results in a smaller increase in sp^2 bonding with no detectable binding energy shift. The fraction of implanted species retained is much smaller for Ar than for N. Differences in behavior between N- and Ar-implanted DLC are consistent with expected chemical reactions. Preliminary results demonstrate the feasibility of mapping the π^* / σ^* intensity (sp^2 / sp^3) ratio by energy-filtered TEM as an alternative to spectrum imaging in STEM mode.

INTRODUCTION

Many of the electrical, mechanical, and electrochemical properties of diamond-like carbon (DLC) coatings are determined by the amount of sp^3 bonding. Correlations of bonding and properties are important in efforts to optimize coatings for specific applications. DLC thin films were grown at LANL by plasma deposition and subsequently ion implanted with nitrogen or argon. Cross-sectioned TEM specimens have been characterized by electron energy-loss spectrometry (EELS), including information on sp^2 / sp^3 bonding ratios from the relative intensities of π^* and σ^* features at the carbon K edge [1]. To minimize beam damage artifacts that can be encountered with intense focussed probes, high-spatial-resolution spectrum lines were acquired in the TEM mode [2] with a Gatan imaging filter (GIF) interfaced to a Philips CM30 at the ORNL SHaRE User Facility. Elemental distribution maps were also acquired by standard core-loss mapping EFTEM methods [3], and preliminary results were obtained for EFTEM mapping of the π^* / σ^* ratio.

EXPERIMENTAL

Appropriate experimental conditions for producing elemental distribution maps and spectrum lines with a GIF interfaced to a Philips CM30 have been refined over several years on a range of materials [4]. Based on this experience, in the present work the following conditions were used: incident beam divergence $\alpha = 2.9$ mrad, collection angle $\beta = 4.8$ mrad, probe current i >100 nA in ~2 μm illuminated area, slit width for elemental mapping $\Delta E = 30$ eV (10 eV for Si-L), exposure times of typically 5 s, 2x-binned 512 x 512 images, and TEM magnification ~2000. The standard 3-window method was used for producing elemental maps with AE^{-t} background extrapolation; 2-window jump-ratio images were also produced [5]. Zero-loss (I_0) and unfiltered (I_T) images were used to produce "thickness" maps of $t/\lambda = \ln(I_T / I_0)$ where λ is the inelastic scattering mean free path (typically ~100nm).

For TEM spectrum lines, 1024 x 256 images were used with dispersions of 0.2 or 0.3 eV/pixel, and exposure times typically <10 s. Image areas were selected with a 2 x 0.5 mm Au slotted washer mounted in the GIF entrance aperture normal to the energy dispersion direction. The substrate-DLC film interface was aligned edge-on and normal to the slot with a tilt-rotation specimen holder. In the TEM spectrum line mode, objective lens chromatic aberration (C_c) limits the spatial resolution ($\delta = C_c.\Delta E/E.\beta$) and introduces an energy-dependent spectral focus. The method is therefore better suited to profiling local changes in fine structure rather than composition. Most image acquisition and processing was accomplished with standard Gatan DigitalMicrograph

(DM) and Image Filter Suite software, but some critical operations used DM custom scripts developed at ORNL. Sub-areas of spectrum lines were extracted into Gatan EL/P for standard spectral processing that included zero-loss peak deconvolution, AE^{-r} power-law background fitting, and Fourier-ratio deconvolution to yield single scattering distributions. Ratios of π^* / σ^* intensities were extracted with a 2 or 3eV window centered on the π^* feature (at ~285eV) and a 10eV window centered on the σ^* feature (at ~295eV) [6]. Attempts to convert intensity ratios to sp^2 / sp^3 bonding ratios were prevented by the unavailability of a suitable calibration specimen; a holey carbon film yielded a lower sp^2 content than the unimplanted DLC films.

RESULTS AND DISCUSSION

Nitrogen-implanted DLC

A little surprisingly, the EFTEM elemental maps (see Fig. 1) indicate an approximately *uniform* distribution of nitrogen throughout the implanted region (the top 160 nm). Quantitative EELS on spectra extracted from spectrum lines (see below) yields N / C = 0.19 ± 0.03. Specimen quality is clearly limited by ion milling: the implanted layer thins preferentially producing analyzable areas with $t/\lambda < 1.0$, whereas for the unimplanted layer for the most part $t/\lambda > 1.0$ (see Fig. 1).

Spectrum lines (see Fig. 2) clearly reveal the presence of nitrogen in the implanted region with no other detectable impurities. Spectra were extracted for ~100nm wide regions of the implanted and unimplanted areas of the film for processing within EL/P. Possible slight misregistry between core-loss and low-loss regions, plus thickness gradients across the selected area, result in slight imperfections in the single scattering distributions (SSDs), especially for the (thicker) unimplanted material at ~50 and 120 eV beyond edge onset (Fig. 2d). However, near-edge fine structure in general and π^* / σ^* intensities in partiular are almost unaffected by multiple scattering.

Nitrogen implantation results in increased sp^2 bonding (19% increase in π^* / σ^* intensity ratio) and a ~1.3 eV shift to higher binding energies (see Fig. 2e). This shift is unambiguous since the TEM spectrum line is recorded "in parallel."

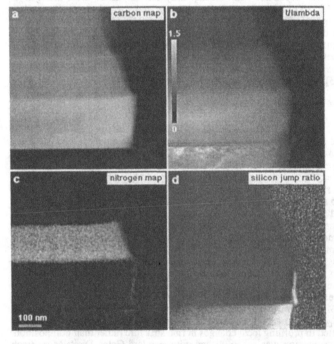

Fig. 1. Cross-sectioned N-implanted DLC film. (a) Carbon, and (c) nitrogen elemental distribution maps (3-window EFTEM net core-loss intensity images). (d) Silicon L-edge jump-ratio image. (b) "Thickness" map (t/λ) with intensites scaled from black ($t/\lambda = 0$) to white ($t/\lambda = 1.5$) as shown on the inset. For the N-implanted layer $t/\lambda < 1.0$ but for the unimplanted layer $t/\lambda > 1.0$.

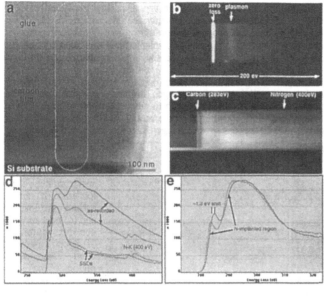

Fig. 2. Cross-sectioned N-implanted DLC film. (a) Unfiltered bright-field image showing position of 2 x 0.5 mm area-selecting aperture. (b,c) TEM spectrum lines of (b) low-loss, and (c) core-loss regions. (d) Core-loss spectra extracted from ~100-nm-wide regions of the implanted (with N K-edge at 400 eV) and unimplanted regions. Single scattering distributions (SSDs) were obtained by Fourier-ratio deconvolution. (e) Expanded C K-edge SSDs normalized to maximum intensity, showing increased π^* intensity and ~1.3 ev shift to higher binding energies for N-implanted material (darker line).

Argon implanted material

The EFTEM elemental maps (Fig. 3) indicate a Gaussian-like depth distribution of argon centered at a depth of ~60 nm below the surface of the DLC film. Argon is not detectable for $t/\lambda <$ 0.4. Quantitative EELS (on spectra extracted from spectrum lines) for a region with $t/\lambda \sim 0.8$ yields Ar / C = 0.006 ± 0.001.

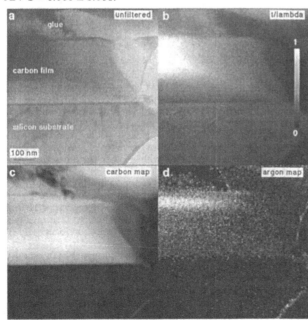

Fig. 3. Cross-sectioned Ar-implanted DLC film. (a) Unfiltered bright-field image (b) "Thickness" map (t/λ) with intensites scaled from black $(t/\lambda = 0)$ to white $(t/\lambda = 1.0)$ as shown on the inset. (c) Carbon, and (d) argon elemental distribution maps (3-window EFTEM net core-loss intensity images).

Spectrum lines (Fig. 4), in addition to the presence of argon in the implanted region, unexpectedly reveal the presence of calcium and oxygen. Quantitative analysis yields Ca / C = 0.042 ± 0.006. The origin of the Ca is uncertain; contamination during ion implantion cannot be ruled out. Spectra were extracted for ~50nm wide regions of implanted and unimplanted film for processing within EL/P. Again, there are clear imperfections in the single scattering distributions (SSDs) shown in Fig 4d,e.

As shown in Fig. 4e, argon implantation results in a much smaller increase in sp^2 bonding (only 5% increase in π^* / σ^* intensity ratio) than for nitrogen implantation, and there is no evidence of a shift in binding energy, which is consistent with the inert character of argon. Even though the specimen quality is clearly less than optimum, no preferential thinning of the implanted region is apparent, which is consistent with the less severe changes in composition and bonding.

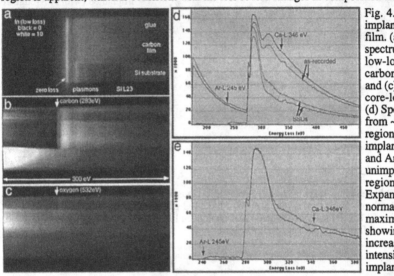

Fig. 4. Ar-implanted DLC film. (a-c) TEM spectrum lines of (a) low-loss, (b) near-carbon core-loss, and (c) near-oxygen core-loss regions. (d) Spectra extracted from ~50-nm-wide regions of the implanted (with Ca and Ar) and unimplanted regions. (e) Expanded SSDs normalized to maximum intensity, showing slight increase of π^* intensity for Ar-implanted material.

EFTEM mapping of π^* / σ^* ratio

Recent work has demonstrated the ability to "map" by EFTEM changes in bonding based on changes in ELNES intensities; e.g., the valence of 3d transition metals (TMs) for different oxides has been mapped by L_3/L_2 white-line intensity ratios [7]. Based partly on this success with the 3d TMs, preliminary attempts were made to map π^* / σ^* intensity ratios for the N-implanted DLC specimen. A series of 25 EFTEM images from 260eV to 308eV, at 2eV spacing, and with ~3eV wide slits, was acquired with 5s exposures (see Fig. 5). The 25 images were aligned with custom DM scripts to correct for drift, and the common area extracted with another custom script. AE-r background subtraction yielded net intensities (Fig. 6a-d), from which a ratio image of net 284eV / net 296eV was produced (Fig. 6e). This image appears to indicate correctly the differences in π^* / σ^* intensity ratio. This success may be because such near-edge net-intensity ratios are largely insensitive to multiple scattering, and thus, to changes in thickness. Extraction of spectra from rectangular image sub-areas was also achieved with custom DM scripts. Although their quality is limited by the poor energy resolution, they are useful for confirming image selection for π^* and σ^* losses.

Although high spatial resolution can be achieved without a field emission gun (FEG), and data for a large number of image pixels can be acquired relatively quickly, compared to spectrum imaging in STEM mode, some important issues for fine-structure mapping by EFTEM remain. Data acquisition is inefficient for a given electron dose to the specimen, energy drift is not presently monitored during series acquisition (but maybe it could be), extracted spectra have poor energy resolution (limited by slit width), chromaticity across the image exceeds 2eV (limited by filter aberrations), and image alignment for drift correction can difficult due to contrast changes or even reversals at ionization edges. Multivariate statistical analysis may be the best way to extract the maximum statistically significant ELNES information from large data sets produced by spectrum imaging in TEM or STEM.

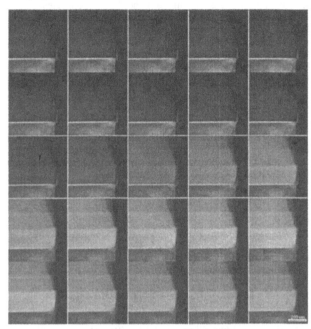

Fig. 5. A series of EFTEM images through the carbon K-edge with slit width ΔE = 3 eV, over an energy-loss range of 260 eV (top left) to 308 eV (bottom right) at 2 eV increments

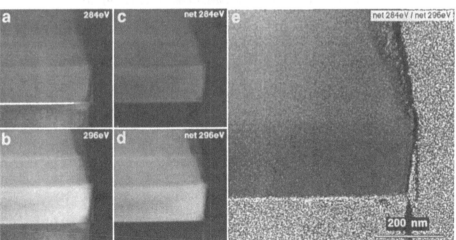

Fig. 6. Energy-filtered images with ΔE = 3 eV at (a) 284 eV loss (π^*) and (b) 296 eV loss (σ^*). Background-subtracted (net) images at (c) 284 eV loss and (d) 296 eV loss. (e) The ratio of (c)/(d) maps the π^* / σ^* intensity ratio and correctly indicates the increased sp^2 content of the implanted layer.

SUMMARY

Ion-implanted diamond-like carbon (DLC) films have been characterized by techniques based on electron energy-loss spectrometry using an imaging energy filter on a 300kV TEM. Nitrogen implantation results in increased sp^2 bonding and a 1.3 eV shift to higher binding energies. Argon implantation results in a smaller increase in sp^2 bonding with no detectable binding energy shift. With the use of a suitable standard of known sp^2 content (such as nanocrystalline graphite), measured π^* / σ^* intensity ratios could yield absolute sp^2 / sp^3 bonding ratios. The fraction of implanted species retained is much smaller for Ar than for N. Differences in behavior between N- and Ar-implanted DLC are consistent with expected chemical reactions. Preliminary results demonstrate the feasibility of mapping the π^* / σ^* intensity (sp^2 / sp^3) ratio by EFTEM as an alternative to spectrum imaging in STEM mode.

ACKNOWLEDGEMENTS

Research at the Oak Ridge National Laboratory SHaRE Collaborative Research Center was sponsored by the Division of Materials Sciences and Engineering, U.S. Department of Energy, under contract DE-AC05-00OR22725 with UT-Battelle, LLC., and through the SHaRE Program under contract DE-AC05-76OR00033 with Oak Ridge Associated Universities.

REFERENCES

1. S.D. Berger, D.R. McKenzie, and P.J. Martin, *Philos. Mag. Lett.* **57** (1988) 285.
2. J. Bentley and I.M. Anderson, in *Proc. Microscopy & Microanalysis '96*, edited by G.W. Bailey et al. (San Francisco Press, San Francisco, 1996) p. 532.
3. A.J. Gubbens and O. Krivanek, Ultramicroscopy **51** (1993) 146.
4. J. Bentley, *Microsc. Microanal.* 4(Suppl 2: Proceedings) (1998) 158.
5. O.L. Krivanek, A.J. Gubbens and N. Dellby, *Microsc. Microanal. Microstruct.* **2** (1991) 315; O.L. Krivanek, A.J. Gubbens, N. Dellby and C.E. Meyer, *Microsc. Microanal. Microstruct.*. **3** (1992) 187.
6. J. Bruley, D.B. Williams, J.J. Cuomo, and D.P. Pappas, *J. Microscopy* **180** (1995) 22.
7. Z.L. Wang, J. Bentley, and N.D. Evans, *J. Phys. Chem. B* **103** (1999) 751.

AUTHOR INDEX

SUBJECT INDEX

Printed in the United States
By Bookmasters